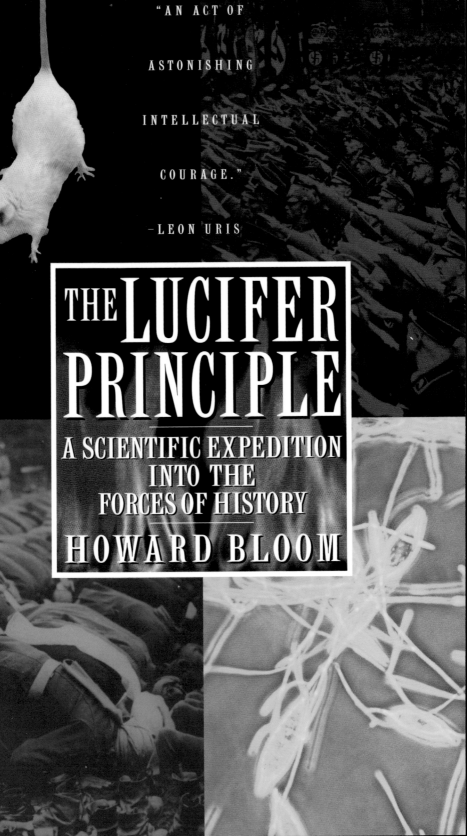

THE LUCIFER PRINCIPLE

A SCIENTIFIC EXPEDITION INTO THE FORCES OF HISTORY

HOWARD BLOOM

THE LUCIFER PRINCIPLE

THE LUCIFER PRINCIPLE

A Scientific Expedition into the Forces of History

HOWARD BLOOM

THE ATLANTIC MONTHLY PRESS
NEW YORK

Grateful acknowledgment is made to the following for permission to reprint previously published material:
Excerpt from *Psychological Care of Infant and Child* by John B. Watson. Copyright © 1928. Published by W. W. Norton. Reprinted by permission of the publisher.
Excerpt from the Squeeze song "Hits of the Year" by Christopher Difford and Glenn Tilbrook. Copyright © 1985 by EMI Virgin Music. Reprinted by permission of EMI Virgin Music.
Excerpts from *Islam and Revolution: Writings and Declarations of Iman Khomeini*. Translated by Hamid Algar. Copyright © 1981. Published by Mizan Press. Reprinted by permission of the publisher.
Excerpts from *The Reign of the Ayatollah: Iran and the Islamic Revolution* by Shaul Bakash. Copyright © 1984 by Basic Books. Reprinted by permission of the publisher.
Excerpt from *The Proud Tower: A Portrait of the World Before the War, 1890–1914* by Barbara Tuchman. Copyright © 1966 by The Macmillan Company. Reprinted by permission of the publisher.

Published simultaneously in Canada
Printed in the United States of America

Library of Congress Cataloging-in-Publication Data

Bloom, Howard K., 1943–
The Lucifer principle : a scientific expedition into the forces of history / by Howard Bloom.
Includes bibliographical references and index.
ISBN 0-87113-664-3 (pbk.)
1. Man. 2. Evolution. 3. History—Philosophy. 4. Culture.
5. Civilization, Modern—20th century. 6. Good and evil. 7. Devil.
I. Title.
BD450.B526 1995   128—dc20   94-11464

DESIGN BY LAURA HAMMOND HOUGH

Atlantic Monthly Press
841 Broadway
New York, NY 10003

01 02 15 14 13 12 11 10 9 8

To Linda Bloom

If only there were evil people somewhere insidiously committing evil deeds and it were necessary only to separate them from the rest of us and destroy them. But the line dividing good and evil cuts through the heart of every human being. And who is willing to destroy a piece of his own heart?

Aleksandr Solzhenitsyn

Only through knowledge will we be able . . . to deliver those with superstitious faith in the omnipotence of violence from their folly.

Fang Lizhi

We have need of history in its entirety, not to fall back into it, but to escape from it.

Ortega y Gasset

CONTENTS

Contents

ONE MAN'S GOD IS ANOTHER MAN'S DEVIL

MAN—INVENTOR OF THE INVISIBLE WORLD

THE MYSTERIES OF THE EVOLUTIONARY LEARNING

MACHINE

Contents

Contents

FOREWORD

⇒ ❦ ❧

The author of this book is an intellectual, originally trained in science, who decided to avoid the limitations of an academic career. Instead of conducting laboratory experiments and competing for federal grant dollars, he put his interest in mass human behavior to work by playing a key role in the careers of rock stars such as Michael Jackson and John Cougar Mellencamp. Meanwhile, he continued to read widely and do what all scientists, as intellectuals, should do: attempt to understand and explain the nature of the world around him. His experience ''at the center of our culture's mythmaking machinery'' may have taught him more about human nature than a university career. Perhaps we should regard him as an anthropologist who has spent many years observing a strange tribe—us.

In the course of his inquiry, Howard Bloom became convinced that evolution could explain the fundamentals of human nature and the broad sweep of human history. He is not alone. It is no longer heretical to study our own species as one of evolution's creations, and many books are appearing on the subject. *The Lucifer Principle,* however, does not merely report on the rapid developments that are taking place within academia. Howard Bloom has his own vision of evolution and human nature that many scientific authorities would dispute. He is a heretic among former heretics.

The bone of contention is the organismic nature of human society. Thomas Hobbes and many others of his time regarded individuals as the cells and organs of a giant social organism—a Leviathan—''which is but an artificial man, though of greater stature and strength than the natural, for whose protection and defence it was intended.'' Today this idea is regarded as no more than a fanciful metaphor. Evolution is thought to produce individuals who are designed to relentlessly pursue their own reproductive success. Society is merely the by-product of individual striving and should not be regarded as an organism in its own right. Even individuals can be decomposed into selfish genes whose only purpose is to replicate themselves.

It is a mark of Howard Bloom's independence of thought that he resisted the extreme reductionism that pervades modern evolutionary biology. He believes that the Leviathan, or society as an organism, is not a fanciful metaphor but an actual product of evolution. The Darwinian struggle for existence has taken place among societies, as well as among individuals within societies. We do strive as individuals, but we are also part of something larger than ourselves, with a complex physiology and mental life that we carry out but only dimly understand. That is the vision of evolution and human behavior found in *The Lucifer Principle,* and at the moment it can be found nowhere else.

When Howard Bloom wrote *The Lucifer Principle,* he studied numerous developments taking place within the halls of academe but was unaware of others. Evolution is increasingly being studied as a process that operates on a hierarchy of units. Even individual organisms are higher-level units, composed of parts that were themselves free-living organisms in the distant past. Truly organismic societies have evolved in insects and even some recently discovered mammal species. As for ourselves, human society may turn out to be far more organismic than the vast majority of evolutionary biologists imagined only a few years ago. These discoveries are unfolding within the scientific community, and many of them have been anticipated by Bloom. Scientists and other academicians might find themselves treading a path forged by an outsider.

As a scientist who has been developing a hierarchical view of evolution from within academia, I have learned from Howard Bloom and value him as a fellow traveler. I do not agree with everything he says, and I sometimes blush at the way he says it—not with the reserve of a scientist but with the brashness of a mass media denizen. Of course, that only makes the book more fun for the average reader. Many of Howard Bloom's ideas must be passed through the scientific-verification machine before they can be accepted. Until then, your motto for *The Lucifer Principle* should be not ''read it and believe'' but ''read it and think.''

DAVID SLOAN WILSON
Binghamton, New York

WHO IS LUCIFER?

How art thou fallen from heaven, O Lucifer, son of the morning! . . .
    *For thou hast said in thine heart, I will ascend into heaven, I
will exalt my throne above the stars of God. . . .*
    *I will ascend above the heights of the clouds; I will be like the
Most High.*

Isaiah 14:12–14

Eighteen hundred years ago in the city of Rome, an influential Christian
heretic named Marcion took a look at the world around him and drew a
conclusion: The god who created our cosmos couldn't possibly be good.
The universe was shot through with appalling threads—violence, slaughter, sickness, and pain. These evils were the Creator's handiwork. Surely
he must be some perversely sadistic force, one who should be banished
from influence over the minds of men.[1]

    More traditional Christians found another way of dealing with the
problem of evil. They created the myth of Lucifer.[2] Lucifer had been a
magnificent angel, a courtier of God, a prime noble in the kingly halls of
heaven. He was trusted, powerful, charming, awesome in his self-posses-
sion. But he had a flaw: he wanted to usurp the seat of heavenly power
and seize the throne of God himself. When the plot was uncovered, Lucifer was hurled from heaven, exiled beneath the earth, and tossed into the
dreary realms of hell. The ancient gods who had been his co-conspirators
were cast into the lightless subterranean caverns with him.

    But Lucifer still bore some of the attributes of his creator and former master. He was an organizer, a would-be crafter of new orders, a creature bent on pulling together forces in his own manner. The fallen angel did not lay facedown long in the muck of the lightless caverns. His first step was to mobilize the squabbling gods trapped with him in hell, regimenting them into a new army.

Then Lucifer set forth to conquer the earth, using as his pawn a fresh, godly invention: an innocent pair Jehovah had placed in a garden— Adam and Eve. The Great Seducer tempted Eve with the apple of knowledge, and she could not resist the Luciferian fruit. Eve's sin against God corrupted all mankind. Ever since that time, man has aspired to the Lord but found himself a victim of the devil.

Marcion the heretic said *God* was responsible for evil. Mainstream Christians absolved the Almighty of responsibility by blaming all that's wrong on the Prince of Darkness and on man. But, in a strange way, Marcion understood the situation better than the more conventional followers of the church, for Lucifer is merely one of the faces of a larger force. Evil is a by-product, a component, of creation. In a world evolving into ever-higher forms, hatred, violence, aggression, and war are a part of the evolutionary plan. But where do they fit? Why do they exist? What possible positive purpose could they serve? These are some of the questions behind the Lucifer Principle.

The Lucifer Principle is a complex of natural rules, each working together to weave a fabric that sometimes frightens and appalls us. Every one of the threads in that tapestry is fascinating, but the big picture is more astonishing still. At its heart, the Lucifer Principle looks something like this: The nature scientists uncover has crafted our viler impulses into us: in fact, these impulses are a part of the process she uses to create. Lucifer is the dark side of cosmic fecundity, the cutting blade of the sculptor's knife. Nature does not abhor evil; she embraces it. She uses it to build. With it, she moves the human world to greater heights of organization, intricacy, and power.

Death, destruction, and fury do not disturb the Mother of our world; they are merely parts of her plan. Only *we* are outraged by the Lucifer Principle's consequences. And we have every right to be. For we are casualties of Nature's callous indifference to life, pawns who suffer and die to live out her schemes.

One result: from our best qualities come our worst. From our urge to pull together comes our tendency to tear each other apart. From our devotion to a higher good comes our propensity to the foulest atrocities. From our commitment to ideals comes our excuse to hate. Since the beginning of history, we have been blinded by evil's ability to don a selfless disguise. We have failed to see that our finest qualities often lead us to the actions we most abhor—murder, torture, genocide, and war.

For millennia, men and women have looked at the ruins of their lost homes and at the precious dead whom they will never see alive again, then have asked that spears be turned to pruning hooks and that mankind be granted the gift of peace; but prayers are not enough. To dismantle the curse that Mother Nature has built into us, we need a new way of looking at man, a new way of reshaping our destiny.

The Lucifer Principle takes fresh data from a variety of sciences and shapes them into a perceptual lens, a tool with which to reinterpret the human experience. It attempts to offer a very different approach to the anatomy of the social organism.

In the process, *The Lucifer Principle* contends that evil is woven into our most basic biological fabric. This argument echoes a very old one. Saint Paul proposed it when he put forth the doctrine of original sin. Thomas Hobbes resurrected it when he called the lot of man brutish and nasty. Anthropologist Raymond Dart brought it to the fore again when he interpreted fossil remains in Africa as evidence that man is a killer ape. Old as it is, the concept has often had revolutionary implications. It has been the thread on which men like Hobbes and Saint Paul have hung dramatic new visions of the world.

I've attempted to employ the subject of man's inborn evil, as have those who turned to the subject in the past, to offer up a restructuring of the way we see the business of being human. I've taken the conclusions of cutting-edge sciences—ethology, biopsychology, psychoneuroimmunology, and the study of complex adaptive systems, among others—to suggest a new way of looking at culture, civilization, and the mysterious emotions of those who live inside the social beast. The goal is to open the path toward a new sociology, one that escapes the narrow boundaries of Durkheimian, Weberian, and Marxist concepts, theories that have proven invaluable to the study of mass human behavior while simultaneously entrapping it in orthodoxy.

We must build a picture of the human soul that works. Not a romantic vision that Nature will take us in her arms and save us from ourselves, but a recognition that the enemy is within us and that Nature has placed it there. We need to stare directly into Nature's bloody face and realize that she has saddled us with evil for a reason. And we must understand that reason to outwit her.

For Lucifer is *almost* everything men like Milton imagined him to be. He is ambitious, an organizer, a force reaching out vigorously to master even the stars of heaven. But he is not a demon separate from Nature's benevolence. He is a part of the creative force itself. Lucifer, in fact, is Mother Nature's alter ego.

THE CLINT EASTWOOD
CONUNDRUM

⇒⋞

We think of ourselves as rugged individuals, cocky Clint Eastwood–like characters capable of making up our own minds no matter what kind of group pressures torpedo the less independent thoughts of people around us. Eric Fromm, the psychoanalytic guru of the sixties, turned the idea that the individual can control his own universe into a rabidly popular notion. Fromm told us that needing other people is a character flaw, a mark of immaturity. Possessiveness in a romantic relationship is an illness. Jealousy is a character defect of the highest magnitude. A mature individual is one who can drift through this world in the self-contained manner of an interstellar transport manufacturing its own oxygen and food. That rare healthy soul, Fromm wished us to believe, had an indestructible sense of his own worth. As a consequence, he had no need for the admiration and reassurance that only the weak long for.*

Fromm was trapped in a scientific fallacy that has become mainstream dogma. Current evolutionary theory, as promoted by scientists like Harvard's E. O. Wilson and the University of Washington's David Barash, says that only the competition between individuals counts; the concept is called "individual selection." Social groups may glare and pos-

---

*To avoid becoming a prose contortionist, I've risked political indiscretion and have used the traditional pronoun "he" to refer to humans (and animals) of indeterminate sex. I apologize to those who find this offensive and assure them that clarity, not sexism, has been my goal.

ture, threaten, connive, and occasionally battle to a grim and bloody death, but none of this really matters. The dogma of the moment declares emphatically that the creature struggling alone, or occasionally helping out a relative, is the only one whose efforts drive the engines of evolution.

However, the accepted view requires a closer look. Among humans, *groups* have all too often been the prime movers. It is their competition that has driven us on the inexorable track toward higher degrees of order. This is one key to the Lucifer Principle.

At first glance, the notion seems elementary, scarcely worth exploring further, but it has revolutionary implications. This book will show how the competition between groups can explain the mystery of our self-destructive emotions—depression, anxiety, and hopelessness—as well as our ferocious addiction to mythology, scientific theory, ideology, and religion, and our one even more disturbing addiction—to hatred.

Group competition solves puzzles of the immune system recently uncovered by researchers in psychoneuroimmunology. It answers perceptual riddles revealed by new studies on endorphins and control. And it even offers solutions to some of our most baffling political dilemmas.

Individualism is a personal credo of great importance. I, for one, am a passionate believer in it. But to scientists, it has been a chimera leading them down a dead-end path. Specifically, individualism has reared its head in science in the form of a simple proposition: If a piece of physiology—a tooth, a claw, an opposable thumb, or the neural circuit underlying an instinct—has emerged from the evolutionary process, it has triumphed for a simple reason: it has helped the individual survive. More specifically, the physiological device has proven useful in the survival of a long line of individuals, each of whom maintained a competitive edge by virtue of the piece of biological equipment in question. The problem: this basic premise is only right up to a point. Individual survival is not the only mechanism of the evolutionary process, as shown in recent research on stress.

The stress response, with its high levels of corticosteroids and its clammy manifestations of anxiety, is usually described as part of a fight-or-flight syndrome, a survival device left over from the days when men were fending off saber-toothed tigers. When our primitive ancestors were confronted with a snarling beast, the stress response supposedly prepared them to engage the brute in battle or to hotfoot it out of the path of danger. But if the stress response is such a marvelous tool for self-defense, why is it so disabling? Why do stress reactions shut down our thought processes, cripple our immune system, and occasionally turn us into stupefied blobs of jelly? How do these impairments help us survive?

The answer is that they don't. Men and animals do not merely struggle to maintain their individual existence; they are members of larger social groups. And, all too often, it is the social unit, not the individual, whose survival comes first.

At first glance, our dependence on our fellow human beings sounds encouragingly angelic, but it is a blessing with a barb. Harvard psychologist Daniel Goleman, paraphrasing Nietzsche, says, "Madness . . . is the exception in individuals, but the rule in groups."[3] A study by social psychologist Bryan Mullen shows that the larger the lynch mob the more brutal the lynching.[4] Freud declares that groups are "impulsive, changeable and irritable." Those caught up within them, he asserts, can become infantile slaves to emotion, "led almost exclusively by the unconscious."[5] Swept up by the emotions of a crowd, humans tend to lose their ethical restraints. As a result, the greatest human evils are not those that individuals perform in private, the tiny transgressions against some arbitrary social standard we call sins. The ultimate evils are the mass murders that occur in revolution and war, the large-scale savageries that arise when one agglomeration of humans tries to dominate another: the deeds of the social group.

The social pack, as we shall see, is a necessary nurturer. It gives us love and sustenance. Without its presence, our mind and body literally switch on an arsenal of interior devices for self-demolition. If we ever save ourselves from the scourge of mass violence, it will be through the efforts of millions of minds, networked together in the collaborative processes of science, philosophy, and movements for social change. In short, only a group effort can save us from the sporadic insanities of the group.

≫ ≪

This book is about the social body in which we are the unwitting cells. It is about the hidden ways in which that social group manipulates our psychology, and even our biology. It is about how a social organism scrambles for survival and works for mastery over other organisms of its kind. It is about how we, without the slightest sense of what the long-term results of

THE WHOLE IS BIGGER THAN
THE SUM OF ITS PARTS

There's a strange concept in the philosophy of science called an "entele-chy." An entelechy is something complex that emerges when you put a large number of simple objects together. Examine one molecule of water in a vacuum, and you'll be utterly bored by the lack of activity in your vacuum tube. Pour a bunch of molecules into a glass, and a new phenom-enon crops up: a ring of ripples on the water's surface. Combine enough glasses of water in a big enough basin, and you'll end up with something entirely different: an ocean. Take the twenty-six letters of the English al-phabet, lay them out in front of you, and you'll have a set of small squig-gles, each of which evokes just one or two specific sounds. String a few million together in precisely the proper order, and you'll have the col-lected works of Shakespeare.[6]

These are entelechies. A city, a town, a culture, a religion, a body of mythology, a hit record, a dirty joke—these, too, are the results of en-telechies. Take one human being, isolate him in a room from the time he's born until the time he dies, and you'll end up with a creature incapa-ble of using language, with little in the way of imagination—an emotional and physical wreck.[7] But put that baby together with fifty other people, and you'll end up with something entirely new—a culture.

Cultures spring into existence only when the crowd is large enough. They are a phenomenon that sweeps across the face of the multitude like a wave. The phenomenon that creates the Beatles, that makes a Hitler, that launches a new philosophy like communism or Christian Fundamental-ism—these are all entelechies at work, waves rolling over the surface of society, incorporating the minor moves of individual human beings into a massive force the way swells of the sea orchestrate insignificant twitches of water molecules into an overwhelming motion.

The perpetual churnings of the waves and the tides are stirred by the gravity of the moon. But what propels the cultural tides of human beings? What causes a horde of barbaric nomads in the desert wastes of the Arab peninsula to suddenly coalesce under the leadership of one man and spill across the known world, carving out an empire? How does an invisible idea preached by an ayatollah pull isolated humans into surging swells of believers ready to die—or kill—for "truth"? Why does a sect originally committed to turning the other cheek flood the world with warriors who literally wade in blood?[8] What makes a country like Victorian England crest in the domination of fully half the planet, then ebb like a spent wave from power and prosperity? What undertow is sucking America into that same path today?

Five simple concepts help explain these human currents. Each section of this book concentrates on one of those ideas and its sometimes startling implications. Together, these concepts are the foundation underlying the Lucifer Principle:

Concept number one: the principle of self-organizing systems—replicators—bits of structure that function as minifactories, assembling raw materials, then churning out intricate products. These natural assembly units (genes are one example) crank out their goods so cheaply that the end results are appallingly expendable. Among those expendable products are you and me.

Concept number two: the superorganism. We are not the rugged individuals we would like to be. We are, instead, disposable parts of a being much larger than ourselves.

Concept number three: the meme, a self-replicating cluster of ideas. Thanks to a handful of biological tricks, these visions become the glue that holds together civilizations, giving each culture its distinctive shape, making some intolerant of dissent and others open to diversity. They are the tools with which we unlock the forces of nature. Our visions bestow the dream of peace, but they also turn us into killers.

Concept number four: the neural net. The group mind whose eccentric mode of operation manipulates our emotions and turns us into components of a massive learning machine.

Concept number five: the pecking order. The naturalist who discovered this dominance hierarchy in a Norwegian farmyard called it the key to despotism. Pecking orders exist among men, monkeys, wasps, and even nations. They help explain why the danger of barbarians is real and why the assumptions of our foreign policies are often wrong.

Five simple ideas. Yet the insights they yield are amazingly rich. They reveal why doctors are not always as powerful as they seem, but why we are compelled to believe in them nonetheless. They explain how Hinduism, the religion of ultimate peace, grew from the greed of a tribe of bloodthirsty killers and why nature disposes of men far more casually than women. They shed light on America's decline, and the dangers that lie ahead of us.

Above all, they illuminate a mystery that has eternally eluded man: the root of the evil that haunts our lives. For within these five small ideas we will pursue, there lurks a force that rules us.

THE CHINESE CULTURAL
REVOLUTION

*The men who are the most honored are the greatest killers. They be-
lieve that they are serving their fellowmen.*

Henry Miller

*It's getting uncommonly easy to kill people in large numbers, and the
first thing a principle does—if it really is a principle—is to kill some-
body.*

Dorothy L. Sayers

In the mid-sixties, Mao Tse-tung tore the fabric of Chinese society apart.
In doing so, he unleashed emotions of the most primitive kind, the true
demons of the human mind. These primordial motivators ripped across
the face of China, bringing death, destruction, and pain. But the frenzy
Mao had freed was not some freak child of Mao's philosophies; it was the
simple product of passions that squirm every day inside you and me.

In 1958, Mao had decided to throw China violently into the future. His
catapult was the Great Leap Forward, an economic plan designed to harness China's manpower in a massive modernization program. Billboards
carried pictures of a Chinese worker astride a rocket. The slogan read,
SURPASS ENGLAND IN FIFTEEN YEARS! Students, senior citizens, intellectuals, and farmers labored ceaselessly to build steel furnaces. They collected

iron pots and tore brass fittings off the ancient doors of their houses to provide the scrap metal required for the construction of those furnaces. Peasants left their homes in mass mobilizations, slogged to communal dining halls, and threw themselves into their work with tremendous enthusiasm. After all, says Gao Yuan, one Chinese schoolboy who lived through it, "people were saying that true communism was just around the corner."[9]

Unfortunately, somewhere along the line the Great Leap Forward stumbled and fell on its face. The communal dining halls closed. Householders who had taken their pots to the furnaces were forced to find new ones. Ration coupons appeared for grain, oil, cloth, and even matches. The little boys who had thrown themselves so enthusiastically into making an economic miracle grew faint from hunger as they sat in school. They learned to catch cicadas on poles with glue-coated tips, then forced themselves to swallow the still-squirming insects. They scoured the hills for edible grass and weeds. Their mothers baked bread with flour augmented by willow and poplar leaves. During three long years of heroic "progress," millions died of starvation.

The Great Leap Forward had crippled the economy, throttling the production of even the simplest things. And the architect of the brave blunder—Mao himself—lost power.[10] He retired into ideological matters, leaving the day-to-day running of the state to a bureaucratic nest of lesser officials. Those officials looked at a citizenry racked with malnutrition and quickly changed gears. They abandoned theoretical rigor and worked to boost production of the household necessities that had all but disappeared. Highest on the priority list was raising food—lots of it. Doctrine took a back seat to the simple task of putting meals on Chinese tables.

The further the new policy proceeded, the more the officials responsible for implementing it felt that they were the real powers controlling China. Their swelling pride told them that they were the new bosses, the men who had taken over the helm of history. Mao was a relic, an antique, a figurehead. When Mao tried to issue orders, his underlings treated him politely but shrugged him off. The commands of China's Great Leader went unheeded.

Mao Tse-tung did not enjoy being led to pasture. And he wasn't the sort of man to take forced retirement lying down. So this demigod of the

Revolution contrived a plan to reassert authority, a plan that would be even more devastating to China than the Great Leap Forward. His scheme would not just starve people, it would torture them, beat them to death, and force them into suicide. It was the Cultural Revolution.

Mao took advantage of a simple peculiarity of human nature: the rebelliousness of adolescents. The defiant attitude of teenage punk rockers and heavy metal head-bangers may seem like a rage spawned by the unique disorders of Western culture, but it is not. Adolescence awakens defiant urges in nearly all primates. In chimpanzees, it inspires a wanderlust that forces some young females to leave the cozy family they've always known and go off to make a new life for themselves among strangers.[11] In langur monkeys, it triggers a restlessness that's much more to the point. Adolescent langur males kick loose the bonds of their childhood family life and cluster in unruly, threatening gangs. Then they go on the prowl, looking for some older, well-established male they can attack. The adolescents' goal is to dislodge the respectable elder from his cushy home and take over everything he owns—his power, his prestige, and his wives.[12]

As we'll see later, humans are driven by many of the same instincts as our primate relatives. Consequently, many adolescents of our species also resent the authority of the adults over their heads. Their hormones have suddenly told them that it is time to assert their individuality and to challenge the prerogatives of the older generation.

Mao didn't address himself to the adults of China. Those comrades saw the good sense of the officials who had shuffled Mao to the side and focused on producing food to fill the stomachs that had ached with emptiness for three long years. Mao turned elsewhere for help in recapturing authority. He turned to the country's teenagers.

Mao started his campaign to regain the reins of China innocently enough. Under his orders, the major papers began a literary debate. They attacked a group of authors who called themselves the "Three Family Village." These essayists were government officials, key figures in the phalanx of bureaucrats resisting Mao's orders. One was vice-mayor of Beijing. Another, the editor of the *Beijing Evening News,* was propaganda director for Beijing's Party Committee. A third was a propagandist for the Beijing city government. Over the years, the articles of these three had been regarded as entertaining diversions, models of witty style. Now official editorial writers "discovered" that the writings of the Three Family Village were hidden cesspools of secret meanings. And what did those meanings amount to? An assault on the sacred precepts of the party.

The attack on the Three Family Village quickly moved from the papers to the schools. Students were encouraged to pen their own excoriations of the traitors, as one newspaper put it, opening "Fire at the Anti-Party Black Line!" Pupils made posters vilifying the scoundrels' names and plastered them over every available wall. Thus they carried out their duty to "hold high the great banner of Mao Tse-tung thought!"

The banner of Mao's thought soon wrapped itself around the necks of more than just the Three Family Village. Schoolchildren were encouraged to find other literary works rotting with revisionism and antirevolutionary notions. The children leapt avidly to their homework assignment. But they became even more enthusiastic a few months later when a new directive came from above: ferret out bourgeois tendencies and reactionary revisionism among your teachers.

The new task was one to which any youngster could apply himself with gusto. That teacher who gave you a poor mark on your last paper? He's a bourgeois revisionist! Humiliate him. The pedagogue who bawled you out for being late for class? A capitalist rotter! Make her feel your wrath. Revenge had nothing to do with it. This was simply an issue of ideological purity.

Students examined everything their teachers had ever written. In the subtlest turns of innocent phrasing, they uncovered the signs of reactionary villainy. At first, they simply tacked up posters reviling the teachers as monsters and demons. Then all classes were suspended so that pupils could work on sniffing out traitors full-time. Instructors who had fought faithfully with Mao's revolutionary forces were suddenly reviled. Others who considered themselves zealots of Maoist thought were pilloried as loathsome rightists. Some couldn't take the humiliation.

Gao Yuan, son of a party official in a small town, was a boarding student at Democracy Street Primary School in Yizhen at the time. At Gao Yuan's school, one teacher attempted to slit his throat. Other pedagogues tried to placate the students. They "exposed" their colleagues and wrote confessions, hoping to get off the hook. It didn't work.

The students at Democracy Street Primary School created a new form of school assembly. Its star attraction was "the jet plane." A teacher was interrogated at great length in private and forced to "admit" his crimes. Then he was taken onstage before a student audience and kicked in the back of the knees until he fell down. One student grabbed him by his hair and pulled back his head. Others lifted his arms and yanked them behind him. Then they held the hapless instructor in this contorted position for hours. When it was over, most teachers couldn't walk. To make the humiliation a bit more lasting, students shaved their erring teachers' heads.

Among their teachers, the diligent students "discovered" the vilest of the vile. Gao Yuan says that they uncovered "hooligans and bad eggs, filthy rich peasants and son-of-a-bitch landlords, bloodsucking capitalists and neobourgeoisie, historical counterrevolutionaries and active counterrevolutionaries, rightists and ultrarightists, alien class elements and degenerate elements, reactionaries and opportunists, counterrevolutionary revisionists, imperialist running dogs, and spies."[13] The students armed themselves with wooden swords and hardware. At night, they imprisoned their teachers in their bedrooms. Another instructor at the Democracy Street Primary School was driven past endurance and hung himself.

Now that they had practiced on their teachers, the students were urged to take their cultural cleansing further and form organized units, Red Guards, to root out revisionism in the towns. Like young monkeys raiding an elder's domain, ten- and fifteen-year-olds rampaged into the cities looking for officials who had strayed from the strict Maoist line. They sniffed out "ox ghosts and snake spirits" among the municipal authorities; subjected magistrates, mayors, and local party heads to interrogations, beatings, and head-shavings; and marched miscreants through the streets wearing dunce caps that were sometimes thirty feet high. Needless to say, the officials under attack had provided the foundation of support for the bureaucratic powers who had ignored Chairman Mao not long ago. The more the Red Guards attacked that foundation, the more the bureaucratic resistance to the Glorious Chairman crumbled.[14]

The Red Guards did not let their enthusiasm stop there. Urged on by Mao's speeches, they went on a campaign against "The Four Olds"— the remnants of prerevolutionary style. The students pulled down store signs, renamed streets, slit the trouser legs of anyone wearing tight pants, stopped women entering the town gates to cut off their braids, pulled down ancient monuments, broke into homes, and smashed everything that carried the hated aura of tradition. Then the Red Guards turned on each other in what started as a debate over the true Maoist line. Behind the argument about Mao's thought, however, was another issue.

Class warfare is a central concept of Maoism. As a result, each citizen of Mao's China was categorized according to the class from which his parents or grandparents came. If your family in the distant past had belonged to an unacceptable social category, you were a pariah. What was acceptable? The poor peasants and soldiers. Middle peasants and intellectuals were beneath contempt. Upper peasants, capitalists, or landlords were beyond the pale. Just to keep things straight, the descendants of these hated social strata were sometimes forced to wear black armbands with white letters broadcasting their status.

In Gao Yuan's school, one student declared categorically that only those whose class background was "pure," those whose parents had come from the Red categories—poor peasants and soldiers—should be allowed in the Red Guard. And what of the children whose parents came from the Black categories—middle-class peasants, wealthy peasants, landlords, and capitalists? Keep them out, said the snobbish student. A parent's class has nothing to do with children, protested Gao Yuan. "All our classmates were born and brought up under the five-star red flag. We all have a socialist education." Not true, snarled the boy determined to keep the Red Guard an exclusive club, "a dragon begets dragons, a phoenix begets phoenixes, and a mouse's children can only dig holes."

In the coming months, belonging to the Red Guard would be a matter of vital importance. The Red Guard would take over the administration of the cities and the schools. If you belonged, you'd have power. If you didn't, every petty grudge against you could be turned into a political charge. And the slightest accusation of ideological sin could be used to make your days worse than your most appalling nightmare. The debate over who should be allowed in and who should be kept out was not an innocent children's game.

Eventually, there would be two different Red Guards in Gao Yuan's school. One would embrace the children of the favored classes. The other would harbor the rejects—the children of the forbidden castes. At first, the two factions were content to squabble over which one upheld Mao's true line. Each accused the other of right-wing revisionism. Both shouted torrents of Mao's quotations, determined to prove the rival faction wrong. Soon, they turned from citations to taunts and insults. Then they graduated to throwing rocks.

The two sides armed themselves. They made slingshots and clubs, then wove helmets from willow twigs soaked in water. The helmets were so hard you could smash them with a hammer and barely make a dent. A few lucky kids found old swords. Others made sabers and daggers out of scrap metal. Everyone in Gao Yuan's town had grown up knowing how to mix gunpowder from scratch, since children traditionally crafted their own firecrackers for annual holidays. Now, the students of Democracy Street Primary School put that skill to a new use: they built arsenals of homemade hand grenades. Some even found ways to get guns.

It wasn't long before the two rival gangs of Red Guards at Gao Yuan's school were engaged in full-scale warfare. Each occupied a separate group of buildings on the campus. And each began a series of raids to unseat the other from its newly established headquarters. In those armed forays, students were wounded with stones, blades, and explosives. The more the blood flowed, the angrier each side became.

One Red Guard faction came across a lone member of the rival gang on campus, dragged him to an empty dormitory room, tied him up, and interrogated him, searching for secrets to their adversaries' weak points. The captured student at first refused to talk. The interrogators beat him with a chair leg. They snared another student and hung him from the ceiling of the room for days, and bludgeoned yet another with a poker. This time, they made a mistake. The poker had a sharp projection at the end that punctured their prisoner's skin every time it struck. When the questioning session was over, the victim's legs were bleeding profusely. He died a few hours later.

Why had the tormentors used so much force? Their captive was a traitor to the precepts of Chairman Mao. The Chairman himself had said that revolution is not a dinner party. Sometimes it was hard to remember that the person hanging from the rafters had sat three chairs away from you in homeroom since the two of you were little kids.

The commitment of students on both sides to the words of Mao was passionate. They spat phrases from the Great Leader like machine-gun bullets, ferocious in their devotion to ''dialectic truth.'' But, in reality, the Maoist ideology, with its noble goal of liberating humanity, was being used by one Red Guard faction to seize power from another. Idealism's rationalizations transformed the rapacity of the students into a sense of selfless zeal.

The Cultural Revolution threw China into chaos. Finally, the military took control of the country and restored order. The Red Guard members were drafted as they came of age. The teenagers who had fought each other went their separate ways. Gao Yuan entered military service, then studied in Beijing. He met an American girl, moved to the United States, and wrote a book about his experience, *Born Red*. Not long after, others who had suffered through the Cultural Revolution would pen memoirs revealing even greater horrors.

Meanwhile, the head of the class-purity-oriented group that had systematically tortured Gao Yuan and his friends for over a year became a member of a trucking company. The leader of Gao Yuan's more liberal Red Guard brigade disappeared for many years. He resurfaced only when China began the modern economic reform that allows a measure of entrepreneurial freedom. Today, the former Red Guard leader once again uses the ability to organize others that helped him marshal his fierce young army of students: he founds successful capitalist enterprises.

Only one person really got what he wanted from the Chinese Cultural Revolution: Mao Tse-tung. When it was all over, he had driven his opponents from their roosts and regained control of China.[15]

But the Chinese Cultural Revolution had unleashed the most primitive and appalling human instincts, providing a clue to the biological machinery that leads us into war and violence. The formerly shy and well-behaved teenagers caught up in the Chinese Cultural Revolution pulled together in tight clusters. The signal that drew them together was the altruism of ideology. Once their groups had been formed, ideology served a second purpose. It became a weapon, an excuse for lashing out at rival groups, a justification for murder, torture, and humiliation. Within their tight gangs, the Chinese teenagers loved each other. Their loyalty to their comrades and to their master, Chairman Mao, was ferocious. But when they turned their attention to outsiders, the folks they labeled as counterrevolutionaries, their feelings were different. Toward those beyond their tiny circle, they radiated only hatred. And they treated those they despised with remorseless brutality.

The Chinese Cultural Revolution was a microcosm of the forces that manipulate human history. It showed how the insubstantial things we call ideas can trigger the loftiest idealism and the basest cruelty. And it demonstrated how under the urge to heroism and the commitment to the elevation of all mankind there often lies something truly grotesque—the impulse to destroy our fellow human beings.

How do mere fragments of thought turn to concepts that kill? Why do groups so readily congeal, face off, and fight? To answer these questions, we have to look at the forces that gave us birth.

BLOODSTAINS
IN PARADISE

MOTHER NATURE, THE
BLOODY BITCH

⇒⋙ ⋘⇐

We do not see, or we forget, that the birds which are idly singing
around us mostly live on insects or seeds, and are thus constantly de-
stroying life.

Charles Darwin, The Origin of the Species

Mankind has always been cutting one another's throats. . . . Do you
not believe . . . that hawks have always preyed upon pigeons? . . .
Then . . . if hawks have always had the same nature, what reason can
you give why mankind should change theirs?

Voltaire, Candide

In 1580, Michel de Montaigne, inspired by the discovery of New World
tribes untouched by Europe's latest complexities, initiated the idea of the
"noble savage." Nearly two hundred years later, Jean-Jacques Rousseau
popularized the concept when he published four works[16] proclaiming that
man is born an innocent wonder, filled with love and generosity, but that
a Luciferian force ensnares him: modern civilization. Rousseau claimed
that without civilization, humans would never know hatred, prejudice, or
cruelty.

Today, the Rousseauesque doctrine seems stronger than ever.
Twentieth-century writers and scientists like Ashley Montagu, Claude
Lévi-Strauss (who hailed Rousseau as the "father of anthropology"),
Erich Jantsch, David Barash, Richard Leakey, and Susan Sontag have re-
worked the notion to condemn current industrial civilization. They have

been joined by numerous feminist,[17] environmentalist, and minority rights extremists. Even such august scientific bodies as the American Anthropological Association, the American Psychological Association, and the Peace and War Section of the American Sociological Association have joined the cause, absolving "natural man" of malevolence by endorsing "The Seville Statement," an international manifesto which declares that "violence is neither in our evolutionary legacy nor in our genes."[18]

As a result, we are told almost daily that modern Western culture—with its consumerism, its capitalism, its violent television shows, its blood-soaked films, and its nature-mangling technologies—"programs" violence into the wide-eyed human mind. Our society is supposedly an incubator for everything that appalls us.

However, culture alone is not responsible for violence, cruelty, and war. Despite the Seville Statement's contentions, our biological legacy weaves evil into the substrate of even the most "unspoiled" society. What's more, organized battle is not restricted to humans. Ants make war and either massacre or enslave a rival swarm. Cichlid fish gang up and attack outsiders.[19] Myxobacteria form "wolf packs" that corner and dismember prey.[20] Groups of lizards pick on a formerly regal member of the clan who has become disfigured by the loss of his tail. Female bees chase an overage queen through the corridors of the hive and lunge, biting over and over until she is dead. And even rival "super coalitions" of a half-dozen male dolphins fight like street gangs, often inflicting serious injuries.[21] Ants do not watch television. Fish seldom go to the movies. Myxobacteria, lizards, dolphins, and bees have not been "programmed" by Western culture.

A host of writers gained attention in the late eighties and early nineties with books that celebrated a return to a mothering earth. They felt that if we scraped away large-scale agriculture, internal-combustion engines, televisions, and air conditioners, nature would return to bless us with her primordial paradise.

Unfortunately, these authors held a distorted view of pre-industrial reality. A pride of lions at their ease enjoys the kind of nature the radical environmentalists dreamed about. You can see the smiles on lions' faces as they lick their paws and stretch out on the ground side by side, clearly pleased with the comfort of each other's warmth. You can see the benevolence with which a mother keeps a cub from playfully tearing her tail apart. She lifts her huge paw and gently shoves the infant aside when his nipping becomes too painful. But nature has given these lion mothers only one way of feeding their children: the hunt. This afternoon, these peaceful creatures will tear a gazelle limb from limb. The panicked beast will try frantically to avoid the felines closing in on her, but they will break her neck and drag her across the plain still alive and kicking. Her eyes will be open and aware as her flesh is gashed and torn.

Suppose for a minute that lions were suddenly stricken with guilt about their feeding habits and swore off meat. What would they accomplish? They would starve themselves and their children. For they have been given only one option: to kill. Killing is an invention not of man but of nature.

Nature's amusements are cruel. A female sea turtle crawls painfully up the beach of a tropical island, dragging her bulk across the sand. Slowly she digs a nest with her hind flippers and lays her eggs. From those eggs come a thousand tiny, irresistible babies, digging out of the sand, blinking at the light for the first time, rapidly gaining their orientation from a genetically preprogrammed internal compass, then taking their first walk, a race toward the sea. As the infants scoot awkwardly across the beach, propelling themselves with flippers built for an entirely different task, sea birds who have been waiting for this feast swoop down to enjoy meal after high-protein meal. Of a thousand hatchlings, perhaps three will make it to the safety of the ocean waves.[22] The birds are not sadistic creatures whose instincts have been twisted by an overdose of television. They're merely engaged in the same effort as the baby turtles—the effort to survive.

Hegel, the nineteenth-century German philosopher, said that true tragedy occurs not when good battles evil, but when one good battles another. Nature has made that form of tragedy a basic law of her universe. She presents her children with a choice between death and death. She offers a carnivore the options of dying by starvation or killing for a meal.

Nature is like a sculptor continually improving upon her work, but to do it she chisels away at living flesh. What's worse, she has built her morally reprehensible modus operandi into our physiology. If you occasionally feel that you are of several minds on one subject, you are probably right. In reality, you have several brains. And those brains don't always agree. Dr. Paul D. MacLean was the researcher who first posited the concept of the "triune brain." According to MacLean, near the base

of a human skull is the stem of the brain, poking up from the spinal col-
umn like the unadorned end of a walking stick. Sitting atop that rudimentary stump is a mass of cerebral tissue bequeathed us by our earliest totally
land-dwelling ancestors, the reptiles.[23] When these beasts turned their
backs on the sea roughly three hundred million years ago and hobbled
inland, their primary focus was simple survival. The new landlubbers
needed to hunt, to find a mate, to carve out territory, and to fight in that
territory's defense. The neural machinery they evolved took care of these
elementary functions. MacLean calls it the "reptile brain." The reptile
brain still sits inside our skull like the pit at the center of a peach. It is a
vigorous participant in our mental affairs, pumping its primitive, instinc-
tual orders to us at all hours of the day and night.

Eons after the first reptiles ambled away from the beach, their great-
great-grandchildren many times removed evolved a few dramatic product
improvements. These upgrades included fur, warm blood, the ability to
nurture eggs inside their own bodies, and the portable supply of baby
food we know as milk. The remodeled creatures were no longer reptiles.
They had become mammals. Mammals' innovative features gave them the
ability to leave the lush tropics and make their way into the chilly north.
Their warm blood allowed them, in fact, to survive the rigors of an occasional ice age, but it exacted its costs. Warm blood demanded that mammal parents not simply lay an egg and wander off. It forced mammal
mothers to brood over their children for weeks, months, or even years.
And it required a tighter social organization to take care of these suckling
clusters of mammal mothers and children.

All this demanded that a few additions be built onto the old reptilian
brain. Nature complied by constructing an envelope of new neural tissue
that surrounded the reptile brain like a peach's juicy fruit enveloping the
pit. MacLean called the add-on the "mammalian brain." The mammalian
brain guided play, maternal behavior, and a host of other emotions. It
kept our furry ancestors knitted together in nurturing gangs.

Far down the winding path of time, a few of our hirsute progenitors
tried something new. They stood on their hind legs, looked around them,
and applied their minds and hands to the exploitation of the world. These
were the early hominids. But protohuman aspirations were impractical
without the construction of another brain accessory. Nature complied,
wrapping a thin layer of fresh neural substance around the two old cortical

standbys—the reptilian and mammalian brains. The new structure, stretched around the old ones like a peach's skin, was the neocortex—the primate brain. This primate brain, which includes the human brain, had awesome powers. It could envision the future. It could weigh a possible action and imagine the consequences. It could support the development of language, reason, and culture.[24] But the neocortex had a drawback: it was merely a thin veneer over the two ancient brains. And those were as active as ever, measuring every bit of input from the eyes and ears, and issuing fresh orders. The thinking human, no matter how exalted his sentiments, was still listening to the voices of a demanding reptile and a chattering ancient mammal. Both were speaking to him from the depths of his own skull.

Richard Leakey, the eminent paleoanthropologist, says war didn't exist until men invented agriculture and began to acquire possessions. In the back of Leakey's mind, one might find a wistful prayer that agriculture would go away so we could rediscover peace. But Leakey is wrong. Violence is not a product of the digging stick and hoe.[25]

In the Kalahari Desert of southern Africa live a people called the !Kung. The !Kung have no agriculture and very little technology. They live off the fruit and plants their women gather and the animals their men hunt. Their way of life is so simple that hordes of anthropologists have studied them, convinced that the !Kung live as our ancestors must have over ten thousand years ago, before the domestication of plants. In the early years of !Kung ethnography, anthropologists became wildly excited. These simple people had no violence, they said. Anthropology had discovered the key to human harmony—abolish the modern world, and return to hunting and gathering.

Richard Leakey used the !Kung as his model of paradisal pre-agriculturists. The !Kung way of life proved that without the plow, men would not have the sword. Yet later studies revealed a blunt and still-underpublicized fact. !Kung men solve the problem of adultery by murder. As a result, among the !Kung the homicide rate is higher than that in New York City.[26]

!Kung violence takes place primarily between individuals. In both humans and animals, however, the greatest violence occurs not between individuals but between groups. It is most appalling in war.

Dian Fossey, who devoted nineteen years to living among and observing the mountain gorillas of central Africa's Virunga Mountains, felt these creatures were among the most peaceful on earth.[27] Yet mountain gorillas become killers when their social groups come face-to-face. Clashes between social units, said Fossey, account for 62 percent of the wounds on gorillas. Seventy-four percent of the males Fossey observed carried the scars of battle, and 80 percent had canine teeth they'd lost or broken while trying to bite the opposition. Fossey actually recovered skulls with canine cusps still embedded in their crests.[28]

One gorilla group will deliberately seek out another and provoke a conflict. The resulting battles between gorilla tribes are furious. One of the bands that Fossey followed was led by a powerful silverback, an enormous male who left a skirmish with his flesh so badly ripped that the head of an arm bone and numerous ligaments stuck out through the broken skin. The old ruling male, whom Fossey called Beethoven, had been supported in the fight by his son, Icarus. Icarus left the battle scene with eight massive wounds where the enemy had bitten him on the head and arms. The site where the conflict had raged was covered with blood, tufts of fur, broken saplings, and diarrhetic dung.[29] Such is the price of prehuman war in the Virunga Mountains.

Gorillas are not the only subhumans to cluster in groups that set off to search for blood. By the early seventies, Jane Goodall had lived fourteen years among the wild chimpanzees of Tanzania's Gombe Reserve. She loved the chimps for their gentle ways, so different from the violence back home among humans. Yes, there were simian muggings, beatings, and rage, but the ultimate horror—war—was absent.

Goodall published a landmark book on chimpanzee behavior—*In the Shadow of Man*—a work that to some proved unequivocally that war was a human creation. After all, the creatures shown by genetic and immunological research to be our nearest cousins in the animal kingdom knew nothing of organized, wholesale violence.[30]

Then, three years after Goodall's book was printed, a series of incidents occurred that horrified her. The tribe of chimps Goodall had been watching became quite large. Food was harder to find. Quarrels broke out. To relieve the pressure, the unit finally split into two separate tribes. One band stayed in the old home territory. The other left to carve out a new life in the forest to the south.

At first, the two groups lived in relative peace. Then the males from the larger band began to make trips south to the patch of land occupied by the splinter unit. The marauders' purpose was simple: to harass and ultimately kill the separatists. They beat their former friends mercilessly, breaking bones, opening massive wounds, and leaving the resultant cripples to die a slow and lingering death. When the raids were over, five males and one elderly female had been murdered. The separatist group had been destroyed; and its sexually active females and part of its territory had been annexed by the males of the band from the home turf.[31] Goodall had discovered war among the chimpanzees, a discovery she had hoped she would never make.[32]

Years later, biological ecologist Michael Ghiglieri traveled to Uganda to see just how widespread chimpanzee warfare really is. He concluded that "the happy-go-lucky chimpanzee has turned out to be the most lethal ape—an organized, cooperative warrior."[33]

So the tendency toward slaughter that manifested itself in the Chinese Cultural Revolution is not the product of agriculture, technology, television, or materialism. It is not an invention of either Western or Eastern civilization. It is not a uniquely human proclivity at all. It comes from something both sub- and superhuman, something we share with apes, fish, and ants—a brutality that speaks to us through the animals in our brain. If man has contributed anything of his own to the equation, it is this: He has learned to dream of peace. But to achieve that dream, he will have to overcome what nature has built into him.

WOMEN—NOT THE PEACEFUL CREATURES YOU THINK

The rivalry of women is visited upon their children to their third and fourth generation.

Gelett Burgess

Males play the greatest role in stirring up bloodbaths. They do most of the killing, and they also do most of the dying. This makes men sound pretty atrocious. And indeed they are. Males by far outdo females in aggression. Remove the testicles from a rooster, and it becomes a peace-loving bird. Sew the testes back into its stomach, and the masculine hormones once again flood the fowl's bloodstream. Now the recently mild-mannered chicken struts off to start a fight.

It's not surprising when pundits declare that if only we had female leaders, war and international aggression would rapidly disappear.[34] Many people are convinced that females are inherently peaceful. Okay, so Margaret Thatcher, the female former prime minister of Britain, won the Falklands War, supplied the British military with nuclear submarines, and packed those subs with atomically tipped ballistic missiles. Indira Gandhi led a military campaign against Pakistan, jailed her opponents, and suspended civil liberties. And Peru's Shining Path guerrilla assassination squads were headed almost entirely by women.[35] But surely these are just aberrations. Or are they? The evidence from the world of our closest relatives in the primate family indicates that the cheerfully idealistic picture of women is a self-delusion. Females, too, are victims of the Lucifer Principle.

Dian Fossey, the chronicler of the central African mountain gorillas, had been following a gorilla band for nine years when she suddenly noticed that one of the tribe's babies had disappeared. This came as a shock. The baby hadn't been sick. Fossey couldn't understand what had happened to it. The naturalist and her aides hunted through the forest, looking for the remains of the body, expecting to find it in one of the spots where the gorilla group had fought with a rival band. But Fossey found no corpse whatsoever at the battle sites.

Finally, acting on a hunch, Fossey and her African helpers collected all of the dung the gorillas had left during the previous days. After their years following this group, the researchers could identify which dung came from which gorilla. For days, the humans painstakingly sifted through the excrement. Finally Fossey found what she'd been looking for: 133 bone and tooth fragments from an infant gorilla—contained in the deposits left by the dominant female and her eight-year-old daughter.[36]

The mother of the dead baby came from a social level these female gorilla aristocrats despised. She was an outcast the well-placed ladies frequently mocked and bullied. Her presence simply could not be tolerated in proper company. Her child was beneath contempt. Fossey concluded that the head female and her daughter had attacked the infant, killed it, and eaten it.[37]

There was more than mere cruelty behind this murder of a helpless baby. Effie, the aristocratic female who had apparently led the infant-killing effort, was in the last stages of pregnancy. Three days after the brutal incident, she gave birth to a baby of her own. Effie had acted like the ambitious wife in a harem who fights to eliminate the children of her rivals. Through infanticide, she had become the only female with four children in the group at one time. She had ensured that she and her young would be the tribe's ruling class. By doing so, she had turned the entire group into a support mechanism for her own offspring.

Effie was very much like Livia, the most powerful woman of Rome in the days a little less than two thousand years ago when that city was reaching the peak of imperial power. According to Robert Graves's careful reconstruction in *I, Claudius*, Livia—like Effie—was one of a number of wives.[38] And like Effie, Livia was mated to the dominant male in the pack. Specifically, Livia had managed to inveigle into matrimony a man named Augustus Caesar, the leader who had grabbed the reins of Rome from his rivals and stabilized the Empire in a time of turbulence. In the process, Augustus had become the most powerful man the world had ever known.

Gorillas manage to keep a whole gaggle of wives trailing behind them simultaneously. Augustus didn't have that privilege. The law forced him to possess his official mates one at a time. Livia was Augustus's third wife. She had won him when she was a tender seventeen. Well, maybe not so tender. According to Graves, the teenage beauty had grown contemptuous of a previous husband because the unlucky gentleman believed in such principles as liberty for Rome's citizens. Livia had no patience with these notions. She was convinced that all power should be centered in the hands of one man—preferably a man under her direct control. So she divorced her soft idealist and sought out a harder husband whose possibilities were more in line with her own aspirations.

At the time, Augustus was married to someone else. In fact, he'd had a number of children by the woman at his side and seemed reasonably pleased with his present wife's ways. But that didn't stop the ambitious young Livia. She managed to tarnish the reputation of his spouse and to drive additional wedges between the unfortunate lady and her husband. Then she inserted herself into the gap, making her presence the only logical consolation for Augustus's distress over his wife's disgrace.

Livia quickly consolidated her hold over Augustus. Soon, he would not make a major decision without her. Like the gorilla Effie, Livia had struggled to become the first lady in the band. And like Effie, Livia was ambitious for more than just herself. She was ambitious for her children. Rome had been run in the past by a democratic senate, but Augustus was shifting power to a one-man emperorship. Livia wanted the newly established imperial throne to go to her own children.

It wasn't that easy. There were rival claimants to the seat of imperial authority. First in line were two of Augustus's old friends and confidants. But, more important, there were Augustus's three grandchildren, born to the daughter of his previous wife. One by one, according to Graves, the rivals died off. Some collapsed mysteriously, others came down with lingering diseases, and still others suffered inconsequential wounds but were given the wrong medical treatment. Surely, Livia's expertise in poisons and her network of murderous helpers—much like the cooperatively cannibalistic cronies in the gorilla Effie's clique—had nothing to do with these deaths.

Finally, only Livia's children were left, as Graves puts it, "to carry on the line . . . Livia's line."[39] Livia, like Effie, had eliminated her children's rivals and guaranteed her offspring a spot on the top of the heap.

One empress in China roughly seventeen hundred years ago took Livia's ambition several giant steps further. To ensure her children power over the empire, she eliminated every single individual in a rival extended family. In all probability, this minor act of manslaughter was not limited to a mere handful of human obstacles. Chinese noble families of the period usually had hundreds or even thousands of members.[40]

Livia, Effie, and the Chinese empress were as bloodthirsty as any male. And the motivation that drove them was distinctly maternal—the desire to give every advantage to their young.

Women are violent. In fact, females are as much a part of the apparatus that triggers male violence as the men themselves. Nobel Prize–winning ethologist Konrad Lorenz described a common behavior in several species of ducks. The female runs out to the edge of her husband's territory and tries to provoke another duck, then runs back to her male, stands next to him, and looks behind her at the enraged rival in the hope that her mate will jump into the fray.[41] Many are the human females who have tried to stir up a similar fight.[42]

Women encourage killers. They do it by falling in love with warriors and heroes. Men know it and respond with enthusiasm. The Crusaders marched off to war with ladies' favors in their helmets. They were not setting out on some mission of gallant gentleness. On their way through Asia Minor, the Crusaders literally roasted Christian babies in cases of mistaken identity. Because the local folk did not speak a language they understood, the chivalrous knights assumed the panicky babblers were heathens. Heathens, of course, deserved no mercy. So the heroes sliced up the adults and baked the infants on spits, all the while thinking of how the damsels back home would admire their bravery.[43]

Technically, this is called sexual selection. The females of a species develop a craving for a certain kind of guy, and all the males compete to live up to the female ideal. Lady peacocks adore hunks with towering blue tails, so peacock gentlemen sport foppish plumes.[44] Lady bowerbirds swoon over bachelors with an architectural flare, so bowerbird males turn sticks and scraps into a Taj Mahal. And what have human females gone for in nearly every society and time? "Courage" and "bravery." In short, violence.

The poetry of the classic sixth-century Arab master Labede is a testament to the feminine ability to bring out the animal in a man. In Labede's lyrical verses, a young fellow jostles along on his camel, dreaming of how he can get his beloved's attention. She, it seems, does not recognize his true worth. He dreams of how he will prove his manliness with feats of daring glory. Now, what is the one feat of daring glory guaranteed to rivet the admiration of a beauty in Labede's tribal desert society? You rush into the nearest village, kill a few of the males, and steal as many camels and old clothes as you can get your hands on. Greatness belongs to the killer. And the young ladies swoon over guys who are great. Labede will tell you, it works every time.[45]

Even T. S. Eliot's erudite "Love Song of J. Alfred Prufrock" is the intellectualized cry of a man who feels women will not look on him with admiration and bear his children unless he wins their attention with a few violent deeds. "I have heard the mermaids singing each to each," mopes the protagonist. "I do not think that they will sing to me." What would get these lovely girls in the sea to give the poet a second look? Well, he could be a bit more like Prince Hamlet—able to finally make a decision and kill. But the poet hesitates. He is not the kind of person to take decisive measures. He imagines himself growing old, a foolish, lonely man, ignored by women all his life. Finally, he consoles himself. "There will be time," he says, "to murder and create."

But females do more than provoke violence among males. They engage in it themselves. Primatologist Jeanne Altman, studying the female baboons of Kenya's Amboseli National Park, noted that when a new baby baboon arrived, the females all rushed over to see it. As it grew older, the baboon ladies came back time and time again. At first glance, their interest looked touchingly affectionate, but, on closer inspection, it was something quite different.

During a typical incident, a mother and baby sat in the savannah grass. A high-ranking female walked haughtily over to the pair. She tugged gently at the baby's arm. When the mother would not give up her child, the socially superior female grew impatient. She tugged the arm more violently. Then she yanked at the baby's leg. The mother reared back, bared her teeth, and made a warning sound. She knew what this meddler was really up to. Given half the chance, the lady of lofty social standing would grab the infant, treat the squealing child like a rag doll, drag it around, pass it back and forth to her friends, and in the end injure it so badly that her "interest" might very well prove fatal.

The chattering anger of the mother did its work. The female from a higher social sphere went back to her clique. The mother was a member of the underclass, looked down on by the haughty and none-too-kind members of the female in-group. The worried nurturer spent the rest of her day clinging tightly to her infant. She could not gather as much food as she needed for herself and her baby, for she was too preoccupied with protecting her offspring from an unforeseen attack. Her child wriggled impatiently in her arms. Further research suggests that the child wanted to go off on its own and have a good time. But this was one baby that would never have the freedom to run and play. It would never be able to wrestle and roll about with the children of the more dominant females. It would never experience that easy social access that leads to self-confidence and a quick mind among baboons. Ultimately, this toddler—like its mother before it—would live its adult years at the bottom of the social pile. The baboon's mother was forced to smother it overprotectively simply to ensure that it survived. For among baboons, the babies of the dispossessed have an omnipresent mortal enemy: the *females* of the tribe.[46]

It is useless for women to blame violence on men, and it would be futile for men to blame violence on women. Violence is built into both of us. When Margaret Thatcher constructed a nuclear navy, she was not behaving in a manner distinctly male, nor was she behaving in a manner distinctly female. She wasn't even obeying a set of impulses that are uniquely human. Thatcher, like Rome's Livia, was in the grip of passions we share with gorillas and baboons, passions implanted in the primordial layers of the triune brain.

FIGHTING FOR THE PRIVILEGE
TO PROCREATE

Why all this savagery? A lot of it springs from a simple biological commandment: be fruitful and multiply. The gorilla Effie led her friends in
the murder of a baby to gain an advantage for her offspring. Livia, mistress
of mighty Rome, did the same to benefit her sons and their sons after
them. Where violence takes place, children crop up over and over again.
Stags fight for the right to have them. Humans declare wars to make the
world safe for them. Strange as it sounds, children—and the genes they
carry—are one key to the mystery of violence.

   An adult male langur who's become the head of the establishment
ensconces himself regally at the center of his group. He has every reason
for sitting pretty. If you take a closer look at the cluster of langurs milling
around him, you'll discover that all of them are either his wives or children. The females do his bidding and offer their bodies only to him. If
they attempt a romance with some dashing bachelor, they are severely
punished; so is the hopeful seducer. No wonder the central male looks so
lordly. He is surrounded by a tribe devoted to one primary purpose—
having and raising his kids.

   As we saw several chapters ago, not every member of the langur society is happy about this state of affairs. In the jungle nearby roams a gang
of postpubertal hooligans who have left home permanently to hang with
toughs their own age. Their newly spurting sexual hormones have triggered the growth of horniness, muscle, and a cocky aggression. Periodically, the gang of youthful thugs advances on the territory where the
well-established elder sits in the midst of his large family. The hoodlums

try to get his attention. They mock and challenge the patriarch. He sometimes sits aloof, refusing to dignify their taunts with a response. On other occasions, he ambles over to the periphery of the harem, then rears up and puts on a display of outrage that chases the young Turks away. But from time to time, the massed delinquents continue their challenge, starting a fight that can be brutal indeed. If they are lucky, the upstarts trounce their dignified superior thoroughly, chasing him from his comfortable home.

Then the newly triumphant members of the younger generation execute an atrocity. They wade into the screaming females, grabbing babies left and right. They swing the infants against the trees, smash them against the ground, bite their heads, and crush their skulls. They kill and kill. When the orgy of bloodlust is over, not an infant remains. Yet the females in their sexual prime are completely unhurt.[47]

The mass murder is anything but random. Like Effie's infanticide, it has a simple goal. This cluster of wives was raising the children of the old man who just fled. As long as the ladies continued to suckle infants, they would be tied to the children of the toppled authority figure. A natural birth-control device called lactational amenorrhea would keep them uninterested in sex, preventing them from entering estrus[48] and blocking the females from carrying the seed of the new conquerors.

When a mother's baby is killed and her suckling stops, however, the whole game changes. Her biochemistry shifts, resurrecting her sexual interest. She becomes an empty womb waiting to have another child. And, this time, the child will not belong to the deposed monarch—it will carry the legacy of one of the invaders.

But surely humans don't indulge in such barbarities. Or do they? In the rain forests near the Amazon live a people called the Yanomamo. Their ethnographer, Napoleon Chagnon, calls them "the fierce people." They pride themselves on their cruelty, glorying in it so enthusiastically that they make a great show of beating their wives. And the wives are as much a part of this viciousness as the husbands. A spouse who does not carry enough scars from her husband's blows feels rejected and complains miserably about her unbruised condition. It is a sign, she is certain, that her husband does not love her.[49]

Yanomamo men have two great preoccupations—hunting and war. The patterns of their warfare bear a strange resemblance to those of the langur. Yanomamo men sneak up on a neighboring village and attack. If they are successful, they kill or chase away the men. They leave the sexually capable young women unharmed, but they move methodically through the lean-to-like homes, grabbing babies from the screaming captives. Like the langurs, the Yanomamo men beat these infants against the ground, bash their brains out on the rocks, and make the footpaths wet with babies' blood. They spear the older children with the sharp ends of their bows, pinning their bodies to the ground. Others they simply throw from the edge of a cliff. To the Yanomamo, this an exhilarating entertainment. They brag and boast as they smash newborns against the stones. When the winning warriors have finished, not a single suckling child remains. Then the triumphant Yanomamo men lead the captured women back to a new life as secondary wives.[50] No wonder the Yanomamo word for marriage means "dragging something away."[51]

What have the Yanomamo victors accomplished? The same thing the langurs did. They have freed the females from the biochemical birth-control device that keeps a suckling woman from bearing new progeny. The Yanomamo fighters have made the wombs of their captured consorts available to carry *their* children.

The Yanomamo are not some strange aberration out of the jungle to illustrate a farfetched point. In the early fourth century, Eusebius, the first historian of the Christian church, summarized what the study of history had focused on until his time: war, slaughters for the sake of country, and *children.*[52] Hugo Grotius in 1625 published *De Jure Bellis ac Pacis,* or *Concerning the Law of War and Peace,* a book that tried to make Christian war more humane. In it, Grotius justified killing children. He cited Psalm 137, which says, "Happy shall he be who takes and dashes your little ones against the rock." Thus, Grotius was well aware of two things: that killing enemy children was common in the days of the Old Testament; and that it remained as common as ever in seventeenth-century Europe.

In fact, the restless effort of human males to find more wombs to carry their seed has been dignified by the forefathers of Western civilization. The rape of the Sabine women, a bit of Roman history anyone with a modest classical education can recount, was a stunt similar to those frequently pulled off by the Yanomamo.[53] The heroes of the tale, a gang of early Romans, invited the neighboring tribesmen and their wives over for dinner and entertainment. The entertainment turned out to be a display of Roman weapons. The hosts pulled their swords, grabbed the girls, then attacked and chased away the husbands. There was a high time among the Roman founding fathers as they indulged merrily in sex with the weeping captives. And nine months later, there was more weeping as the kidnapped ladies gave birth to a fresh crop of infants—the babies of the banquet hosts.[54]

The Trojan War also ended with a scene that any Yanomamo warrior would have understood. It started as a battle over one woman, a lovely creature who behaved very much like Konrad Lorenz's female duck: the aquatic female who provoked a fight, then ran back to her mate and tried to get him to join in. The instigator, in the case of the human conflict, was Helen of Troy. When the fighting was over, the winning Greeks were rewarded with a Yanomamo-esque bonanza—captured plunder and conquered Trojan females. The warriors took the women home and ravished them, but they didn't bother to carry Trojan infants on the trip back across the Aegean Sea. (As Troy was going down in defeat, Andromache, one of the Trojan wives, told her baby that the odds were good that "some Achaian will take you by the hand and hurl you from the tower into horrible death.")[55] Less than a year later, the fresh crop of babies from the Trojan captives fattened the Greek bloodline.[56]

The Yanomamo, the langurs, the Romans, and the Greeks were all driven by the same force. They were hungry for sex, and that hunger translated into something else—the desire to populate the world with their own offspring. But the men were not alone; Effie the cannibalistic gorilla and Livia the Roman schemer were out for the same thing. Under the impulse toward violence often lies the simple urge to have kids.[57] Which leads us to one of the fundamental forces behind the Lucifer Principle: the greed of genes.

THE GREED OF GENES

꩜

According to current cosmological theory, the universe was born in an explosion that goes by the quaint name of the big bang. In its first second of existence, the newborn cosmos began a habit it has never overcome: it started evolving higher forms. What began as powerful, inchoate energies soon coalesced into elementary particles. Those particles were attracted to each other and banded together in tight microsystems called atoms. From nothingness and energy, matter in its simplest form had been born. Obeying the rules of a magnetic square dance, some atoms linked arms and do-si-doed into the void as molecules. The universe had taken another quantum leap up the stairway of complexity.

Molecules spinning through the emptiness were drawn together by gravity into suns and planets.[58] *Voilà*—the universe lunged once more up the ladder of intricacy. In its beautifully mindless way, nature was disgorging whole fresh batches of inventions.

Then on the face of at least one of the new planets, an assembly mechanism that used something even more wondrous than the power of gravitational or electromagnetic attraction arose for the first time.

꩜

In the beginning, says Oxford University zoologist Richard Dawkins in *The Selfish Gene,* the face of the earth was washed by primitive seas.[59] On the surface of those waters, lightning and sunlight knit together molecules of ammonia, water, carbon dioxide, and methane to form the first organic substances. These substances sloshed inertly beneath the waves, a slowly accumulating, murky sludge. One day a miracle occurred. Some accident twisted a few of the organic clumps of atoms together into a new shape, giving them a property the universe had never seen. The molecular pretzel could make copies of itself. It mindlessly attracted scraps of muck to its surface and—quite accidentally—snapped the molecules it was embracing together like pop-em beads. When the pretzel let the finished product go again, it had unwittingly made a mirror image of itself.

The replica had the same property as its pretzel-like parent. Molecules of sludge were attracted to its surface. Each segment of surface would pull toward it a very specific atomic shape, so the replica's exterior acted like a paint-by-numbers canvas, drawing precisely the correct component to exactly the right spot. Once all the new molecules were lined up in order, they'd snap together. The result was yet another spanking new copy, ready to unpeel from its parent and drift away. The fresh-born copy, in its turn, would attract other wandering molecules to its face, where they would line up, pop together, then uncouple to be carried off by the currents of the sludge-filled early seas. The molecules with the peculiar ability to make copies of themselves are called replicators.[60] These replicators, like the innovations that had preceded them, would move the universe one more step up the ladder of complexity.

For eons, replicators drifted through the chemical soup of the early earth, casually copying themselves. But eventually, the population of molecular Xerox machines grew overwhelming, and the supplies of untouched organic sludge began to run short. That's when the replicator that could do more than merely reproduce itself had an edge. The replicators that could do more, says Dawkins, were those that "learned" to make copies from more than just raw sludge. They could take apart their competitors and reassemble the components for their own purposes. Other replicators arose that could defend themselves. The first defense was probably a simple chemical armored shell, like those that protect some bacteria. But over time, the armored suits became more intricate, developing muscular whips to provide speed, movable fins for steering, and, far into the future, hands, feet, and brains. The descendants of the early replicators are the genes of today. And the latest versions of those first primitive protective suits are you and me.

There's another aspect to Dawkins replicators that helps explain some of nature's more reprehensible habits. Imagine a day in the future when some clever engineer invents an entirely new industrial process, a manufacturing technique that makes factories and workers obsolete. Under the new system, management committees that sit around anxiously pondering the next profitable move are as useless as last week's donuts. The enormous stamping machines, pressing devices, and even welding robots are unnecessary artifacts to be tossed into museums and gawked at from time to time. What has replaced them?

An ultraminiaturized factory complete with a built-in blueprint for its finished product. The device is so small that you can fit millions of them on a flyspeck and so inexpensive that a penny will buy you more than you can count. These little wonders have another advantage. You can scatter them at random. They take care of the rest. There's no more need to spend billions digging metals out of the earth or cracking chemicals from oil and turning them into plastics. The automated minifactories find what they need without help, sensing the presence of unprocessed industrial materials in a pile of garbage, a whiff of air, or a lump of dirt. If they run across the necessary substances, they immediately go to work assembling the finished goods. If they don't discover what they need, the mechanisms simply fail to activate. A deactivated minifactory is no great loss. After all, millions of the microconstruction units can be turned out for the price of a stick of gum.

When the new system becomes popular, however, it turns out to have a glitch. The scheme is *too* successful. Each product is cranked out in a world overrun with other gizmos stamped out by the same system. What's more, each product of the minifactories is programmed to go out and gather the raw materials to make more copies of itself. Suddenly the fields, streets, kitchens, and garbage heaps are awash in bright new contraptions stumbling over each other in an effort to grab the stuff more contraptions are made of, and there simply isn't enough raw material to go around.

It doesn't take long before some bright designer endows his gadgets with a clever twist. The improved models speed up the raw-material gathering process by pouncing on finished products from rival microplants, stripping them to pieces, and using the parts as prepackaged parcels of raw industrial stuff. The insidious new models spread fast, stalking unwary gizmos left and right, strewing the planet with discarded components that others snatch in seconds. Then the updated models increase their efficiency even further by working in packs. Sometimes millions of complex final products are lost at a time. But in the grand scheme of the new economy, the loss is not that great. Gadgets abducted for scrap can be replaced at a price that even the Koreans would find hard to believe. This autoconstruction technique, based on Dawkins's replicators, is analogous to the genetic system. The genetic method's sheer efficiency is one of the primary reasons Mother Nature isn't nice. She doesn't have to be.

For over three billion years, raw materials have grown so scarce and finished goods so numerous that gizmos have scrambled through a mad dash to grab and disassemble each other; but that doesn't disturb Mother Nature at all. In fact, she's discovered that research and development is *aided* by the finished products' competition. Just toss them out there and watch them try to outwit each other. Keep the clever and flip the unsuccessful into the circular file. From a million failures will spring the breakthroughs of tomorrow.

The generative power of the genetic process helps explain why we are so appallingly expendable in the eyes of an indifferent cosmos. Our prehistoric cousin the Neanderthal was a clever contraption. Numerous anthropologists believe the Neanderthal was capable of philosophy, religion, and language. Unfortunately, once *Homo sapiens* evolved, Neanderthal became an obsolete scrap on the garbage pile of history.

In human terms, obsolescence means suffering and mass death. Each individual product of the minifactory system is built with a vast array of sensory devices designed to protect it from damage. It's regrettable that those sensory mechanisms happen to be quick and efficient, because they produce what we humans call pain. But, then, no product is perfect. The Neanderthal woman who had lost her children and mate was probably plunged into emotional agony.[61] She may well have lost her family to rampaging *Homo sapiens* tribes, glorying in their victories, gloating over their superiority, and reveling in their conquests.[62]

The ability to give speeches on glory, conquest, superiority, and nobility may well have been one of the features built into the latest model, our "heroic" ancestor *Homo sapiens*. The new *Homo sapiens* concepts of nobility and heroism would have been great leaps forward. Why? They suited the hunger of genes.

Dawkins's *The Selfish Gene* rearranged the way many of those who deal with animal and human behavior see the world around them. Dawkins said that we tend to see ourselves as masters of our genetic endowment, but in reality we are merely servants. We are not using genes to achieve our own ends. Our genes are using us. (The idea had been anticipated by the seventeenth-century English poet and satirist Samuel Butler, who quipped, "A hen is just an egg's way of making more eggs.")

If Dawkins is right, humans and their social groups originated as mere puppets, complex tools of tiny molecules. You and I were fashioned as if we were cranes, dump trucks, and tanks, designed to be driven by a set of replicators. We are gatherers of raw materials, operating at the behest of microscopic minifactories seated at the center of our cells. For genes are infected with an overweening ambition: their ultimate goal is to reproduce, and in the process, to overrun this world.

Despite the opinions of Montaigne, Rousseau, and their contemporary followers, modern civilization is not the generator of violence. Nor is brutality limited to the "patriarchal" male. The creator of human savagery is Nature, who works her ways through brain segments bequeathed to both men and women by our animal ancestors. Ironically, it is female aggression that gives the greatest clue as to why nature has found conflict so indispensable. Creatures of every species fight for the privilege of procreation. They battle to immortalize the replicators at their core.

No wonder the wives of ancient emperors and high-ranking ladies of the gorilla clan have been out to corner the world for their children. No wonder Greek heroes, Yanomamo warriors, and rampaging Romans have risked their lives in the hunt for new wombs to sow. Every time a sperm and egg spew a fresh creature into the world, the victor is a gene.

WHY HUMANS
SELF-DESTRUCT

THE THEORY OF INDIVIDUAL
SELECTION AND ITS FLAWS

≫€

Richard Dawkins's theory is a powerful tool for cracking the mysteries of the cosmos, but it has a limitation. In reality, genes were never the loners that Dawkins makes them out to be. Though he labels them "selfish," even he is forced to admit that genes were compelled to coagulate in teams, just as their minions—from termites to humans—would later be.

Current evolutionary theory, known as "neo-Darwinism," says that preservation of his genes is the first priority of the individual: preservation for himself, his children, and for his remaining relatives. And, as the examples in previous chapters show, when it comes to children, at least, that view has much evidence to support it. Yet it is missing something vital in the human experience. When Rudolph Valentino died, numerous women committed suicide.[63] Survival for themselves and their immediate families was the last thing on their minds.

Underlying the notion of genetic selfishness is another, even more basic assumption: the theory of individual selection. According to the theory of individual selection, when it comes to picking and pruning, evolution sorts creatures one at a time. Hence, the most potent impulse in the makeup of every micro- and macrobeast is the drive for personal survival.

But, somewhere deep inside, each of us knows that individual survival is not his only raison d'être. So thoroughly is that fact built into us that we find it in our physical structure. We come complete at birth with an arsenal of survival weapons, but we're also equipped with devices that can *negate* our existence. These are our self-destruct mechanisms.

In 1945, the Japanese had been fighting American soldiers for three and a half years. They had known they could not lose. Their gods had made them a superior people.[64] They had swept through China and the Pacific Islands in the thirties like an avenging wind, taking vast territories, conquering hordes of "inferior" peoples, showing the heaven-given supremacy of their race. The enemy who faced them was a contemptible lot, unblessed by the divinity that buoyed Japan, and crippled by racial impurity.

Yet the mongrels from the West were accomplishing the unthinkable. They were beating down the warriors of Japan. By the time the Americans reached Okinawa, the Japanese could see that heaven had deserted them. The shame was unendurable. Four thousand Japanese killed themselves in Okinawa's underground naval headquarters. Another thirty thousand military men and civilians threw themselves from a nearby cliff.[65]

On the Japanese mainland, pilots volunteered to keep the American marines in Okinawa from getting supplies. Those flyers were promised honor and death. Their mission was to guide their planes to the enemy and stay at the controls as the explosive-laden aircraft slammed into the enemy's ships. "I will be doing my duty by dying," they wrote in final letters to their families. Fifteen thousand of them fulfilled that fatal obligation. One commentator, describing the kamikaze experience forty years later, explained that "Japan is a society of groups, not individuals."

To us the kamikazes' ultimate devotion seems baffling, alien, something that could never happen here; but it *has* happened here. Patrick Henry was declaring his loyalty to his fellow revolutionaries and their cause when he said, "Give me liberty or give me death." He was confessing that the social organism of which he was a part was more important than his own existence.

Suicides in 1929, the year of the Great Crash, tended to be flamboyant and highly publicized. There was the head of the Rochester Gas and Electric Company who asphyxiated himself with his firm's chief product; the two stock speculators who flung themselves from a New York City roof hand in hand; and the investor who poured flammable liquid on himself and lit a match.[66] But these were exceptions, not the rule. However, from 1930–33, once the Depression hit its stride, the number of suicides skyrocketed. In 1932 alone, it tripled.[67] The men and women who killed themselves contributed very little to their own survival or to that of their closest relatives.

Back in 1897, the seminal French sociologist Emile Durkheim compiled a set of statistics that demonstrated the rise of self-inflicted deaths after the market crashes of 1873 and 1882, and coined the term "altruistic suicide." Durkheim seemed to sense that beneath the surface, the suicide was destroying himself to rid the wider social group of a burden.[68] Sociologist and ethnologist Marcel Mauss, a relative and follower of Durkheim, was even more specific. He noted an occasional "violent negation of the instinct for self-preservation by the social instinct."[69]

If our actions are geared to increasing the odds that our personal genes or those of our near relatives will make it into the next generation, what is the reason for suicide's existence? And what about the other bits of death-in-life built into the human psyche? Why do humans get depressed? Why do they sometimes feel like crawling off into a corner and dying? There is an answer, but it doesn't quite square with the notion of genes fighting for themselves no matter what. We are parts of a larger organism and occasionally find ourselves expendable in its interests.

This idea is not very fashionable at the moment. Evolutionists, myself included, believe that competition is vital to the creation of new species. The beast with the bigger brain, the sharper claw, or the cleverer way of building a nest outdoes his or her clumsier rival and has more children. His offspring inherit his advantageous cranial capacity, natural weaponry, or architectural skill and in turn have plentiful broods of their own. Within a few hundred thousand generations, the creatures with the anatomical or mental advantages have outbred their dull-witted or blunt-pawed rivals. Less favored creatures may easily find themselves extinct. According to the current evolutionary party line, this competition takes place between individuals. The idea that it could occur between *groups* has been resoundingly dismissed because of a chain of arbitrary twists in the history of evolutionary theory.

The concept of evolving life dawned long before the publication of Charles Darwin's theories. In roughly 580 B.C., the Greek philosopher Thales of Miletus declared that life had not been created by the gods but had emerged by natural means from water.[70] Twenty-three hundred years later, Enlightenment thinkers like France's Georges-Louis Buffon reinterpreted petrified oddities formerly dismissed as stone tongues and dragon's teeth. The objects, the audacious naturalists said, were parts of fossilized creatures from a previous era. Using the latest theories of geology, Buffon and his fellow iconoclasts demonstrated that placement of fossils in the rocky strata suggested primitive creatures had occupied the earth far before the supposed biblical date of Creation and had progressed to increasing levels of complexity as they'd moved from their birthplace in the seas to footholds on dry land.

Meanwhile, Pierre Louis Moreau de Maupertuis, another scholar who preceded Darwin by a hundred years, worked out a remarkably prescient theory explaining how advances from one species to another might occur. Even Darwin's grandfather Erasmus Darwin anticipated his younger relative by more than half a century, putting forth an evolutionary overview in his *Zoonomia* (1796). But it was Darwin's meticulous fact gathering,[71] his family connections, and his methodical campaign to win over the scientific community that finally reoriented the thinking of specialists and laymen alike. (Darwin kept a checklist of influential thinkers, then used his social ties to bring them on board one by one.) As a result, Darwin's 1859 *Origin of the Species* created a splash so great that its propositions were even the subject of newspaper cartoons.

Twenty-seven years earlier, Darwin's evolutionary thinking had been thrown into high gear when a book called *An Essay on the Principle of Population* brought the young naturalist's attention to the hyperactive output of the replicatory system. The essay was the work of Thomas Robert Malthus, a pessimistic English clergyman who had proposed in 1798 that food supplies increase at a sluggish arithmetic rate, while population explodes in a geometric progression, making mass death through starvation inevitable. Population excess of this magnitude, Darwin concluded, would create competition for survival. And the creatures best suited to get the most out of a hostile environment would be the contestants who survived.

Hence Nature would prune her flock like the breeders of sheep near the Kentish country home where Darwin did most of his writing. These careful squires selected for reproduction only the animals that were the hardiest and produced the most wool. A culling of this sort performed by nature, if continued over eons, would produce radical changes in a species. Because of the similarities between the methods of gentleman farmers and the less tender mechanisms of competition in the wild, Darwin dubbed the results of the battle for survival ''natural selection.''

Darwin saw competition taking place at several levels, including that which occurs between individuals and that which occurs between groups.

When discussing ants, he acknowledged that evolution could easily induce individuals to sacrifice their self-interest to that of the larger social unit.[72] In his later writings, he proposed that a similar process occurs among human beings.[73]

In the 1930s, a new school of "population geneticists" led by men like J. B. S. Haldane and Sewall Wright cranked out theories whose arcane mathematical formulas gave evolutionists the sense that they were making the climb from Darwin's mere observation and speculation to the higher scientific ground normally occupied primarily by those most envied practitioners of the discipline, physicists. The popularity of Haldane and Wright's algebraic hypotheses grew despite a substantial flaw: they were not strongly supported by empirical evidence. Equally disturbing, each of the formulations seemed based in large part on the premise that the individual is the basic unit of evolutionary change. Competition between groups had been hustled off the stage.

Then, in 1962, the Scottish ecologist V. C. Wynne-Edwards, a careful observer of his country's native red grouse, concluded that these birds sometimes sacrificed their reproductive privileges to keep their flock from starvation. The grouse, Wynne-Edwards contended, gauged the amount of food the moors could provide each year and adjusted their behavior accordingly, delaying breeding when supplies looked meager or even opting for total chastity.[74] The interests of the group, concluded Wynne-Edwards, overrode those of the individual.

The backlash to the University of Aberdeen professor's heresy was immediate and intense. Scientists like G. C. Williams and David Lack declared that group selection was "all but impossible."[75] And august theorists like W. D. Hamilton and R. L. Trivers explained away the "altruistic" tendencies Wynne-Edwards had discerned by generating a new mathematical system, the theory of kin selection, that said that individuals would only sacrifice their own interests in favor of others if the others in question were relatives, creatures who possessed similar sets of genes.[76] In other words, self-sacrifice represented an individualistic gene selfishly protecting a copy of itself.

The newly consolidated theories of individual and kin selection were hailed as major achievements and became biological dogma. Wynne-Edwards's carefully reasoned theory, based on decades of fact gathering in the field, was tossed aside as a disreputable aberration. The Scotsman spent fourteen years in the heather gathering fresh information, tabulated the resulting statistics, and published the conclusions in his 1986 work *Evolution through Group Selection.* The book was virtually ignored.[77]

However, in the late eighties, an uneasy sense that evolution may not be limited to the level of the individual organism or gene showed signs of inching toward science's peripheral vision. Stephen Jay Gould puzzled over the fact that there's too great a variety of seemingly useless genes lurking in the chromosome—more than one would statistically expect if each gene were subjected to extermination by natural selection.[78] Some genes, he concluded, seem to be "invisible to selective pressures." Competition between groups could account for the conundrum, since the group preserves a wide variety of individuals incapable of surviving on their own; however, Gould sidestepped considering this option. Though he was forced to acknowledge that not all selection takes place on the individual level, he contended that selection transpires below that level between gene fragments and above that level between species. Gould assiduously avoided mentioning the possible importance of social groups.

California State University's David J. Depew and Bruce H. Weber, on the other hand, asserted forthrightly that "a group . . . can be considered as an 'individual' " and that "the population level remains primary" as a unit of selection. But their brief observation, buried in a work on an unrelated subject, went unnoticed.[79]

E. O. Wilson, in his keystone book *Sociobiology,* cites numerous examples of behavior in which individuals sacrifice themselves for the good of the larger whole. But current theory continues to explain these away by claiming that members of the group who give up their lives do it to protect brothers, sisters, and cousins who share bits of the same genetic legacy.

Much of the enthusiasm over the theory of kin selection comes from W. D. Hamilton's brilliant mathematical demonstration of how genetic relatedness might account for the cohesion of bees, wasps, and other Hymenoptera in a hive. However, recent evidence shows that Hamilton's 1964 notion doesn't always mesh with the real world. Tropical wasps live together in cooperative colonies and function as a social unit. Most of the females become workers and give up on having offspring of their own, working not in their own interests or that of their kin, but in the interests of the group. Yet they do *not* show the high degree of family relationship—i.e., genetic similarity—predicted by Hamilton.[80]

In many cases, human beings who willingly form squadrons, march off, and fight to the death have no genes in common at all. In fact, during the American Civil War, relatives squaring off on opposite sides did not protect those who shared their genes; instead, they threatened to destroy them.

Even more damning, women in a murderous frame of mind usually do away with their own *children*.[81] Researcher Donald T. Lunde says, "[A]lmost all infants who are killed are killed by their mothers."[82] These mothers wipe out the very offspring who would carry their genes into the next generation. (The next favorite target of a married woman is her husband or lover.) And these grim facts of life are not restricted to the United States. Murderers in the former Soviet Union, Hong Kong, and Britain also show a predilection for killing those who share their genes.[83]

Kin selectionists have had a difficult time explaining yet another mystery: why among some social animals a few members of the herd will stand up and shriek when a predator approaches, even at the risk of making themselves obvious to the predator and becoming his first meal. For example, a herd of Thompson's gazelles is grazing quietly in an open East African field. A hungry leopard approaches quietly from downwind, holding its body low in the tall grass. Suddenly, a gazelle raises her head, cocks her ear, and freezes. A snapping sound has aroused her suspicions. Looking around, she spots the silhouette of the leopard's head. What does she do? To enhance her own survival and that of her genes, her best strategy would be to move to the center of the herd, making herself as unobtrusive as possible. The leopard would then pick off some unsuspecting and unrelated creature on the herd's periphery. The gazelle's worst move, on the other hand, would be to draw attention. Research shows that predators almost invariably go for a herd animal that is acting differently from the rest.[84]

But the gazelle who has just spotted the clawed creature does not quietly blend into the bunch. She breaks into a strange run punctuated by abrupt jumps into the air. Her behavior alerts her herd-mates to the prowling cat. One after another, they join the running and jumping. The leopard, thrown off by the commotion, eventually gives up and walks away.[85]

The Thompson's gazelle is not alone. Social animals of all kinds—mammals and birds alike—shriek, thump, or jump to warn their companions of an impending attack. Every one of the shriekers takes the chance that its warning gesture will make it the first victim of the hunter's assault. The theory of kin selection says the jumping or thumping animal is protecting its relatives. In a small number of cases, this hypothesis has worked out brilliantly, but in many others, it has been a failure. Large groups of animals do not just consist of brothers, sisters, and cousins. In fact, mobs like the flocks of birds that migrate thousands of miles each spring and fall seem to contain very few close relatives at all.[86] Yet members of the flock still shriek a warning when a hungry raider approaches. Why do these creatures choose to make themselves conspicuous?

A stealthy meat eater will have an easy time creeping up on a group whose members dare not act as lookouts for their neighbors. That social band's days on the savannah are numbered. But the aggregation whose participants court destruction by shrieking is primed for self-defense. An occasional individual may suffer, but the group will live to face another day.[87]

Individual selectionists have made a heroic effort to deal with the problem of altruism via the concept of kin selection. But there is a more subtle challenge to the primacy of personal survival that they haven't yet dared tackle: intropunitive behavior. In the 1950s, psychologist Harry Harlow at the University of Wisconsin wanted to see how necessary the love of a mother and friends were to humans. He couldn't wrench newborn babies from their mothers' arms and raise them in isolation cages, but he could do the next best thing. He tried the experiment on newborn monkeys. The simians raised without social contact frequently sat in a corner of their cage, curled into a ball, their eyes staring emptily into space, and chewed at their own skin, gouging themselves until they bled. That is intropunitive behavior.[88]

When you feel like kicking yourself around the block, you are in the grip of the intropunitive force. At times, whole herds of humans have unleashed this impulse in an orgy of self-punishment. Once a year, during the festival of Muharram, the Shiite men of the Middle East parade through the streets pulling out their hair, lacerating their scalps with swords, covering themselves in their own blood, and even injuring themselves with wounds that kill.[89]

Occasionally, the imagination can cooperate with the intropunitive emotions to make the mind a living hell. Some extreme Christian fundamentalists see demons and Satan lurking in every shadow. Their imaginations have created creatures that constantly threaten to torment them. The slightest slip from the true path, they feel, can send Satan's minions writhing through every vein of their body. These visions of dancing demons do little to enhance an individual's survival or that of his genes. In fact, during the first thousand years of Christianity, many of the devout swore there was one sure way to avoid Satan's seductive embrace: chastity. A few castrated themselves; others withdrew to the cloister and swore off sex forever. Most of these died childless.[90]

In a sense, there *are* demons lurking in the human flesh ready to explode with activity. They are biologically built-in self-destruct mechanisms. A talented advertising executive in New York was suffering from an unusual problem: phobia of cancer. He didn't *have* cancer, but his fear of it had practically incapacitated him. One night, he called all his friends at three o'clock in the morning in a panic, convinced that numerous vessels in his brain had hemorrhaged and that blood was filling his sinus passages to the point of a fatal explosion. Finally, one friend took him to the emergency room of a local hospital, where he was diagnosed as having a minor virus that gave him a stuffed nose. The next day, the executive fell into a panic again, certain that his nose was about to burst and kill him.

Something was running rampant through his psyche, tormenting this man; but it wasn't a physical disease in the standard sense of the term. The tormentor was a set of self-destruct processes that wait within us for their day of use. In the case of the man with the cancer phobia, the day had arrived, in part, because his career as a successful executive had come to a sudden halt a year earlier when the company for which he'd been working shut down.

Wynne-Edwards has demonstrated that red grouse on the moors of Scotland compete with each other at the beginning of the season for territory. The winners end up comfortably fed and mated, but the losers usually die of predation or disease. The deaths, says Wynne-Edwards, are "the aftereffects of social exclusion."[91]

In the body, each cell comes equipped with a mechanism for what scientists call "apoptosis," "programmed cell death," "an intrinsic cell suicide program" that researchers at University College in London say must be actively restrained from going into action by positive feedback indicating the cell is necessary to the larger organism.[92] When a hospital patient is forced to spend months in bed, seldom using his legs, many of the legs' constituent cells, sensing that they are no longer needed, dwindle to mere shadows of their former selves. Others simply disappear.[93] When a human spends weeks or months in space, his heart no longer has to labor mightily, pumping blood upward in defiance of gravity's force. The heart shrinks dramatically[94] as the cells that no longer deem themselves of value scale down to an existence just one step removed from death. The individual is a cell in the social superorganism. When he feels he is no longer necessary to the larger group, he, too, begins to wither away.

As we'll see more clearly in later chapters, the demons driving the advertising executive mad were the circuits of social disposal, "intrinsic suicide programs" similar to those that remove cells whose lives are no longer needed by the larger social beast. If our instincts were geared solely to the survival of ourselves and our relatives, such internal demons could not exist.[95]

SUPERORGANISM

※

Over a hundred and fifty years ago, the German botanist Matthius Schleiden was pondering the recently discovered fact that organisms as simple as water fleas and as complex as human beings are made up of individual cells. Each of those cells has all the apparatus necessary to lead a life of its own. It is walled off in its own miniworld by a surrounding hedge of membrane, carries its own metabolic power plants, and seems quite capable of going about its own business ruggedly declaring its independence. Yet the individual cells, in pursuing their own goals, cooperate to create an entity much larger than themselves. Schleiden declared that each cell has an individual existence and that the life of an organism comes from the way in which the cells work together.[96]

In 1858, pathologist Rudolph Virchow took Schleiden's observation a step further. He declared that "the composition of the major organism, the so-called individual, must be likened to a kind of social arrangement or society, in which a number of separate existencies are dependent upon one another, in such a way, however, that each element possesses its own peculiar activity and carries out its own task by its own powers." A creature like you and me, said Virchow, is actually a society of separate cells.[97]

As we've already seen, the reasoning also works in reverse—a society acts like an organism. Half a century after Virchow, entomologist William Morton Wheeler was observing the lives of ants and found that no ant is an island. Wheeler saw the tiny beasts maintaining constant contact, greeting each other as they passed on their walkways, swapping bits of regurgitated food, adopting social roles that ranged from warrior or royal handmaiden to garbage handler and file clerk. (Yes, at the heart of many ant colonies is a room to which all incoming workers bring their discoveries. Seated at the chamber's center is a staff of insect bureaucrats who examine the new find, determine where it is needed in the colony, and send it off to the queen's chamber if it is a prized morsel, to the nursery if it is ordinary nourishment, to the construction crews if it would make good mortar, or to the garbage heap kept just outside the nest.)[98] Viewed from the human perspective, the activities of the individual ants seemed to matter far less than the behavior of the colony as a whole. In fact, the colony acted as if it were an independent creature, feeding itself, expelling its wastes, defending itself, and looking out for its future. It was Wheeler who dubbed a group of individuals acting collectively like one beast a superorganism.[99]

The term superorganism slid into obscurity until it was revived by Sloan-Kettering head Lewis Thomas in his influential 1974 book *Lives of a Cell*.[100] Superorganisms exist even on the very lowest rungs of the evolutionary ladder. Slime molds are seemingly independent amoebas, microscopic living blobs that race about on the moist surface of a decaying tree or rotting leaf cheerfully oblivious to each other when times are good. They feast for days on bacteria and other delicacies, attending to nothing but their own appetites. But when the food runs out, famine descends upon the slime mold world. Suddenly, the formerly flippant amoebas lose their sense of boisterous individualism. They rush toward each other as if in a panic, sticking together for all they're worth.

Gradually, the clump of huddled microbeasts grows to something you can see quite clearly with the naked eye. It looks like a slimy plant. And that plant, a tightly packed mass of former freedom lovers, executes an emergency public works project. Like halftime marchers forming a pattern at a football game, some of the amoebas line up to form a stalk that pokes high into the passing currents of air. Then the creatures at the head cooperate to manufacture spores, and those seeds of life drift off into the breeze.

If the spores land on a heap of rotting grass or slab of decomposing bark, they quickly multiply, filling the slippery refuge with a horde of newly born amoebas. Like their parents, they race off to the far corners of their new home in a cheerful hunt for dinner. They never stop to think that they may be part of a community whose corporate life is as critical as their own. They are unaware that someday they, like their parents, will have to cluster with their fellows in a desperate cooperative measure on which the future of their children will depend.[101]

Another creature enlisted in a superorganism is the citizen of a society called the sponge. To you and me, a sponge is quite clearly a single clump of squeezable stuff. But that singularity is an illusion. Take a living sponge, run it through a sieve into a bucket, and the sponge breaks up into a muddy liquid that clouds the water into which it falls. That cloud is a mob of self-sufficient cells, wrenched from their comfortably settled life between familiar neighbors and set adrift in a chaotic world. Each of those cells has theoretically got everything it takes to handle life on its own; but something inside the newly liberated sponge cell tells it, "You either live in a group or you cannot live at all." The microbeasts search frantically for their old companions, then labor to reconstruct the social system that bound them together. Within a few hours, the water in your bucket grows clear, and sitting at the bottom is a complete, reconstituted sponge.

Like the sponge cells and the slime mold amoebas, you and I are parts of a vast population whose pooled efforts move some larger creature on its path through life. Like the sponge cells, we cannot live in total separation from the human clump. We are components of a superorganism.

ISOLATION—THE ULTIMATE POISON

≫❦≪

Remove the sponge cell from the sponge, prevent it from finding its way back to its brethren, and it dies. Scrape a liver cell from the liver, and in its isolation it too will shrivel and give up life. But what happens if you remove a human from his social bonds, wrenching him from the superorganism of which he is a part?

In the 1940s, the psychologist Rene Spitz studied human babies isolated from their mothers. These were the infants of women too poor to care for their children, infants who had been placed permanently in a foundling home. There, the children were kept in what Spitz called "solitary confinement," placed in cribs with sheets hung from the sides so that the only thing the babies could see was the ceiling. Nurses seldom looked in on them more than a few times a day. And even when feeding time came, the babies were left alone with just the companionship of a bottle. Hygiene in the homes was impeccable, but without being held, loved, and woven into the fabric of a social web, the babies could not maintain their resistance. Thirty-four out of ninety-one died. In other foundling homes, the death rate was even higher. In some, it climbed to a devastating 90 percent.[102] A host of other studies have shown the same thing: babies can be given food, shelter, warmth, and hygiene, but if they are not held and stroked, they have an abnormal tendency to die.

Two means have been discovered to produce depression in laboratory animals: uncontrollable punishment and *isolation*. Put an animal in a cage by himself, separated from his nest-mates, and he will lose interest in food and sex, have trouble sleeping, and undergo a muddling of the brain.[103]

Tampering with an individual's bonds to the larger social organism can have powerful consequences.[104] In humans, feeling you're unwanted can stunt your growth. The flow of growth hormones, according to recent research, is affected strongly by "psychosocial factors." Monkeys taken away from their families and friends experience blockage of the arteries and heart disease.[105] On the other hand, rabbits who are petted and hugged live 60 percent longer than those who have not received such attention.

When their mates die, male hamsters stop eating and sleeping, and often succumb to death themselves. They are not alone. A British study indicated that in the first year after a wife dies, a widower has a 40 percent greater risk of death. In another study at New York's Mount Sinai School of Medicine, men who had lost wives to breast cancer experienced a sharp drop in the activity of their immune system one to two months after the loss.[106] A survey of seven thousand inhabitants of Alameda County, California, showed that "isolation and the lack of social and community ties" opened the door to illness and an early demise.

An even broader investigation by James J. Lynch of actuarial and statistical data on victims of cardiovascular disease indicated that an astonishing percentage of the million or so Americans killed by heart problems each year have an underlying difficulty that seems to trigger their sickness: "lack of warmth and meaningful relationships with others."[107] On the other hand, research in Europe suggested that kissing on a regular basis provides additional oxygen and stimulates the output of antibodies. Closeness to others can heal. Separation can kill.

The cutting of the ties that bind can be fatal even in the wild. Jane Goodall, who has studied chimpanzees in the Gombe Stream preserve of Africa since 1960, saw the principle at work in a young animal named Flint. When Flint was born, his mother adored him. And he, in turn, doted on her. She hugged him, played with him, and tickled him until his tiny, wrinkled face broke out in the broad equivalent of a chimpanzee smile. The two were inseparable.

When Flint reached the age of three, the time came for his mother to wean him. However, since Flo, the mother, was old and weak and Flint was young and strong, this proved difficult. Flo turned her back and tried to keep her son away from the nipple, but Flint flew into wild tantrums, lashed about violently on the ground, and ran off screaming. Finally, a worried Flo was forced to calm her son by offering him her breast.

Later, Flint developed even more aggressive techniques for ensuring his supply of mother's milk. If Flo tried to shrug him off, Flint struck her with his fists and punctuated the pummeling with sharp bites. At an age when other chimps have freed themselves from parental apron strings, Flint was still acting like a baby. Though he was a strapping young lad and his mother was increasingly feeble, Flint insisted that she carry him everywhere. If Flo stopped to rest and Flint was anxious to taste the fruit of the trees at their next destination, the hulking child would push, prod, and whimper to get her moving again. Then he'd climb on her back and enjoy the ride. When shoves and whines didn't motivate his mother to pick him up and cart him where he wanted to go, Flint would occasionally give the exhausted lady a strong kick. Although Flint was old enough to build a sleeping nest of his own, at night he insisted on climbing into bed with his mother. In the course of normal development, Flint should have turned his attention from Flo to the other chimps his age, forging ties to the superorganism—the chimpanzee tribe—of which he was a part. But he did not, and the consequence would be devastating.[108]

Flint's mother died. Theoretically, Flint's instincts should have urged him to survive. But three weeks later, he went back to the spot where his mother had breathed her last and curled up in a fetal ball. Within a few days, he, too, was dead.

An autopsy revealed that there was nothing physically wrong with Flint: no infection, no disease, no handicap.[109] In all probability, the youngster's death had been caused by the simian equivalent of the voice that tells humans experiencing a similar loss that there's nothing left to live for. Flint had been cut loose from his single bond to the superorganism, and that separation had killed him.

Social attachment is just as vital to human beings as it is to simians. Research psychiatrist Dr. George Engel collected 275 newspaper accounts of sudden death. He discovered that 156 had been caused by severe damage to social ties. One hundred and thirty-five deaths had been triggered by "a traumatic event in a close human relationship." Another twenty-one had been brought on by "loss of status, humiliation, failure or defeat." In one instance, the president of a college had been forced to retire by the board of trustees at the age of fifty-nine. As he delivered his final speech, he collapsed from a heart attack. One of his closest friends, a doctor, rushed to the stage to save him. But the strain of losing his companion was too much for the physician, and he, too, fell to the floor of the platform, dying of heart failure.[110]

Our need for each other is not only built into the foundation of our biological structure, it is also the cornerstone of our psyche. Humans are so uncontrollably social that when we're wandering around at home where no one can see us, we talk to ourselves. When we smash our thumb with a hammer, we curse at no one in particular. In a universe whose heavens seem devoid of living matter, we address ourselves skyward to gods, angels, and the occasional extraterrestrial.[111]

Our need for other people shapes even the minor details of our lives. In the early 1980s, a group of architects decided to study the use of public spaces outside modern office buildings. For over twenty years, architects had assumed that people long for moments of quiet contemplation, walled off from the bustle of the world. As a consequence, they had planned their buildings with solitary courtyards separated from the street. What the architects discovered, to their astonishment, was that people shunned these secluded spots. Instead, they parked themselves on low walls and steps near the packed sidewalks. Humans, it seemed, had an inordinate desire to gawk at others of their kind.[112]

Even mere distortions in the bonds of social connectedness can affect health. According to a study by J. Stephen Heisel of the Charles River Hospital in Boston, the activity of natural killer cells—the body's defenders from disease—is low for people who, on the Minnesota Multiphasic Personality Test, demonstrate depression, social withdrawal, guilt, low self-esteem, pessimism, and maladjustment. Those who withdraw have pulled away from the embrace of their fellows. Those with guilt are certain that their sins have marked them for social rejection. The maladjusted have failed to mesh with those around them. And those with low self-esteem are convinced that others have good reason to shun them. In the study, low natural killer-cell activity wasn't linked to use of medication, alcohol, marijuana, or recent medical treatment—just to measures of impaired social connection.[113]

Meyer Friedman, the doctor who delineated the Type A and Type B personalities and their relationship to heart disease, says, "If you don't think what you do is very important, and if you feel that if you died, nobody's going to mourn, you're asking for illness."[114] Even the well-being of the men we would imagine to be most invulnerable to social forces depends on the sense that the superorganism needs them. When President Dwight Eisenhower had his heart attack on September 24, 1955, mail came in by the sackful from all over the world. Ike said, "It really does something for you to know that people all over the world are praying for you." Eisenhower's doctor sensed that the president's position in the social network could heal him. He insisted that Ike's aides continue to discuss business with the recuperating president, making him feel he was still important. Eventually, Ike went to Camp David for five weeks of rest. It was the worst thing he could have done. Stripped of his sense of social purpose, he went into severe depression, the first setback Eisenhower had experienced since his heart attack. The ailing chief executive recovered only when he was allowed to go back to work.[115]

Finding himself necessary to the social organism had a similar impact on another warrior—Colonel T. E. Lawrence, "Lawrence of Arabia." In the Middle East, Lawrence had been a dashing, energetic figure. He had dressed as an Arab and worked hard to win the respect of tribal leaders. He had taught himself to jump nine feet onto the back of a camel, something few Arabs could accomplish. He had steeled himself to ride across the desert for days without food. He had stretched his limits until he'd gained an endurance far beyond that of the average desert dweller, and he was admired greatly for it.

At the same time, Lawrence convinced the British that he could successfully mobilize the Arab nomads into a unified fighting force. With that force, Lawrence argued, he could help defeat the Germans and Turks in the First World War. The success of his argument boosted his power. When he rode into a circle of bedouin tents, his camels frequently carried several million dollars' worth of gold, a gift to cement his negotiations with the desert chieftains.

Using bribery and the force of his personal reputation, Lawrence drew together the widely scattered Arab tribes to storm Akaba. His force took the city despite seemingly impossible odds, defeating a small Turkish army in the process. After riding the desert for days and leading the charge in two successful battles, Lawrence was totally exhausted. Yet when he realized his troops in Akaba were starving, he mounted his camel and rode three days and three nights, covering 250 miles, eating and drinking on his camel's back, to reach the Gulf of Suez and summon help from a British ship.

The sense that he was critical to the success of the social organism had given the young British officer an almost unbelievable physical endurance. When at last the war was over, Lawrence rode into the city of Damascus in a Rolls-Royce as one of the conquerors of the massive Turkish Empire.[116]

However, once the fighting ended and Lawrence was forced to pack his Arab robes away and return to England, he felt totally out of place. True, he had friends in high places—Winston Churchill and George Bernard Shaw, among others[117]—but he felt wrenched from the social body into which he had welded himself. He was bereft of purpose, unneeded by the larger social beast.[118] Lawrence went back to live in his parents' home. His mother said that the former war hero would come down to breakfast in the morning and would remain sitting at the table until lunchtime, staring vacantly at the same object that had occupied his gaze hours earlier, unmoving, unmotivated. Eventually, at the age of forty-seven, Lawrence died on a lonely country road, the victim of a motorcycle accident.[119] Or was he really the victim of something far more subtle?

Not long before his death, Lawrence wrote to Eric Kennington, "You wonder what I am doing? Well, so do I, in truth. Days seem to dawn, suns to shine, evenings to follow, and then I sleep. What I have done, what I am doing, what I am going to do, puzzle me and bewilder me. Have you ever been a leaf and fallen from your tree in autumn and been really puzzled about it? That's the feeling."[120] Experts on suicide explain that vehicular accidents often occur to those who are depressed and *courting* death.[121] Was it mere chance, then, that T. E. Lawrence, a man of almost superhuman physical skills, was killed by a bit of sloppy driving on a vehicle he had used for years? Or had the inner calculators of the Arabs' former leader come to the conclusion that, like the unneeded cell in a complex organism, it was time for him to simply slip away?

EVEN HEROES ARE INSECURE

From none but self expect applause
He noblest lives and noblest dies
Who makes and keeps his self-made laws.
                    Sir Richard Burton

The man mindful of his reputation
Does not reveal his sadness.
                    early Anglo-Saxon author

If we are so desperately dependent on our connection to other human beings, why are we plagued with the peculiar notion that we *should* be detached, aloof, dignified, and independent? Why does the modern ideal of
self-sufficiency appeal to us so strongly?

The theme of self-sufficiency crops up in pop psychology with nag-
ging consistency. Marilyn Machlowitz, author of a 1985 book called *The
Whiz Kids,* for example, profiled entrepreneurs who achieved success
before the age of forty, then criticized them for being insecure. Often,
Machlowitz said, these young businessmen and women don't feel worthy
of their success. Implicit in her complaint was the notion that healthy humans would never be plagued by such doubts.[122]

But do the supremely confident and self-contained individuals to
whom you and I are constantly compared exist? Apparently not. Ernle
Bradford, the military historian, portrays both the legendary Carthaginian
general Hannibal and the Roman who defeated him, Scipio Africanus, as

men who were easily able to exist without emotional dependence on others. In reality, these men were anything but self-contained. Their fortunes from one day to the next relied on the loyalty of tens of thousands of troops, and on their ability to maintain the confidence of figures even farther from them in the grand web of the superorganism: the powers of the state back home. Without the financial backing of the Roman senate and the council of Carthage, both men's efforts would have been doomed.[123]

When Rome's senators accused Scipio of pilfering money from the public treasuries, the general was far from indifferent to the charge. He burst into the senatorial hall with his account books, tore them up in front of the lawmakers, and stormed out of Rome, never to return—not exactly the gesture of a man untouched by the opinions of others.[124]

Bradford, convinced that Scipio was, indeed, a master of self-reliance, had probably been deceived by a trick of aristocracy, a theatrical charade used by those who wish to exert power over others. It's a device even ambitious chimpanzees employ to maintain authority. The ruse goes something like this: The dominant male sits in the center of a noisy multitude looking utterly indifferent to what goes on around him. Lower-ranking apes nervously glance left, right, and behind them for clues as to what they should do next. They cast frequent, furtive glances at the master chimp to see if it is time for them to honor him with a deferentially downcast gaze or to discover if he has turned aside. For when his back is toward them, the underlings can get away with some forbidden gesture. However, the lofty head of the chimpanzee clan seems to look at no one and gives the impression that he need take his cues from no mere earthly beast.[125]

But even the chimpanzee at the top of the hierarchy who looks so impressively aloof is boiling with social emotions he doesn't dare show. Ethologist Frans de Waal made a six-year study of chimps in Belgium's Arnhem Zoo and published the results in a brilliant and illuminating volume called *Chimpanzee Politics*. In it, de Waal describes two males competing for top position in the group. The combatants confront each other with all the dignity of chivalric knights. Each stands erect, his hair raised in a magnificent mantle about him, looking massive and heroic. The pair stare into each other's eyes without flinching.

The manly stoicism with which the duo square off is a pose maintained only by an extreme exertion of self-control. After the confrontation is over, both chimps march away. When one is certain he is out of sight of gawkers, all the emotions he's been holding back suddenly rampage across his face. His upper lip flies up, leaving his teeth bare—the ultimate chimpanzee sign of nervousness. Realizing that another member of the tribe might spot him and note his delayed terror, the recent combatant tries again and again to pull the rebellious lip down over his teeth and regain his dignified appearance, but the stress-filled grimace simply will not leave his face.[126] Under the dignity and confidence of a few moments before was a seething cauldron of insecurity.

Hitler used to go through something similar at the height of his power. He would bully an opposing head of state, shouting, fuming, seemingly invulnerable to the inhibitions that weaken other men. Then, when he was alone in his room, the indomitable leader would collapse into a screaming nervous wreck. To make sure no one he wanted to intimidate ever saw that side of his personality, the führer would carefully rehearse for major meetings in front of a mirror.[127]

T. E. Lawrence even saw the ruse of imperturbability at work among the desert aristocrats of Arabia. Lawrence described a group of Arab chiefs summoned to the tent of their leader, who had just discovered that he and his vassals were about to be given two thousand camels—a veritable fortune in livestock—by the British. The chieftains scrambled excitedly toward the dwelling where their meeting was about to take place, then quickly stopped themselves outside the tent flap to compose their faces and postures into an appearance of hauteur, so that they would not march in with the silly looks of glee that the news of the unexpected fortune had stamped all over them.[128]

The lofty independence projected by Scipio, Hannibal, Hitler, and the Arab chiefs was nothing but an old reflex left over from prehuman times—a masquerade designed to further the impression of stature, magnificence, and strength. But under the mask of independence, even the most fearsome leaders are vulnerable to the views of others. Men of power pull off their disguise of aloofness well enough to fool even the psychological experts. Yet, the success of their sham is our undoing, saddling us with a false ideal of self-sufficiency, imbuing us with guilt for our dependence on the superorganism.

LOVING THE CHILD WITHIN IS
NOT ENOUGH

Current psychology is often blind to the existence of the larger being of which we are a part. Because of that blindness, many highly regarded therapeutic experts have a habit of dictating impossible cures. A group of prestigious doctors, for example, appeared on television's Phil Donahue show determined to show how you can confront stress as one human in isolation.[129] If you're a housewife and society seems convinced that your daily chores are on a par with garbage collection, the specialists gave the impression that all you have to do is sit at the kitchen table and talk yourself into self-respect. One homemaker stood up in the audience and put the prevailing notion in two short sentences. "It all depends on the image you project," she said. "If you think well of yourself, other people will think well of you." Thinking well of himself was not enough to save Scipio Africanus from the contempt of the Roman senate. Nor will it be enough to magically alter the degree of respect the Western world accords those who clean house, cook, and raise children.

One of Donahue's guest authorities declared in no uncertain terms that "you don't have to be a victim of what society or what anyone else thinks of you." But according to psychologist Sol Gordon, founder of the Institute for Family Research and Education, projecting a positive attitude works, not because of changes it makes on the individual psyche, but because of improvements it generates in relationships with others.

The best way to turn off the self-destruct mechanism is not to weep over childhood traumas until we can finally love the child within. It is to realize that the self-destruct devices are controlled by social forces: our sense of how we measure up to the standards of those we respect and our relationships with friends, husbands, wives, and even our dogs and cats. (The idea that relationships with animals can protect our physical and emotional health is not whimsy. Studies of heart attack victims have shown that owning a dog or a cat diminishes the odds of a second attack.)

Science would be well served to retain individual and kin selection's insights, admit their limitations, and move on. The fact is, if individual selection's survival instinct is our ruling force, then self-destruct mechanisms should not exist. Or, at best, their action should be limited to aiding those who carry genes nearly identical to our own. But animals of all kinds are born with a virtual arsenal of built-in poison pills. And a range of evidence even wider than that which I've had the space to present here indicates that these biological circuits are linked to the interests of a coagulation of fellow creatures who are *not* necessarily kin.

The concept of the social organism is more than a mere metaphor. It is closer to the comparison between a wave of water and a ray of light, two radically different phenomena subject to many of the same natural laws. For example, cyclic AMP acts as an intracellular messenger nearly everywhere in your body, but it is also the alarm substance that slime mold amoebas sensing starvation send to rally their solitary fellows into a consolidated, sluglike beast.[130] The gene NM-23, which controls the clustering or dispersion of cancerous cells in humans, also handles congregation and dispersion among the cells of the slime mold.[131] Contrary to contemporary theory, evolution is not built solely on competition between self-interested loners. It also relies on contests between teams of individuals striving for *group* survival. As a result, physiological feedback loops often call upon the individual to sacrifice his health—or even his life—for the sake of a larger whole. We have inherited much of our biology, including that involved in behavior, from the cellular ancestors who first learned to form communities. As a consequence, innumerable organismic mechanisms operate within assemblies of human beings. In coming chapters, we'll encounter quite a few of them.

ONE MAN'S GOD IS ANOTHER MAN'S DEVIL

US VERSUS THEM

⟫ ⟪

Nature uses only the longest threads to weave her patterns, so each small piece of her fabric reveals the organization of the entire tapestry.

Richard Feynman

White blood cells in the immune system function as soldiers on patrol. They are constantly scouring the corridors of the body, prowling for intruders. As they move through the veins and capillaries, they encounter billions of friendly cells, and myriad scraps of flotsam and jetsam that belong to the body itself. Should they make a mistake and attack these compatriots, the body would be in severe trouble.

How does the immune system manage to avoid cases of mistaken identity? The cells of the body have the equivalent of a uniform—a chemical combination as unique as a human face or a fingerprint. What's more, invading viruses also have a distinctive chemical costume. When a white blood cell detects the markings of the virus, it goes on the attack and sends out signals summoning its legion of confederates to the assault.[132]

Uniforms are necessary on the cellular level. They also prove indispensable to human society. Margaret Mead says every human group makes a simple rule: thou shalt not kill members of our gang, but everyone else is fair game. According to Mead, each group says that all *humans* are brothers and declares that murdering humans is out of the question. Most groups, however, have very strange means of defining who is human. A tribal member, in most primitive societies, is a full-fledged human being. A citizen of some other tribe, on the other hand, is usually not. Most primitive tribes, says Mead, feel that if you run across one of these subhumans from a rival group in the forest, the most appropriate thing to do is bludgeon him to death.[133]

Like white blood cells harmlessly passing each other in the body's corridors but destroying "foreign" intruders, humans of the same tribe recognize each other as parts of the same flesh and avoid hostilities. A body is a collective of cells that have to get along to survive; a society is a collective of individuals that have to do the same.

Humans ranging from the most primitive to the most sophisticated form cozy in-groups that assault outsiders, clumping together as competitive superorganisms.[134] This tendency, as we've already seen, is not limited to human beings. Lewis Thomas has pointed out that even lethargic-looking sea anemones engage in cold wars. Two seemingly identical patches of anemones on a rock may appear to live in peaceful harmony. In reality, however, the colonies edge against each other, aggressively trying to dislodge the rival community from the rock both call home. Like white blood cells and the denizens of the primitive tribe, the members of each anemone clump know who is one of "us" and who is one of "them."[135]

According to Harvard's E. O. Wilson, xenophobia, the fear and hatred of interlopers, is universal in higher animals. Wilson explains that squabbling *within* a group is minor compared to the snarling, spitting, and raking of claws that occur when group members encounter an outsider.

Much of the animal communication observed by ethologists, in fact, seems to have evolved to help one animal tell his killer companions: "Hey, I'm one of *us*." For example, calls allow one bird to tell others he's a part of their flock.[136] Some bird societies have even evolved their own dialects to aid in this purpose.[137] Animal markings and specialized scents also help beasts tell who is part of their own pack and who is not.[138]

Humans, too, need ways of identifying who they're supposed to take care of and who they're allowed to oppose. Some of those identifiers include how you hold your fork, what language you speak, what kind of clothes you wear, how close you stand to someone,[139] how you say hello, what haircut you choose, and what color you paint your face.

Leaders fashioning new social organisms seem instinctively to recognize the fact that they will have to find ways to differentiate their followers from everyone else. Moses coined a slogan: "Hear, O Israel, the Lord our God, the Lord is one.'' He enjoined his followers to inscribe the sentence on their doorposts, where they would see it every time they went out and came back in, and to tie scraps of parchment marked with the phrase to their arms every morning and every night.[140] To make sure that Jews would be marked as separate from the members of any other group, he even gave them a distinctive diet.

Lenin was born into a supremely middle-class family, but he dressed in the unmistakable clothing of a worker, adopted workers' slang,[141] and encouraged his followers to do the same. Then he pointed to an enemy with a radically different style of couture and speech—the well-coiffed bourgeoisie.

Mohammed gave his followers a set of prayers and ritual washings to execute five times a day, then instructed the faithful to signal their identity with the whiskers on their chins—shunning the razor and glorying in the growth of their beards.[142]

Each leader gave his followers a set of markings to identify them as one of us, and he set up the signals that would make it easy to spot the unbeliever. The quick identification of us and them is necessary because the competition between superorganisms—whether they're cultures or subcultures—can get very serious indeed. The Israelites Moses had gathered would soon be battling Canaanites. In those campaigns, the Hebrews could easily have been wiped out. Mohammed's followers were about to take on the superpowers of the entire Western world. To an informed handicapper, the odds of Moslem survival would have seemed astonishingly slim. Lenin's converts would soon kill the czar, his wife, and his children; exile the aristocrats who had tyrannized Russia for centuries; and wipe out the entrepreneurs and successful farmers who had fueled the country's rapid economic expansion. Meanwhile, in 1917, *counter*revolutionaries would mount a bloody civil war against the Marxists. Had they been able to, these adherents of the old order would have presided over the Bolsheviks' extermination. The battle between social groups is no mere pantomime. Being mistaken for a member of the wrong team can be fatal.

The battles between groups in a society at peace may be far less bloody, but they are no less persistent. Welfare families want to raise their payments; middle-class folks would like to avoid upping the taxes that provide the welfare checks. Landlords want to raise rents; tenant groups want to lower them. Rockers want to start a club on the corner; the older couples in the neighborhood want to protect their peace and quiet. Conservatives want to seize more power; liberals want to render them powerless. Men want to avoid housework; women want them to do more cleaning and mopping. All are clashes between clusters of humans who feel you're either with us or against us. They are battles for turf, like the slow struggles between competing clumps of anemones on a rock.

Within the group of those who wear the markings of the correct superorganism, all may be cozy and humane. But if your markings are wrong, watch out! Bertha Krupp, heiress of the German industrial family that armed Hitler's Third Reich, regularly visited Krupp factory workers who were ill. She generously comforted those in need. Bertha saw herself as warm, compassionate, and giving.[143] She had no compunctions, however, about the fact that her son ran slave camps in which people were beaten from the time they got up in the morning until they crawled into a lice-ridden bunk at night, deliberately underfed until they had worked themselves to death. Bertha was not even concerned that her family maintained gas ovens on the factory grounds to eliminate those forced laborers who might prove recalcitrant. In her own eyes, Bertha Krupp was a good and charitable person. Her kindness extended to those she considered human. The Slavs and Jews from whose bones she ground her fortune, on the other hand, were of a distinctly different subspecies. Bertha and other Germans of the time referred to these subservient beings with one simple word: *Stücke*—"livestock."[144]

As Margaret Mead said, killing real people is forbidden; but folks beyond the boundaries of our own superorganism aren't really people, are they.

THE VALUE OF HAVING AN ENEMY

⇒❦❦

True to the principles enunciated by Margaret Mead, every culture hand-cuffs hostility within the group. But in exchange for this imprisonment of anger, the culture offers a set of outsiders that it's acceptable to loathe and sometimes kill. These are the folks we call enemies.

A charismatic leader's invocation of the enemy's image is frequently what draws the social organism together.[145] Orville Faubus, governor of Arkansas from 1955 to 1967, knew how to pull together the social beast. He did it by creating an enemy that didn't exist. In 1957, Faubus was facing an uphill battle for reelection. His popularity was down, for he had upset liberals by allowing utilities and railroads to raise their rates, and had stepped on the toes of conservatives by increasing taxes. But Faubus had an ace up his sleeve—the creation of a bogeyman. Three years earlier, the Supreme Court had ruled that schools must be integrated. The South was outraged. Numerous Southern politicians were trying to capitalize on the issue, but only Faubus knew how to turn it into a full-scale, headline-grabbing drama with himself at the center.

The governor's opening move was simple. He phoned the deputy attorney general in Washington to ask what the Federal government planned to do to head off violence when Little Rock's schools opened their doors in September to both black and white kids. The attorney general's staff was perplexed. As far as they could tell, Little Rock seemed remarkably peaceful, so the Feds flew an official in to find out what the governor knew that they didn't. Faubus played it cagey. He had evidence that violence would indeed erupt, he said, but it was "too vague and in-definite" to turn over to anyone else.

"Violence," said the mayor of Little Rock in astonishment, "there was no indication whatever [of it]. We had no reason to believe there would be violence." Governor Faubus declared that the mayor was wrong. Little Rock's stores, he said, were running out of knives, and most of those newly purchased blades were in the pockets of blacks. The FBI checked on Faubus's claims. In one hundred stores, the sale of guns and knives was actually down. But there *were* weapons brandished openly in the streets. They belonged to the National Guard troops the governor had called in to "defend" the citizens of his state against the assaults of the supposedly well-armed enemy. When nine black children showed up outside Central High ready to enroll on the first day of school, those guardsmen raised their rifles and turned the kids away. Again, the FBI checked for signs of the budding black insurrection Faubus kept implying was about to begin, clues to the existence of the horde who threatened to slit the throats of innumerable innocent whites. This time, the authorities in Washington issued a five-hundred-page report. They couldn't find a shred of evidence to prove Faubus right.

The Federal investigators should have checked the home of Orville Faubus's close friend Jimmy ("the Flash") Karam. Karam was Arkansas's state athletic commissioner, a man who could whip together a squad of oversized brutes on a minute's notice. One day in mid-September, Faubus slipped away to a southern governors' conference in Sea Island, Georgia, leaving the Flash to carry out a delicate mission. Early in the predawn hours outside of Central High, Karam positioned a squad of thugs recruited from local sports teams. When the first class bell rang at 8:45 A.M., four African-American reporters showed up to cover the black students' attempts to approach their school. One of Karam's heavies let out a cry: "Here come the niggers." The black journalists beat a hasty retreat but not hasty enough. Twenty of the whites planted by the Flash cornered the reporters and started punching hard. As the police moved in, Karam roared, "The niggers started it!" Radio newscasts reported the melee, and, soon, the dregs of white Little Rock, spoiling to defend Caucasian honor, began to pour in. Hundreds of them. When they couldn't find enough blacks to beat, they turned on northerners. They mercilessly pounded three reporters from *Life* magazine. The violence that Orville Faubus had predicted for Little Rock had arrived.

Jimmy Karam darted to a gas station phone booth to fill the governor in on the situation. Faubus threw a press conference at the Sea Island governors' convention and declared soberly that "the trouble in Little Rock vindicates my good judgement."

President Eisenhower was forced to send in the 327th Battle Group of the 101st Airborne Division to try to restore calm and enforce the Supreme Court's desegregation order. Now Faubus had two enemies threatening his virtuous citizens: the blacks and the Yankee government, the same government that had humiliated the South in the Civil War. Faubus strutted, preened, and protested. He got national network news time on ABC, lambasting the president for having stripped the South of its freedom. He declared that the FBI had taken innocent southern girls into custody, grilling them for hours, and condemned the Federal soldiers for putting "bayonets in the backs of schoolgirls . . . the warm, red blood of patriotic Americans staining the cold, naked, unsheathed knives." He accused the soldiers of invading the girls' locker rooms to leer at the helpless young women pinned there by brute force. Investigation showed that the events Faubus described so vividly had never occurred, but at the time that scarcely mattered. In the minds of Arkansas's white citizens, only one man was standing up to these northern attacks—Governor Orville Faubus.

The result was simple. Faubus had been in danger of losing the election. Instead, he outstripped his closest opponent by nearly five-to-one and won every election after that until he finally retired. By creating an enemy, Faubus had galvanized Arkansas behind him, turning a cloud of disorganized citizens into a social mass.[146]

Fidel Castro found the existence of enemies equally indispensable, but he had a stroke of luck that Faubus had lacked. His enemy actually existed. The foe Fidel used to achieve social cohesion was the United States, the massive, imperialist monster that over the years, he said, had stripped Cuba of her sovereignty. Fidel needed to distract his constituency from a mind-boggling string of broken campaign promises. In the days before Castro seized power, the bearded leader had stirred up popular support for his revolution by posing as a political moderate, a champion of democracy and of an open society. The guise was a deception. Castro had been studying Lenin, Marx, and the Argentinean despot Juan Perón for years. His real goal was a dictatorship that would put every last scrap of power in his own hands.

Selling the idea of totalitarianism to the Cuban population, however, might have been a difficult proposition. Instead, Fidel played on the
island's dreams of freedom. While still fighting his guerrilla war in the
mountains, Castro appointed as president of his provisional government a
patriotic, well-meaning, prodemocratic judge from the Cuban city of San-
tiago named Manuel Urrutia Lleo. Urrutia's presence gave Fidel's movement an unmistakably democratic flavor. Fidel went even further. In an
interview with *Look* magazine and an article he penned for *Coronet* magazine, Castro declared passionately that his goal for Cuba was the liberty of
its people, civil rights, free enterprise, and the privilege of electing officials.[147] These statements were a sham. Now that he'd toppled Batista,
the leader who had sold his followers on a freely elected government
would get away with imposing a despotic political system through the
adroit use of an enemy.

This technique was embodied in Castro's manipulation of an unsuspecting pawn, President Urrutia. When Fidel first came down from the
mountains in triumph and took over the reins of authority, he filled his
cabinet with figures of indisputably moderate credentials. Then he secretly established a shadow government. In this clandestine group were
the "real revolutionaries," committed Marxist-Leninists like himself, figures determined to implement a program of "social justice" that would
entail snatching land from the peasants, establishing state-controlled collective farms, drafting the populace into militias, seizing all businesses
(eventually even the hot dog carts owned by scuffling members of the
lower class), shutting down the free press, and shifting all control into the
hands of a bureaucracy answerable directly to Castro himself. As for elections, they would be totally out of the question.

Gradually, the moderate figurehead President Urrutia began to
sense what was going on, and he, like many other Cubans, was not
pleased. Urrutia resisted the power grab in the only ways he knew. He
refused to attend cabinet meetings when Fidel was present, and made
anti-Communist speeches and television statements. He warned the
Cuban people that something sinister was taking place. Many Cubans
were alarmed, but Fidel stayed firmly on his course of secretly collectiviz-
ing society.

Finally, in despair, Urrutia volunteered to go on a leave of absence
from his official duties and not return. This, however, would not suit

Fidel's purposes. It could be read too easily by the people as precisely what it was, an act of protest. Castro had a better idea. He persuaded Urrutia to remain in office, then he embarked on a campaign to smear Urrutia's name. First, Fidel appeared on television, implied that he himself was a staunch anti-Communist, and called Urrutia's public statements dishonorable. Next Fidel arranged to have hundreds of thousands of Cuban peasants shipped to Havana for a celebration. As the peasants streamed into the Cuban capital, the bearded national savior went on television and delivered a two-hour speech announcing his resignation as prime minister. Who had forced this hero of the people to leave office? President Urrutia. Castro claimed that Urrutia had concocted the phony specter of a Communist menace and used this appalling lie to sabotage the revolution. Urrutia, explained Castro, was an American patsy, paid to spread the poison of Yankee propaganda. His duplicity, said Fidel, moved to the very ''brink of treason!''

The mood of the crowds gathered in the streets of Havana grew ugly. They chanted for Urrutia's resignation. The earnest president who had tried to warn his countrymen signed his resignation while Fidel was still in mid-oration. Then Fidel piled the fantasy of an American-inspired conspiracy to even larger heights. A former Cuban air force commander flew a light plane over Havana dropping anti-Castro leaflets. Fidel's officials said that the plane had machine-gunned the city. One of Fidel's top revolutionary fighters resigned his post in disgust over the pro-Communist direction events were taking. Fidel had him arrested and sentenced to a twenty-year prison term. Fidel gathered a crowd of a million Cubans in Havana and announced that these gestures of protest had been part of a massive counterrevolutionary plot engineered by the fiendish Americans. The answer? The creation of massive militias, the restoration of the death penalty, and the reinstatement of ''revolutionary tribunals'' to ferret out conspirators and send them to a string of concentration camps. The ultimate result of these measures, of course, was to help Fidel banish the old ideal of democracy and to move the country firmly into the grip of the one man capable of ''defending'' it from its hulking enemy to the north.[148]

Later, America played into Fidel's hands by plotting his assassination and launching the pathetically bungled invasion of the Bay of Pigs. These U.S. maneuvers provided the Maximum Leader with an indispensable po-

THE PERCEPTUAL TRICK THAT
MANUFACTURES DEVILS

Perception is a highly selective process. We see and vividly remember some things that pass before our eyes. We ignore many others. And still others we work to actively deny. What happens to those realities that consciousness shuns? They become part of the process that makes the notion of an enemy click.

We struggle for position on the hierarchical ladder, trying to get as close to the top as we possibly can. Very few of us arrive there. When the wriggling and kicking is over, most of us find ourselves somewhere in the middle. We are excluded from the realm of the beautiful people, barred from the inmost circles of power and prestige, and never quite reach the utopias of love and fame toward which our fantasies beckon us.

How do we live with the daily humiliations built into our middling roles? The mind is replete with gentle anesthetics that soothe these pains. The greatest among them is a perceptual process of cosmetic surgery. The mind covers up the harsher facts of our existence, and it focuses our attention on those few things that can give us self-esteem.

From 1972 until 1977, University of Utah psychologist Marigold Linton kept detailed, daily records of every aspect of her life. Periodically, she reviewed those records to see what she could remember and what she had forgotten. The review was a painful experience. Linton "warmed up" for each of these sessions by trying to recall the events of the last year without her notes. What came to mind were the high points—the good times with her friends, the research successes, the new things she'd been able to buy. But examining her three-by-five cards brought back a welter of details her mind had thoughtfully buried: the helpless feeling when her car had broken down and there had been no one to help; the fight she'd had with a lover; the feeling of dejection when a professional journal had turned down one of her papers.[150]

Linton is not alone. The mind also rewrites reality for you and me. Elizabeth Loftus, the pioneering University of Washington memory researcher and author of the superb book *Memory,* points out that people remember themselves as much harder workers than office attendance records show they actually have been. They recall past salaries as being much higher than their old paycheck stubs indicate. They recollect buying fewer alcoholic drinks than they actually did and are certain that they gave much more to charity than they ever have. They remember, in short, the glory, the positive accomplishments. What's more, they exaggerate those triumphs. But their minds erase the tiny, daily shames—the manifold humiliations of which life below the top of the hierarchical ladder consists. Instead, says Loftus, the mind erects a comfortingly false picture of the self and of the past.[151]

Where do the ugly events and the aspects of ourselves we need to forget go? We imagine them as parts of our enemy. When World War II was at its peak, the American Jewish Committee commissioned a psychological research project to determine the causes of the fascist horrors. Under this program, a team of behavioral scientists at the University of California at Berkeley developed a test to probe for the kind of tendencies that may have helped a Hitler or Mussolini gain power. That test, called the F (for fascism) scale, became one of the most widely used research tools in the history of modern psychology.

Literally thousands of studies revealed a profile of what the researchers called "the authoritarian personality." Generally, this was an individual raised in a strict home where the father was the clear holder of power. The parents had shown a stern disapproval of hostile outbursts on the part of their children. They had also rigidly prohibited the acknowledgment of any form of sexuality.[152]

But hostility and sexuality are both unavoidable aspects of human life. How had the authoritarian personalities coped with their unwelcome aggressive and sexual impulses? Through a technique that Freudians call projection. Like researcher Marigold Linton, who forgot most of the distressing things that happened to her in everyday life, the authoritarian

personalities excluded their own aggression and sexuality from their consciousness. Like the subjects who had misremembered their own past pay
and work habits, the authoritarian types pictured themselves as people in
whom sexual and aggressive tendencies did not exist. Aggression and sexuality, they were convinced, boiled up only in the minds of some *enemy*.

And here's the real trick. The authoritarians thought frequently of
that enemy and his loathsome preoccupation with lechery and hate. They
could actually feel the smarmy sexual sensations and livid hostility that
coursed through their enemies' veins. Why could they sense this so vividly? Because they had projected their *own* set of forbidden emotions onto
a faceless opponent like a ventriloquist projecting his voice into the mouth
of a dummy. By seeing their unacceptable impulses in some unsuspecting
outsider, they managed to dwell on those impulses and deny them at the
same time!

Here's how the principle works in real life. During the early eighties, a group of women in Orange County, California, were convinced that
the dark forces of "secular humanism" were using elementary school
textbooks to destroy their children's minds. The women's group was certain that the godless foes of true religion were trying to swamp their
youngsters with brain-crippling pornography. To find out if their suspicions were true, the women's group members examined the illustrations
in the local school's textbooks through a microscope. Sure enough, they
discovered minuscule pictures hidden subliminally on the pages. These
microscopic images portrayed women with naked, nippled breasts and
men with enormous erections. The outraged mothers actually succeeded
in getting some of the textbook illustrations changed on the basis of their
"discovery."

But where, in reality, did the pictures of the naked men and women
exist? Not in the printed pages, but in the minds of the women peering
through the microscopes. Like Marigold Linton, the psychologist who
"forgot" the events in her own life that had humiliated her, the micro-
scope-wielding religionists threw a mental sheet over their sexual impulses, hiding them from view. Then they conceived an enemy—the
secular humanist—roiling with forbidden sexuality and working in devi-
ous ways to insert sex into their innocent children's lives. The only way to
prevent this intrusion was to be eternally vigilant, perpetually on the
lookout for the humanists' sexual invasion. By searching for the humanist

danger, the clubwomen of Orange County managed to keep their thoughts focused on sex, but this sexual obsession, they could now tell themselves, was not *their* fault. It was the fault of their sex-crazed enemy, without whom sexuality would never have crossed their minds. Of such rejected pieces of ourselves are our devils made.

The perceptual flimflam practiced by the Orange County women contributed mightily to a war between two subcultures. The concept of a secular humanist enemy undermining American youth is currently preached by television and radio evangelists on the national Christian Broadcasting Network, on the nation's 221 fundamentalist TV stations, and on 1,370 fundamentalist radio stations.[153] It has been aggressively promoted by figures like Jimmy Swaggart, Pat Robertson, and the Reverend Donald Wildmon, who among them have had annual access to hundreds of millions of dollars' worth of media time. And it is being bolstered by a series of multimillion-dollar mailing campaigns whose sophistication makes the Reader's Digest Sweepstakes look amateurish by comparison.[154] The prophets of religious rigidity have made full use of the psychological ploy with which we project our own worst tendencies onto some distant figure. According to fundamentalist leaders, the secular humanists are a massive, well-organized "religious" group that has taken over our radio stations, our television networks, our newspapers, and our schools. These diabolical subversives have allegedly turned everything from television sitcoms to classroom syllabi into anti-Christian weapons designed to enslave the minds of decent American children to a godless, immoral creed.

Like Orville Faubus and Fidel Castro, the fundamentalists have used a spectral enemy to forge a flock that is hungry for power. Some fundamentalist leaders declare that God has made a covenant with his true followers. According to that covenant, the Almighty will deliver the political power of the presidency and the congress into the hands of his believers.[155] Other fundamentalist power brokers declare that this is a Christian country, erected on Christian principles by Christian founding fathers. As such, the government of our Constitution is designed to be placed in Christian control. Non-Christians should be excluded, some of these leaders say, from running for public office. After all, non-Christians, in Pat Robertson's words, are "termites" undermining the foundations of this country.

And who are non-Christians? Nonfundamentalists. According to Jimmy Swaggart, for example, Catholics and traditional Protestants are not Christians but the mistaken followers of a "monstrous lie." Some fundamentalist leaders are more blunt. Catholics, traditional Protestants, followers of EST, dabblers in Buddhism, New Age fans, and others, they say, are followers of Satan.[156] In the name of God, power must be wrested from the grip of the Satanists and placed in the palms of the godly.[157]

A simple perceptual device designed to anesthetize us from the nastier aspects of our inner reality has given the fundamentalist movement much of its power. From the sexuality their followers reject within themselves, the leaders have conjured up the lechery of a satanic enemy. From the hostility the faithful hide from themselves, the leaders have built a fantasy of an adversary obsessed with violence. They have crafted the indispensable tool that pulls together a superorganism. With the illusion of that demon, they have relieved the women of Orange County of their sins and welded them into a political force.

HOW HATRED BUILDS THE WALLS OF SOCIETY'S BUNGALOW

Frustration turns into hate.

José Napoléon Duarte

Politics, as a practice, whatever its professions, has always been the systematic organization of hatreds.

Henry Adams

One more ingredient is necessary to make the notion of the enemy click: hatred. The persistence with which societies offer permission to hate is astonishing. Jesus gave permission to frown upon the rich.[158] Medieval Christianity gave permission to hate heathens. Islam gives permission to hate infidels. Marxism gives the have-nots permission to hate the haves. Unions give permission to hate bosses. Peace groups give permission to hate militarists. Conservatives give permission to hate liberals. Each culture chooses an enemy on which to blame a goodly portion of the earth's evil and turns hatred of that group into a virtue. But from what raw substance is this group adhesive distilled?

A vast number of studies by clinical psychologists like N. H. Azrin, R. R. Hutchinson, and D. F. Drake show that frustration generates rage. Train a rat to run down a straight tunnel toward a piece of food. When it reaches the tunnel's end, it gets to eat. Then, place a plexiglass barrier in the rodent's path. When it reaches the transparent obstacle, the rat can still see its food but can no longer reach it. This is frustration. How does the animal respond? By going into a fury. If you place a bottlebrush next to the plastic barrier, the enraged rat will tear the object to pieces. Give him a smaller cousin, and he will abuse the diminutive beast terribly.[159] From these observations and a host of human studies we'll go into later in this book, came one of psychology's classic formulations—that frustration and aggression go hand in hand.

But frustration is an experience we cannot dodge. Fashionable as it is to think that none of us can be happy until we fulfill our potential, fulfilling our potential to its limit is an absolute impossibility. If a bacterium were allowed to fulfill its potential, within only four days it could produce more progeny than there are protons in the universe.[160] Fortunately, reality's constraints have kept bacteria from acting out their full reproductive possibilities.

Human males, like microbes, have reproductive capacities that could swamp the solar system. During his lifetime, each man produces enough sperm to inseminate every woman on the planet many times over. (A single ejaculation contains between 100 and 300 million sperm—a supply sufficient to impregnate almost every woman in North America.)[161] But the individual's sexual potential is something he can never fulfill. Men everywhere from Sri Lanka to Savannah, Georgia, watch women go by and fantasize possessing them sexually. In his imagination, each male couples with thousands of females by the time he dies. Yet in real life, the average man has mated with only a few.

Even chimpanzees endure this ignominious fate. The dominant chimp in a group hoards all the ladies for himself. Subordinate males are starved for sex. When the head chimp is not looking, the lesser beasts sneak up to friendly looking females and make supplicating, seductive gestures. If the lady seems willing, the furtive Romeo tries to lead her off to some secluded spot where the couple can catch a moment of forbidden love. But, all too frequently, as the pair skulks into the shadows, the lordly top-ranking animal catches a glimpse of their departure and punishes the impertinent commoner brutally for poaching on his harem.[162] The result for humans and chimpanzees alike is a frustration that is inevitable and inescapable. If the psychologists are correct, the upshot should be an ever-growing buildup of aggression. In human groups, how can this hostility be channeled to keep from blowing society apart? It can be aimed safely away from ''us'' and used to blast away at ''them.''

There are innumerable causes for frustration in human life. But the inability to fulfill one's potential is way up there on the list, and that inability is by no means restricted to sex. An ant is born with all the genetic equipment it needs to take on any role in society.[163] If it is fed one mixture of food by the colony's nurses, it turns into a soldier—a powerful beast far larger than normal size, equipped with savage jaws, and designed to defend the colony against attack. If it is fed another mixture, it becomes a worker, a small but nimble creature capable of carrying loads many times its own weight. And if it is fed a rare potion reserved for the select few, it can blossom into a queen—the one central creature toward whose preservation all efforts are directed, the only ant who gets the privilege of having children.[164]

Though an ant may have been hatched as a lowly worker, the blueprints for soldier and queen still lie dormant within her. In fact, if the colony's ruler is killed by marauders or disease, a worker may suddenly blossom with the reproductive powers fate has denied her all her life. She will begin to lay eggs. Leaving her food gathering and housecleaning chores behind, she will exude a substance that instills in passing ants the urge to feed and serve her, for she has ascended to royalty.[165]

Humans also on occasion go through transformations that dramatize the possibilities buried within them. During 1860, the citizens of Point Pleasant, Ohio, were distressed by the daily sight of a slightly rundown-looking store clerk in his late thirties. The man had been in the military, resigned under suspicious circumstances, failed at farming, failed at real estate, and ended up working in his father's leather-goods store. He was a poor excuse for a salesman, a worse-than-incompetent bill collector, and didn't even seem to know the establishment's stock. What's more, there were rumors that he had a problem with the bottle. Then, in 1861, the Civil War broke out, and the town failure enlisted in a regiment of volunteers. Less than two years later, he was promoted to major general. Eventually, he became president. His name was Ulysses S. Grant.[166]

Like ants, each one of us is built with all the equipment necessary to be a master or a slave, a beggar or a king. Most of us, however, will be only one of these. We will dream of the higher fortunes that could have befallen us, but, for the most part, we will never taste those possibilities in real life. And, as we grow older, many of us will carry an increasing burden of resentment for the fates we failed to have.

In some ways, it is the social organism and its needs that determine the role each of us will play and the many more roles that each of us will never be given the power to act out.[167] How the demands of the larger social beast determine our fate is hinted at by another aspect of the life of ants. Some of these Hymenoptera are lazy and sit around all day doing very little; others work their tails off in the interest of the community. But try separating the ne'er-do-wells from the industrious and setting them up as two new colonies—one composed exclusively of layabouts and the other made up entirely of nose-to-the-grindstone types. A strange thing happens. In the community of laggards, a large proportion of the lazy little beasts suddenly become imbued with a furious sense of industry. They turn into workers. On the other hand, in the community composed completely of workers, a small portion of the formerly zealous toilers seem overcome with boredom and settle down to spend their days doing nothing. They become the new leisure lovers. Each new colony takes on the shape of the old one.[168]

An individual ant behaves very much like a cell in a developing embryo. Any embryonic cell could just as easily become part of a liver, an eyeball, or a toe. What determines which of those things that cell turns into? Its *position* in the rapidly unfolding body. The embryo is "striving" to develop in a certain form. The individual cell behaves almost as if it were looking at a blueprint, figuring out where it is, and determining what it has to be to make the embryo come out according to plan. In a chick, you can take a cell that was about to develop into a wing feather and move it to the location that's destined to be a foot. If you perform the maneuver in time, the former wing-feather cell will turn into a perfectly normal piece of claw. The process is called cellular differentiation.[169]

The same thing happens in the all-worker and all-drone ant colonies. They undergo differentiation. There seems an implicit sketch for the contours of the community. A lone ant, in some peculiar way, looks around and sees where it sits in the social matrix, then becomes what it has to be to make the community fit the master plan.

Human groups go through a similar process. Researcher Richard Savin-Williams spent a season watching summer campers interact. In June, the bunk-mates met for the first time. For roughly an hour, the campers felt each other out, probing each other's strengths and weaknesses, deciding who would be friends with whom. Then they quickly sorted themselves into a superorganism with a head, limbs, and a tail.

One camper became the "alpha male," the dominant individual, the group leader. Another became the "bully," a big, strong brute nobody particularly liked. A third became the "joker," everybody's good-natured sidekick. And one became the "nerd," the unathletic, overly eager sort that everyone else felt free to kick around. Like the ants and the embryonic cells, each boy had taken his place in a kind of preordained social blueprint.

Just how preordained that blueprint was and how much of his potential each boy had to sacrifice to assume his role became clear when another researcher tried an experiment. The scientist assembled a cabin composed entirely of "leaders," boys who had been dominant, "alpha males" in their old groups. Very quickly, the new cluster sorted itself out according to the familiar pattern. One of the leaders took charge. Another became the bully. A third became the group joker. And one of the formerly commanding lads even became the new group's nerd.

When the researchers went through the scientific literature to find other data related to their work, they discovered that studies of Chicago gangs in the 1920s had shown these long-gone groups arranging themselves according to an almost identical unconscious plan. The gang members of a bygone era also had their leaders, bullies, jokers, and nerds. Each individual had taken up a position in the superorganism's unfolding structure. And each had shaped his personality to fit the spot in which he landed.[170]

For the ant, the possible roles the insect discards may never come back to haunt her. In humans, however, the personalities that could have been are always there, always uncomfortable in their imprisonment. And periodically they scream from the dungeons of the mind, demanding their freedom.

The novelist Hermann Hesse said that we each have a thousand personalities hidden in a mental closet. The circle of our consciousness centers on one, but the others are in the darkness waiting to come out. Implicit in each of us is the whole society, the dominant individual, the outcast, and all the variations in between. Novelists, more than the rest of us, realize how many possible people inhabit our minds. When these authors sit down to their typewriters, whole casts of characters come parading into the light of awareness, each ready to live out a new life. And each of these fictional humans is disgorged by the brain of just one writer, who in real life has settled on the single personality and fate he will call his own.

The buried personalities may be erased from the surface of consciousness, but they still wriggle toward the light—in anger, frustration, and jealousy. Every male is built with the same neuronal networks that compelled Genghis Khan to conquer an empire twice the size of Rome's,[171] the same set of circuits that motivated some of Genghis's descendants to accumulate hundreds of wives and even more concubines, the same instincts that impelled Turkish sultans to have attractive women from all over their domains shipped in for a few nights of physical glory.[172] But in most cases, those circuits will never unfurl their ambitions in the real world. We have thousands of mental mechanisms crying out vainly for a moment of triumph, thousands of potential personalities that will never be allowed to live.

Frustration, as the researchers have demonstrated, breeds rage. Hatred is a despicable by-product of the human condition. Nature, however, frequently utilizes such garbage as building material. We will see how she has employed the psychological detritus of hatred in a minute, but first, let's take a close look at another example of how the natural world often turns a poisonous excrescence to good use. The instance I have in mind keeps each of us from turning to an oozy blob.

Communities of cells living in the seas roughly 600 million years ago had a chemical disposal problem. From the surrounding waters, they took in large quantities of calcium, a substance that could in megadoses poison them. To function effectively, the cells had to constantly filter the calcium out of the water and deposit it outside the cellular backdoor, where the mineral wouldn't interfere with the cell's internal functions.

Somewhere along the line, a cellular community evolved a clever way of getting rid of its unwanted calcium contaminants. The collective compacted its discarded calcium sludge into safe cylinders and laid these solid slivers of deactivated toxic waste along the interior corridors between the huddled cells. The disposal technique produced a surprising benefit. The discarded calcium rods became structural beams that gave added strength and power to the cellular cooperative. They were bones, devices that made revolutionary forms of movement possible, and eventually enabled cellular superorganisms to lift themselves out of the water onto the land.[173]

In human society, another kind of garbage, this one psychological, is used for similar structural purposes. The waste product, in this case, is the frustration from which hatred is distilled. The frustration of humans collects much as calcium accumulated in the space between cells of the early ocean-living, cellular communities. To avoid damage within the group much of it is directed somewhere else, at outsiders. Envy and fear are turned from a source of disruption to a creator of cohesion. Nature has compacted mankind's frustrations to build the superorganism's bones.

The demon one society wants to eradicate is all too frequently the god of some rival group.[174] Baal, the god of the Canaanites, was a false idol to the Jews. The former Soviet Union's longtime gods—Marx and Lenin—are our devils. Our revered middle class was the former Soviet Union's hated bourgeoisie. Social organisms—like clusters of anemones on a rock—face off and fight. From that struggle, they frequently derive their identity. Battle draws a set of straggling individuals together into a firmly consolidated social clump. It gives the formerly quarreling separatists a powerful common bond.

Leaders like Orville Faubus and Fidel Castro have skillfully manipulated a few basic rules of human nature: that every tribe regards outsiders as fair game; that every society gives permission to hate; that each culture dresses the demon of its hatred in the garb of righteousness; and that the man who channels this hatred can rouse the superorganism and lead it around by the nose.

MAN—INVENTOR
OF THE
INVISIBLE WORLD

FROM GENES TO MEMES

※ ←

How noiseless is thought . . . it will not rule over, but in all heads,
and with . . . solitary combinations of ideas, as with magic formulas
bend the world to its will.

Thomas Carlyle

Let's return for a moment to Richard Dawkins's concept of the replicator. Dawkins's proposal squares brilliantly with the reality of upwardly
spiraling life, but the eminent zoologist's individual selectionist contention that self-reproducers operate on their own, as I've mentioned
before, fails to fit the facts. Despite the high degree of competition between individuals in evolving systems, every form of replicator is nestled
in a team. Genes that fail to work effectively with their partners on the
chromosomal string are doomed. The self-replicating wonder we are
about to meet is also forced to fit into a constellation of its fellows, or it,
too, disappears.

What is this relative newcomer in the field of autoduplication? It has
no physical substance, and cannot be studied under a microscope or kept
in a jar. The new replicator, like its predecessor the gene, is capable of
assembling vast amounts of matter. Like genes it can pull together products the earth has never seen, but unlike genes it can manufacture forms
of order that mere genetic stuff could never dream of.

Pinpointing the date of the new device's birth is difficult. Its primitive precursors may have begun to wriggle across the planetary face
roughly twenty-two million years ago with the emergence of protochim-

panzees, whose modern bands devise widely varying dialects, ways of using tools, and methods of cooperative hunting.[175] Yet its first fully identifiable forms may not have gone to work swallowing substances and churning out copies of themselves until the last thirty-five thousand years. Dawkins calls these new replicators memes.

Genes, says Dawkins, swam through the protoplasmic soup of the early earth, nourishing themselves on organic sludge. Memes float through another kind of sea—a sea of human brains. Memes are ideas, the snatches of nothingness that leap from mind to mind. A melody wells up in the reveries of a solitary songwriter. It seizes the brain of a singer. Then it infects the consciousness of millions. That melody is a meme. A scientific concept starts as a vague glimmer in one researcher's thoughts. It ends up with whole schools of adherents. That concept is a meme. Each flips from the puddle of one brain to another, crazily copying itself in the new environment. But the memes that count the most are the ones that assemble vast arrays of resources in startling new forms. They are the memes that construct social superorganisms.

Genes sit at the center of each cellular blob, dictating the construction of a multibillion-celled body like yours and mine. As genes are to the organism, so memes are to the superorganism, pulling together millions of individuals into a collective creature of awesome size. Memes stretch their tendrils through the fabric of each human brain, driving us to coagulate in the cooperative masses of family, tribe, and nation. And memes—working together in theories, worldviews, or cultures—can make a superorganism very hungry.

From 1852 to 1864, Karl Marx sat alone nearly every day in a corner of the library of the British Museum, going through books and assembling his theories.[176] Little did he realize it, but the bearded writer was simply the tool of fragmentary memes. Those ideas had been floating in the zeitgeist, waiting for a receptive human mind to come along and function as an enzyme functions in human metabolism—splicing together molecules destined for each other. Marx pasted together the ideas of his time and came up with the ideology named after him.

At its birth, the new ideological meme was vulnerable and powerless. The only small batch of matter over which it had any control was the body and mind of Karl Marx, a hundred and fifty pounds of isolated humanity.[177] Marx was not a promising person in which an idea would wish to start its life. Though he occasionally made money as a newspaper correspondent, Marx's work was definitely not in demand. He was so foul tempered, so cantankerous, so subject to turning even the tiniest discussion into a quarrel that he had few friends and almost no followers. One of his college professors said young Marx was always waving his fists in the air in a fury, "as if a thousand devils gripped his hair." And one of the many would-be friends Marx alienated recalled that "the sarcasms with which he assailed his adversaries had the cold penetration of the executioner's axe."[178]

It's little wonder that for the next fifty years, the meme that had assembled itself in Marx's brain barely stayed alive. The conceptual tangle leaped tenuously from one mind to another, constantly searching for the opportunity to expand its power. Marx muttered his ideas to his acquaintances, a few of whom passed them on to others, and he preserved them in his book, *Das Kapital* (1867). But that book, too, seemed like a poor vessel in which to keep the struggling meme alive. Russian censors found the volume utterly incomprehensible. As a consequence, these normally repressive servants of czarist autocracy blithely allowed the importation of the murky work into their land.[179] Most Russians had the same difficulty understanding the new ideas that the censors had. The result: though *Das Kapital* was occasionally passed from hand to hand, Marx's meme still barely clung to life.

Then, Marx's mental progeny had a modest stroke of luck. It found its way into the cerebral substance of a few men capable of something Marx hadn't been—organization and the recruitment of followers. These were Lenin, Stalin, and their friends,[180] but even they, at first, seemed wretched prospects for the multiplication of a meme. Lenin, like Marx, spent seventeen years sitting in libraries, poring over the records of the Paris Commune and researching the techniques of street fighting. He too was an exile from his mother country, forced to while away the hours between articles and books with a handful of other radical Russians whose extremist views had also gotten them tossed out of the land of their birth.

Lenin frequently took time out from his research and writing to lecture at European meetings and issue orders to underground cells back in the motherland; but even these efforts, ironically, underscored his isolation. At times, Lenin's communication with his co-conspirators back home was so tenuous that, as Harrison Salisbury puts it, "he was almost completely cut off from Russia."[181] Lenin attempted to spread the idea of which he was the incubator through a series of newspapers—one of those singularly useful devices with which a meme reaches out and acquires power over new minds—but circulation was abysmal. For example, during the First World War, Lenin put out *Sotsial-Demokrat*. Only five or six copies of the debut issue straggled into Russia. Later, distribution improved, and Lenin managed to have twenty copies smuggled in a pair of shoes.[182]

A curious accident occurred in 1905. Czar Nicholas went to war with Japan—a nation that had just climbed out of feudal backwardness and built its first modern military machine. Astonishing as it seems, the czar lost nearly his entire Baltic fleet to the postfeudal upstarts.[183] A moment of defeat is a great time for an ambitious idea to seize minds that are fleeing from the precepts of a luckless leader. The result was a revolution. Angry crowds rampaged in the streets of Saint Petersburg calling for the czar's befuddled head.

But Marx's fragile idea was trapped, unable to seize the opportunity, because Lenin and his friends were too busy squabbling to join the revolution, much less lead it. Fortunately for the meme lodged in their minds, fate would give the quarrelers a second chance. After the revolution of 1905, the czar regained control over the country, but in 1917, Nicholas was losing once again. This time the occasion was the First World War. Millions of Russians died on the eastern front. Soldiers struggled through the snow without ammunition, winter clothing, or food. Back home, the railway system broke down. Grain and meat could not reach the towns. Men and women of modest means starved in the boulevards of the Russian empire's most magnificent cities.

After a new revolution had spontaneously begun to heave through the streets of Petrograd (formerly Saint Petersburg) and Moscow, Lenin finally got his act together. He took a train from Zurich to the Russian capital. The minute he arrived at the Petrograd railroad station, the bearded book-lover began haranguing the crowds with slogans, and the meme that had first been born in the mind of Karl Marx finally fell like a scrap of wandering bacteria into nutritive jelly, spreading with explosive force.

By the mid-eighties, the ideas brought together by one man's brain in the corner of a lonely library, ideas that looked from year to year as if their disappearance from the planet was all but inevitable, had gone from controlling one 150-pound man to mastering millions of tons of this planet's matter. These memes were alive in the minds and the social mechanisms of over 1.8 billion human beings, spreading their influence over the lands, the minerals, the machines, and the domesticated animals that those human beings controlled.[184] The new replicator took up even less space and mass than the tangle of atoms required for a strand of DNA. That replicator—like the ones incubated by Jefferson and Madison, and Saint Paul—had assembled under its rule more of the planet Earth than any gene had ever brought together, any gene, that is, save one: the human gene.

The new replicator, the meme, is a vast upward step on the ladder of creation. The old genetic system could take ten thousand years to effect a product improvement in a large and complex beast,[185] but memes can rearrange sprawling networks of outrageously intricate creatures in only a few centuries or less. The meme of Christianity restructured the Roman Empire a mere three hundred years after Jesus completed his Sermon on the Mount. Similarly, Marxism radically altered the shape of Russian society a startling sixty years after Karl the cantankerous ambled out of the British Museum's library with the final manuscript of *Das Kapital* tucked tightly under his arm.

The meme has done its work by assembling massive social systems, the new rulers of this earth. Together, the meme and the human superorganism have become the universe's latest device for creating fresh forms of order. They are the newest innovation in a climb toward complexity that started with the big bang.

THE NOSE OF A RAT AND THE HUMAN MIND—A BRIEF HISTORY OF THE RISE OF MEMES

How many gods can there be in one sky?
—from "Hits of the Year"
by Christopher Difford and Glenn Tilbrook

Memes probably arose by accident. At least that's the implication of one clue to their origin—a clue from the world of the rat.

Rats are obsessed with those who share their genes. They absolutely love their relatives. In their nests, fathers, mothers, aunts, uncles, and children crawl over, under, and around each other, seeking constant body contact. And they watch out for one another. For example, if a rat discovers that a tempting hors d'oeuvre is actually laced with poison, he defecates and urinates on the insidious morsel to make sure none of his loved ones is duped into tasting the fatal snack.

But the kindness of rats only extends to family. Rats will mercilessly hunt down members of a rival clan. And if a nonrelative accidentally stumbles into their nest, the homey little creatures who a moment before were hugging one another will turn on the guest with the foreign genes and tear him limb from limb.[186]

How do rats know who's kin and who's not? How can they tell who shares their genes? Rodents don't have a compound microscope with which to examine the replicators at the center of a visitor's cells, so they're stuck with making genetic guesstimates, guesstimates based on smell.

Each rat household has its own telltale odor. The smells of scent glands; excreta; food; the soft, warm stuff that lines the nest; and the wood, straw, or earth of the chamber's walls—all blend into an unmistakable olfactory brew. Every inhabitant wears that living-room perfume. Chances are that if two rats are sporting the same aroma, they're carrying the same genes, since the pair were raised in the same spot—probably by the same mother and father. Unless some experimenter decides to muck things up.

That is exactly what one scientist did. He removed a rat from his nest, washed the complaining creature off, then rubbed it thoroughly in the shavings of another nest, giving it the smell of a stranger. Then the experimenter put the innocent beast back in its own home, where it should have been safe among its brothers and sisters. Unfortunately for the furry victim, he'd returned home wearing the wrong cologne. His loving family, blind to his familiar physical appearance, bared their teeth and lunged. When the experiment was over, the unwitting animal was dead, killed by those who had previously hugged and nuzzled him. Smell had told the brood that their brother was carrying the wrong set of genes.[187] In this case, the noses of the rats misled them.[188]

Early human groups were stuck with the same problem. How do you tell who's family and who's not? How do you know who shares your genes? Like rats, primitive humans turned to external signs. Fortunately, they didn't rely on their noses. Instead, the inventive *Homo sapiens* used ideas, manners, morals, and peculiarities of clothing. The Children of Israel were typical of the tribal nations of the time. To belong, you had to have the right genetic stuff. How could an early Hebrew tell if you were entitled to insider treatment? Your god, your mannerisms, and your ideas were the outward labels of your genetic contents. Memes were the equivalents to the rat's perfume.

But a strange thing happened as human groups grew larger. Memes became detached from genes. In the days of the Old Testament, memes seldom made the effort to leap from one gene pool to another. The ancient Hebrews, for example, made no effort to convert the heathens. It wouldn't have made sense. Unbelievers weren't family. If the meme was to retain its role as a genetic marker, only those who shared the same genes could share the same god. That was the tribal concept held by primitive peoples the world over.

But as Leo W. Buss, a biologist at Yale University, says, "at each stage in the history of life in which a new self-replicating unit arose—the rules regarding the operation of natural selection changed utterly.''[189] One result was that two to three thousand years ago, the gods who had been mere labels for a genetic stock detached themselves and took on a new purpose.

You can see the gene-free god unfolding in the days of the New Testament. Jesus was a Jew, and all available evidence indicates that, like the other Jews of his time, he felt his god was a genetic one. The only people to whom Jesus preached were other Jews, and they alone were the folks to whom his god and his DNA were attached. When Jesus was crucified, most of his disciples followed in his footsteps, trying to convince Jews that Jesus was, indeed, the Messiah.[190] Like the tribal god he served, a Messiah would aid only the people who carried the chosen genes.[191]

Then, after his death, Jesus acquired a new kind of apostle. The original followers of the carpenter from Nazareth had been simple people from the hills of Galilee, poor backwoods folk with only the most rudimentary education. The figure who would transform Christianity was a city sophisticate with a university education.

His name was Saul. He knew aspects of the world the original disciples had never dreamed of. He had grown up in the cosmopolis of Tarsus, a bustling center of trade where men from all over the vast Roman Empire did business. Saul's father, though Jewish, had been a Roman citizen. Saul had been educated in Israel's greatest urban center, Jerusalem, and he spoke the language of international high culture—Greek.

Saul was a Johnny-come-lately to the teachings of Jesus. He didn't even get involved until after Christ was dead. When Saul first heard of the redneck sect, he was so infuriated that he organized squads of vigilantes, broke into the homes of Christian Jerusalemites, and hauled the inhabitants off to prison. Then Saul volunteered to bust up a community of Christians in Damascus. But on the road to the northern city, Saul had a strange experience. He felt enveloped in light. He heard the voice of Jesus, the deceased leader whose views he so deplored. Saul became Saint Paul and dubbed himself the newest of Jesus' apostles; then the freshly minted holy man went off to win others to his idiosyncratic notions of what Jesus' teachings were all about.

The community of Jesus' followers does not seem to have welcomed Paul's posthumous reinterpretation of their leader's ideas with open arms. They probably regarded him with suspicion. With his big city ways and complex ideas, he was anything but their rustic kind. Finally, the self-styled apostle, in exasperation, decided that if he couldn't dig up followers among the Jews, he'd turn elsewhere.[192] Thus, Paul began a vigorous campaign to win over "the gentiles"—citified Greeks, Romans, Anatolians, Sicilians, Spaniards, and others whose urbane views were more congenial to his own.[193] In the process, Paul was one of the early innovators of a new concept: transferable religion. He broke free of the old notion that a god was an emblem of tribal heritage and sliced the ties that bound divinity to genes.

Paul was not the first to free gods from chromosomal components. Buddha had done the same over five hundred years earlier. But Paul was among the most influential ever to apply the idea. Thanks to Paul, the Christian meme would eventually sweep together an awesome jumble of genes. Dark-haired Greek and Roman genes; blue-eyed, blond Scandinavian genes; red-headed Irish genes; russet-skinned American Indian genes; black-complected African genes; and even the occasional Chinese and Japanese genes. Folks whose genetic coils were dramatically distinct would find themselves yoked together by a common thread. That intangible tie was a meme.

Under Paul,[194] beliefs became the focal points for movements that, freed of genetic anchors, could sweep across the face of the world, gathering humans of all kinds within their grasp. For when Paul separated genes and gods, he helped unleash a force that would bring together superorganismic groupings on a scale the world had never seen. He helped make the meme the world's most powerful form of replicator.

Nineteenth-century psychologist William James once said that Saint Paul was a biological "failure because he was beheaded." Paul had no children. His genes simply ceased to be. "Yet," said James, Saint Paul "was magnificently adapted to the larger environment of history." For history is no longer the sole province of the gene. History is the environment of the meme.

HOW WRONG IDEAS CAN BE RIGHT

~><~

In the 1950s, three Midwestern scientists—Leon Festinger, Henry Riecken, and Stanley Schachter—heard of a small cult in Chicago dedicated to the proposition that the earth was about to come to an end. To learn all they could about how such a belief system works, they joined the group. They were astonished by the resulting experience.

The cult was led by a middle-aged couple. Dr. Thomas Armstrong (a pseudonym the researchers used in their writings to spare the real leader embarrassment) was a well-educated doctor who worked for a college health service. His co-leader, Mrs. Marian Keech (another pseudonym), claimed to be receiving messages from spiritual beings on other planets, beings she called "the Guardians." These kindly souls did not bother to send space-suited messengers. Instead, they delivered their truths through automatic writing.

Mrs. Keech would sit at a desk, a pencil in her hand. Suddenly, the pencil would move. Jerking spasmodically across the page, it would fill the paper with words. When the "possession" was over and her hand at last came to rest, Mrs. Keech would lift her pad and read. The words her fingers had formed "unknowingly" were messages from a distant world.

For a while, the creatures from another planet were content to issue Mrs. Keech a set of pronouncements remarkably similar to Christianity. Then they changed their tone. They predicted that a massive flood would soon erupt in the Western Hemisphere and drown the entire world; but there was hope. Just before the inundation, the Guardians would send flying saucers to rescue those believers who had been wise enough to follow the teachings conveyed through the intergalactic counselors' spokes-woman on earth, Mrs. Keech. For several weeks, the thirty followers of the Guardians obsessively carried out the departure instructions their ex-traplanetary advisers had provided. They memorized the passwords that would get them through saucer security. ("I left my hat at home." "What is your question?" and "I am my own porter.") They took all the metal out of their clothes. Some gave away their personal belongings, quit their jobs, abandoned their college classes, and got ready to say good-bye to the earthly life they had known.

Neighbors and relatives did not always look kindly on these activi-ties. Dr. Armstrong's sister filed a lawsuit to have his children taken away. He was fired from the college health service. Neighbors and family members of his followers threatened to have the space-crazed believers clapped into an asylum. However, few of the Guardians' little flock were deterred, for they believed that their very survival depended on faithfully following extraterrestrial orders.

When the day of the great flood arrived, reporters appeared to in-terview the followers. Not anxious to share the good fortune of their im-minent survival, the believers in the Guardians preferred to keep the wisdom that could save their fellow earthlings to themselves. So they turned away journalists with a simple "No comment."

As the critical hour drew nigh, the believers shooed a cluster of re-porters away from Dr. Armstrong's house and made ready for their mid-night escape from this doomed earth. They donned the slashed clothing from which they had ripped zippers, fasteners, eyelets, and every other metal device. The men held up their pants with ropes. Gathered in the living room, the flock received last minute instructions from the Guard-ians through the automatic scribbling of Mrs. Keech and methodically drilled the use of its passwords, chanting in unison phrases like "I am my own porter, I am my own pointer."

As midnight approached, the followers fell silent. They sat with their coats in their laps, holding themselves so still that the air seemed hammered by the ticking of the room's two clocks. One minute before the saucer was due to arrive, Mrs. Keech exclaimed proudly in a strained, high-pitched voice, "And not a plan has gone astray!"

But twelve o'clock passed, and there was no whirring of antigravity engines, no great rush of air from above, no luminescence lighting up the lawn outside the windows. The believers sat stock still, their faces frozen. The researchers noted that "it became clear later that they had been hit hard."

Finally, the faithful stirred. In desperation they went over the messages and the original prediction to see if they had made an error in calculation. They argued one explanation for the delay after another. Then, at 4 A.M., Mrs. Keech began to cry. Reported the researchers, "She knew, she sobbed, that there were some who were beginning to doubt but that the group must beam light to those who needed it most and that the group must hold together." The catalyst that might, indeed, save them from dissolution arrived at 4:45 A.M. Mrs. Keech's hand suddenly jerked across the page of her writing pad with a new message from the Guardians. The faith of the believers, it said, "had spread so much light that God had saved the world from destruction."

Then the Guardians delivered an even more significant directive: publicize the fact that the believers had saved humanity! Mrs. Keech leaped to the phone to call the very newspapers whose reporters she and the others had chased away just the night before. When she hung up, others rushed to the phone, seized with a sudden sense of purpose, and dialed more media outlets whose inquiries they had shunned in the previous weeks. The saucer group had found a new mission: to increase the size of their congregation by winning new converts. New converts to a belief that had been proven false.[195]

Why did the failure of a prediction trigger a spasm of new activity? Because the measure of the success of a web of memes—a myth, a hypothesis, or a dogma—is not its truth but how well it serves as social glue. If a belief system performs that function well enough, it can trigger the growth of a superorganism of massive size, even if its most basic tenets prove dead wrong.

In the early 1840s, a wealthy American farmer named William Miller claimed to have figured out the date of the world's end. The great time of reckoning, when fire would scourge the earth and usher in a thousand years of peace and righteousness, would arrive, said Miller, in 1843. Miller's passionate preaching inspired fifty thousand converts to wait patiently for the moment when their belief in the agriculturist's words would put them at the center of a new world order. Many closed their shops and gave away their farms in anticipation of the happy day. But

1843 came and went, and this earthly orb remained intact, unsinged by any cosmic blowtorch. Miller changed the date of his prediction to October 22, 1844. When the newly revised moment of death and glory rolled around, Miller's claim once again proved empty. But none of that prevented Miller's ideas from welding together a massive social group. The creed his disciples founded is best known today as Seventh Day Adventism. As of 1981, the movement could claim 3,668,000 adherents in 184 countries.[196]

Marxism also attracted its following with predictions that failed to come true. The dictatorship of the proletariat, believers said, would be only a temporary phase. Freed from the oppression of capitalism, citizens would lose their greed, their aggression, and their desire to lord it over their neighbors.[197] The New Communist Man would emerge—a creature of infinite goodwill, eager to work and help his comrades. Once the society consisted entirely of these blissfully transformed mortals, government would simply melt away. Unfortunately, when the shackles of capitalism were removed in the Russian Revolution of 1917, men and women remained as selfish, lazy, and argumentative as ever. A rapidly bloating Soviet state was bolstered by armies of secret police whose efforts were designed to restrain the foibles of those who had failed to undergo transformation. But for generations, the flaws of Marx's idealistic prophecies no more stopped the march of his ideas than the failure of William Miller's forecasts slowed the growth of Adventism.

Even today, when the Marxist states have supposedly crumbled, Western academics and Russian protesters bearing posters of Stalin are convinced that the old ideology's woes are merely a temporary rest stop on the long march to triumph, for Marxism has been a marvelous social glue.

If you can convince enough people of your worldview, no matter how wrong you are, you're right! The real significance of a meme is its power to pull together a superorganism.[198]

THE VILLAGE OF THE
SORCERERS AND THE RIDDLE
OF CONTROL

All diseases of Christians are to be ascribed to demons.

Saint Augustine

Why are humans drawn to ideas like filings to a magnet? Why do memes have the power to create, elevate, pacify, and kill? One answer is a Rube Goldberg machine built into the human frame, a physiological widget with some rather remarkable tricks.

In the Nilgiri hills of India, shortly before the Europeans came, there lived four tribes. One tribe, the Badaga, were farmers. Another, the Kota, were craftsmen. A third, the Toda, were herdsmen. And the fourth, the Kurumba, made and raised almost nothing at all. The Badaga, Kota, Toda, and Kurumba lived together in a delicate harmony, each supplying a vital something that the other three needed and paying for the indispensable products of its neighbors with its own handiwork. But there was one form of merchandise for which the Badaga, the Kota, and the Toda were willing to pay far more than for any of the others. Sometimes their need for this single good mounted to the level of hysterical panic; yet to us this commodity might seem the least essential of them all.

After the yearly harvest, the women of the Kota craftspeople fashioned pots and carried them, along with iron hoes, forks, and plowshares forged by their husbands, to the village of the Badaga farmers. There, the two tribes held a feast. When the celebration was over, the Kota craftspeople left, carrying a gift from their Badaga farmer hosts—a year's supply of wheat.

The Kota craftsmen not only made household goods, they were accomplished musicians. To obtain their meat, they played their tunes at the rituals of their herdsmen neighbors—the Toda. In exchange for music that soothed the gods, the Toda cattle raisers gave the Kota artisan-musicians meat from the sacrifice of a buffalo. So the Kota lived in cozy economic interdependence with the Toda livestock breeders and Badaga farmers, swapping handicrafts and music for steaks and bread.

But of all four tribes, the one with the greatest economic power was the Kurumba. Living in the jungle, the Kurumba did not raise wheat, did not make household utensils, and did not provide meat. They never even set forth to sell their wares; yet the work they offered brought the Kota craft folk trekking through the dense foliage to the Kurumba village, begging for a service that was totally intangible, one whose value cannot even be proven to exist. The Kurumba were sorcerers.

The Kota utensil makers paid regular insurance to these forest magicians. After all, the Kurumba spell weavers controlled the dark forces that could snatch you in the middle of the night and bring you down with dropsy, epilepsy, or sleeping sickness. In his classic book *Economic Anthropology,* Melville J. Herskovits writes:

> The Kurumba exacted all the market would bear, and on occasion their demands were anything but modest. When a Kota fell ill, for example, his relatives, indicating how they had been regular and generous in sending gifts to their Kurumba worker of magic, would complain that he had not fulfilled his part of the agreement to keep them from harm. The customary reply would be that some especially powerful Kurumba sorcerer had been insulted by a Kota, or had become envious of their good fortune, and was therefore sending unusually strong magic against his victim. Only sustained effort, to be called forth by the giving of extra gifts, might counteract this influence; and since there was no other recourse, the Kota would have to give more and more lavishly.[199]

Since nature endows the body with vast arsenals for self-defense, the majority of the Kurumba necromancers' clients recovered. Occasionally, however, one succumbed. When a relative died, the furious Kota artisan family did not ask for their money back. Nor did they complain that the sorcery of their jungle neighbors was a fraud. Far from it. The Kurumba 'protector' was offered sympathy for having had to grapple with so powerful an adversary.[200]

The members of the Kota tribe did not know that their Kurumba spell makers provided them with about as much protection against the evil spirits as a Kmart Superman costume would offer against a drive-by shooter's bullets. They clung to the belief in their "medical" protectors with a passion that went far beyond the borders of reason.

But why? To the Kota tribesmen, illness and death were forces they could not touch, manipulate, slow down, or shoo away. The sorcerers sold the notion that through their intercession, mere mortals could stop the unstoppable, bringing death to heel like a well-trained dog. They offered a fantasy that is more important to us than the reality of our daily bread, an illusion that actually can make the difference between life and death: the illusion of control.

Control is extremely powerful stuff. Pioneering anthropologist Bronislaw Malinowski, researching the Trobriand Islanders from 1914 to 1918, demonstrated that these South Sea fishermen used ritual to generate a false sense of mastery when they were about to set out on deep-sea expeditions—highly unpredictable ventures in which finding prey was more a matter of luck than skill and the chances of being drowned in a storm were great. On the other hand, the islanders didn't bother with mumbo jumbo when harvesting sea life in their predictable lagoons. Ritual, Malinowski concluded, was a means of creating a false sense of control when reality was intolerably slippery. In his 1927 *The Future of an Illusion,* Sigmund Freud went a step further and declared that man will cling to religion's fantasy of control as long as science fails to give him actual power over his destiny.[201]

But the extent to which the need for control is stitched into our biology did not become apparent until roughly forty years later when it was discovered that lab animals who are given control live longer, and have higher antibody counts and fewer ulcers.[202] Jay Weiss, at New York's Rockefeller University, put two groups of rats in cages with electrically wired floors. When the current was turned on, the rats' unprotected pink feet would receive a painful shock. One group of rats could neither stop nor escape the onset of the electrical charge. A second group was granted a modest privilege. Its cages were outfitted with a switch that allowed these rodents to turn off the electricity. The group with the switch, in short, had control.

The rats who had been able to control their punishment emerged from the experiment in relative health—despite their heavy dose of electrical torment. But the group that had no control came out of their ordeal with a high incidence of stomach ulcers. The difference occurred despite the fact that the two groups of rats received precisely the same amount of electrical shock at precisely the same time. What's more, in similar studies at the University of Colorado, after only a short exposure to uncontrollable pain, the immune defenses of helpless rats crumbled, leaving the creatures highly vulnerable to infection and to cancer-causing drugs.[203] Control protected the health of the rats with the switch, and the lack of control stripped their hapless brethren of even their own internal protectors. Numerous additional experiments have confirmed control's benefits to physical health.[204]

Control is also the magic ingredient that keeps us alert in the face of danger. It does so by suppressing the output of endorphins. Endorphins are chemicals produced by the body to soothe our pain. They are similar in molecular construction to morphine,[205] and they have morphine's ability to smother suffering. Endorphins have been glorified in popular literature as blessed biological benefactors. In reality, however, they are seductive poisons. Endorphins' power to anesthetize is great, but they cripple us in the bargain, shuttering our perceptions and cutting our resistance to disease.[206]

Two groups of rats were placed in shock cages. One group could escape the shock by jumping to an unelectrified platform; a second group could not. In short, the rats who could make the leap had a primitive form of control. The systems of the rats who could not control their shock were flooded with endorphins. The rats who did have control, on the other hand, avoided an endorphin surge. The rats without control were afflicted with endorphin's dulling of the senses and the mind, while the rats with a handle on their fate remained perceptive and alert.[207] Other experiments indicate something equally ominous: the lack of control disables the ability for long-term potentiation of neurons—in other words, it wreaks havoc on the ability to retain and act on vital information.[208]

Control, in humans and rats, energizes the mind. A lack of control can cripple mental powers. Two groups of human subjects were given a set of complicated puzzles and a proofreading chore. Both groups had to carry out their tasks while an irritating noise grated away in the background. There was one major difference between the two experimental contingents: one had control and one did not. The tables at which one group of subjects sat had a button. With that button, they could turn off the wretched sound. The second group's members had no control at all. Without a button, they simply had to grin and bear it. The group with the control buttons on their desks sailed through the puzzles and made only a modest number of errors in their proofreading. The group with no control did miserably. They mastered five times fewer puzzles, and their proofreading was atrocious. The deprivation of control had clouded their minds. Strangest of all, the group with the button never actually pushed that button once. It wasn't the noise or lack of it that affected their performance; it was the mere *idea* that if they'd wanted to, they could shut it off. It was the thought of control.[209]

The Kota came crawling to their sorcerer neighbors begging for an invisible form of help because forest witch doctors sold the impression that through their services men could hold their fingers above the unseen button that turns off death and disease. This belief in magic is one clue to our need for memes. Religious and scientific schemes—clusters of guesswork that sometimes seem like a madman's dreams—offer the feeling of control, an indispensable fuel for the physiological powerhouses of life.

THE MODERN MEDICAL SHAMAN

How backward these Kota villagers were, you may very well say. What strange nonsense preyed on their primitive minds. How fortunate we are that in our modern age, few of us are this gullible. But we are.

Like the Kurumba sorcerers, modern doctors sell the illusion of control. Often when you describe your symptoms to your M.D., he gives you an indifferent look, as if no such problem exists. You are not the only one your doctor treats this way.

Norman Cousins, in *Human Options,* described a case that he considered all too typical, one "in which the hospital staff was contented with a half truth. The investigation of the patient was decidedly unscientific in that it stopped short of even an attempt to determine the real cause of the symptoms. As soon as organic disease could be excluded, the whole problem was given up. But the symptoms persisted. The case was a medical failure in note of the fact that the patient went home with the assurance that there was nothing the matter with her."[210]

According to the National Center for Health Statistics, Americans make *six hundred million* office visits per year. Doctors conclude that the patients in over half these cases have no real problem.[211]

Why would a man selling his ability to deal with disease pretend that your affliction is a whim? After all, the symptoms one generation swears are "in your head" are often shown by research to be real a few decades later. But a doctor does not generally confess ignorance. He is selling the illusion of omnipotence: the illusion that through consulting him you gain control over your body, the same illusion sold by the sorcerers of India.

Occasionally, your physician changes tactics. He gives you a name for your problem but no cure. The name alone—like a magic talisman—makes you feel you have a problem your doctor can control. Or the doctor gives you a prescription. Often it's simply for an anti-inflammatory drug, an imitation aspirin. In most cases, an anti-inflammatory agent will not cure you; but it may provide momentary relief from a symptom, and it could produce the placebo effect, another benefit of the illusion of control.

The doctor may send you to the hospital for tests, but try asking him what treatments are available once the tests are over. In many cases, there are none—or, at most, one or two. The tests—and the elaborate equipment used to administer them—are frequently part of a show designed to enhance the illusion of a vast technology providing the doctor with control.

Take, for example, one of the current age's most common afflictions—the back problem. If you come down with back pain and are sent for tests, you can have myelograms, X rays, CAT scans, or nuclear magnetic resonance scans. But ask first what the doctor can do to treat you when the tests are over. The range of remedies includes a laminectomy (surgical removal of a disk, after which two vertebrae are generally fused);[212] an anti-inflammatory drug; a psychiatrist; physiotherapy; or the injection of papain. Generally, the test results do not influence the doctor's choice of cure. So why the tests? To cover the doctor against potential malpractice claims and to sell the illusion of control.

Medical science has made massive progress since the days of leeches and bleedings but is still helpless in the face of many of mankind's ailments. In fact, there is a good chance that the doctor's art has had far less to do with the advance of modern health than any of us would like to think. Several medical authors in 1987 and 1988 produced illuminating historical critiques of mortality statistics. These analyses demonstrated a puzzling fact: that mankind's lengthening lifespan owes less than we might imagine to modern drugs, diagnostic techniques, surgery, hospitals, or any of the other tools of contemporary health care. The investigators pointed out that devastating illnesses like typhoid, cholera, measles, smallpox, and tuberculosis began their decline in the mid-1800s. Over time, these sicknesses dwindled to a tiny fraction of their previous levels. The wonder drugs usually credited with eradicating the diseases—antibiotics—were not invented until nearly one hundred years *after* the illnesses started to disappear. Tuberculosis, for example, decreased by a startling 97 percent from 1800 to 1945. Only then was streptomycin finally introduced to mop up the meager fraction still remaining. Apparently, it was not just the set of bottles in the doctor's bag or the miraculous pills dispensed by his prescriptions that virtually destroyed the deadly scourge.[213]

Exactly what did produce the dramatic improvement in contemporary health still eludes the experts. Some say it was better nutrition, the introduction of clean water supplies, and the improvement of sanitation. Others, like California epidemiologist Leonard A. Sagan, suggest that it was the greater freedom—and hence degree of control—that became available to the average citizen.

Maybe it's not so surprising, then, that a vast variety of the symptoms that afflict mankind puzzle even the most erudite doctor. Over 50 percent of the patients who troop into medical offices are sent home with the assurance that there's nothing really wrong. Even "well-understood" illnesses are far less controllable by medical techniques than physicians are willing to admit. However, the medical profession deals with this dilemma by hiding its ignorance. According to a 1987 study by psychologist Dan Bar-On of Israel's Ben Gurion University,[214] patients are usually better able to predict the impact of their malady than their doctors. What doctors are selling, then, is not necessarily the ability to cure us but the illusion of control.

While medical practitioners make their living from the human hunger for control, they themselves are also frequently victims of that hunger. Like their patients, doctors need desperately to believe that they can, indeed, dominate the forces of disease and healing. Psychological research suggests that people tend to block out what they can't control and focus on what they can.[215] The medical profession's eyes are frequently blinded by this phenomenon.[216]

One good example is the medical community's former refusal to acknowledge the existence of adolescent depression. Frederick K. Goodwin, scientific director of the National Institute of Mental Health, explains that the psychotherapeutic community only recognized the existence of severe depression in teenagers and children in the mid-1970s. Goodwin says, "Until maybe 10 years ago, we believed that severe depression was solely an illness of adults. Adolescents didn't develop 'real' depression—they just had 'adolescent adjustment problems,' so most psychiatrists didn't and still don't think to look for it in kids. Now, however, we know that idea is dead wrong.''[217]

In the fifties or sixties, you may have had a fourteen-year-old who couldn't sleep, who'd stopped eating, who was spending a good deal of his time weeping, who may have been a good student—bright, cheerful, and sociable—but now seemed cut off from other people. If you'd taken that child to a psychiatrist or psychologist, the doctor would almost certainly have dismissed the symptoms as merely a phase.

Meanwhile, in 1951, medical researchers developed a new drug, iproniazid, to treat tuberculosis. When the substance produced strange side effects—making patients euphoric and unusually energetic—it was slated for the medical trash can. Then, in 1957, a few psychiatrists discovered that iproniazid could be used to relieve depression in patients who hadn't responded to any other form of therapy. But it wasn't until the seventies that the use of these miraculous new pills spread through the medical community.[218]

When antidepressants began to spill from the pharmaceutical factories, a phenomenon that hadn't existed in the minds of doctors emerged from the shadows of denial. The physician was suddenly willing to see a set of symptoms that a few years earlier he had airily dismissed. Today, the good doctor sagely pronounces the new diagnosis he's picked up from the medical journals. And he centers his pen over his pad to jot a prescription.

One old expression says, ''A man whose only tool is a hammer analyzes every problem in terms of nails.'' Without the hammer of antidepressants, the doctor once denied the existence of adolescent depression. Now that the physician has a tool that provides control, he is ready to open his eyes. The doctor, it seems, craves control as deeply as his patient.

CONTROL AND THE URGE
TO PRAY

Religious leaders in the distant past capitalized far more avidly than doctors on man's addiction to control. During the eleventh century, a battle raged over who would hold supreme authority in Europe. The leading contenders were the Holy Roman emperor and the pope. The emperor commanded massive armies and treasuries overflowing with wealth, and he had the loyalty of noblemen spread over a vast distance. The pope, too, had his treasury and armies, but he possessed a tactical weapon of a kind no emperor could match: the illusion of control.

In the two hundred years since Charlemagne, the Holy Roman emperors had lorded it over the popes. To ascend the papal throne, a candidate was forced to obtain the emperor's permission. And just to make his subservience crystal clear, the newly elected pope underwent a solemn ceremony in which he was handed his symbols of power by the emperor himself. The ceremony telegraphed the message that in heaven the pope may derive his authority from God, but on earth, he receives it from the Holy Roman emperor.

In 1073, a gentleman won the papal election who was not willing to take this procedure lying down. His name was Hildebrand (later known as Pope Gregory VII). Hildebrand wanted to make the Church answerable only to the dictates of God. In other words, he was dead set on elevating the Church's power over that of the emperor.

Hildebrand had a reputation for never sidestepping a good fight. The new pope started his reign with a virtual slap in the face to the Holy Roman Emperor Henry IV. The pontiff never bothered to arrange the elaborate ceremony in which he would humbly accept his crown from Emperor Henry. Instead, the new Holy Father simply took the throne himself. Then he sent a curt message to the emperor informing his abashed majesty of this fait accompli. Two years later, the pugnacious pope added insult to injury. Rulers had long enjoyed the privilege of appointing the bishops within the boundaries of their own realm. Hildebrand declared that this practice would immediately come to a halt. From now on, the church in Rome would make these appointments.

The gall of the new pronouncement was startling. Under Pope Hildebrand's new system, religious officials would no longer be men the secular authority could count on for loyalty or cooperation. They would no longer be extensions of the king's own bureaucracy. Instead, these powerful local figures would be virtual foreign agents. As if this weren't bad enough, Hildebrand went out of his way to tweak the emperor's nose. He excommunicated a few of the sovereign's closest advisers. Then he demanded imperiously that His Majesty show up in Rome to "defend himself against charges of misconduct."

This was about as much as Henry IV could take. The ruler finally decided to show the upstart pope just who was boss. Henry convoked a church synod within his territory—a synod of clerical figures loyal to himself. Under Henry's guidance, the cooperative churchmen declared in no uncertain terms that Pope Hildebrand was fired. Then Henry sat back, smugly confident he had won. After all, how could an ecclesiastical prayer juggler stand up to a sovereign who commanded the greatest armies in Europe and could crush whole countries on the merest whim? But Henry had overlooked something.

The Vicar of Christ demonstrated that he, like the inhabitants of the village of the sorcerers, had a stranglehold over a weapon no mere king could command. Hildebrand excommunicated the population of Germany. In response, the German citizenry, fearing that their souls would be cast into eternal torment, so pressured their ruler that he was forced to travel to Canossa and stand barefoot in a snowy courtyard for three days begging the pontiff's forgiveness.[219]

Like the Kurumba sorcerers, the pope claimed command over forces that were invisible. He maintained that his priests had power over the hidden gateways that led to an unseen heaven and hell. With his influence over an unprovable realm, the pope claimed the right to control the uncontrollable.

The Medieval Church made substantial sums by selling its illusions.[220] And indeed the phantasms pardoners peddled—control and hope—were vital to individual survival. (A study of 2,832 subjects by Robert Anda of the Centers for Disease Control showed that adults deprived of all sense of hope are four times more likely to die of heart disease.)[221] In real life, the average serf was nailed to the land as Christ had been nailed to the cross. He was subject to famine and plague. The major decisions that affected his days were made by the lord of the manor. Periodically, a serf's cottage was ravaged, his crops destroyed, his cattle taken, and his wife raped by the troops of a neighboring noble, a passing foreign military group, a band of brigands, or sometimes by the troops of his own king.[222] The serf had no hope and he had no control.

But hope and control, as we've seen, are biologically necessary to both the immune system and the brain. The Church said this earthly life would be but a short period of torment followed by a stretch of time that counted infinitely more—an eternal afterlife. Only one in a thousand, said the Church, would make it to the golden gates of paradise. But that, at least, was hope. And there were ways to guarantee that you would be among those bound for heaven's glory: you could show repentance for your sins; you could purchase pardons that brought you immunity to the results of your evil acts; you could share in the body and blood of Christ through the Communion offered by your local priest on a regular basis for a small fee; and you could go on a pilgrimage. No wonder the citizens of Germany found the pope's excommunication intolerable. He was excluding them from hope, snatching away the only thing that made life tolerable. More to the point, he was depriving them of the fantasy of control, a necessary device for tricking the body into survival.

Religion continues to offer its minions that vital slice of delusion. Fundamentalists Anonymous—a group of former religious extremists who have bolted from the reactionary Christian movement—says that to the fundamentalist, Christ is a device for resolving every puzzle. Submission to religious authority, the fundamentalist believes, will give him control over the vagaries of life.

As we've seen before, modern fundamentalist leaders are following in the footsteps of Pope Hildebrand and attempting to translate their grip over the illusion of control into political power, using organizational entities like the Christian Coalition, the Christian Voice, the American Coalition for Traditional Values, the American Family Association, and the

POWER AND THE INVISIBLE WORLD

⇒❦⇐

"Fetch me a fruit of the Banyan tree."
"Here is one, sir." "Break it."
"I have broken it, sir."
"What do you see?"
"Very tiny seeds, sir."
"Break one."
"I have broken it, sir."
"What do you see now?"
"Nothing, sir."
"My son," the father said, "what you do not perceive is the essence,
and in that essence the mighty banyan tree exists.
Believe me, my son, in that essence is the self of all that is. That is the True."
                                          Chandogya Upanishad, vi, 13

The fundamentalist submits to the authority of his minister. Holy Roman Emperor Henry IV bent his knee in the snow to the pope. And we submit to the authority of the doctor. Control is power, says the dictionary, and the dictionary is quite right. But how is the illusion of control turned into power over you and me? Witch doctors, prophets, priests, scientists, and medical doctors gain our confidence by giving us the feeling that they have tapped into an invisible truth, a truth hidden behind the surface world we see. The keepers of the mysteries exude a certainty that through their contact with this invisible world, they are able to solve the problems that to us seem baffling.

As a result, we give these figures almost anything they want. Americans are currently pumping money into the pockets of physicians and their attendants at a rate that is truly astonishing. Expenditure for medical care, to everyone's dismay, is the fastest growing segment of the American economy. Yet, on one of the simplest measures of overall health—infant mortality—we ranked a dismal twentieth. The babies of those nations that have not indulged in a medical spending spree actually are more likely to stay alive.[223] Said Senator Lawton Chiles of Florida, ''If your child was born in Singapore or Hong Kong, it would have a better chance of reaching the age of one than if it was born in the United States.''[224]

Over five hundred years ago, folks offered up the same kind of frantic financial sacrifice to their priests. The result: roughly a third of the land of England was in the hands of the Church until Henry VIII took it away. What's more, the income of the Church in Henry's day was 300,000 pounds per year. The revenue of the English government, on the other hand, was a mere 100,000.[225] Outside the Western world, the power of holy men to pull in earthly goods like a supermagnet lasted much longer. Before China's unconscionable invasion of Tibet in 1950, priests ran the country and controlled an unbelievable percentage of the Himalayan nation's wealth.[226] In America, television evangelists are resurrecting the priestly phenomenon, sucking in the dollars of believers by the sackful. But how do priests, scientists, and physicians manage to cement their power?

The rise of Isaac Newton in the early 1700s allows us to see one such power structure being crafted. Newton established the absolute authority of science in the minds of Western men by implying that he could see into the very forces of the cosmos. With his mathematical theories, Newton accomplished something that had eluded all the wise men of his age: he explained the motions of the moon and planets.

Newton's followers claimed that Sir Isaac had generated a method that would penetrate *all* of the workings of the universe.[227] When the Newtonian scheme successfully mastered the intricacies of the solar system, it seemed these enthusiasts must be right. In reality, Newton's system predicted very little about the universe in which men actually lived. It gave no explanation whatsoever for men's depressions and despairs, for their desires and their greed. It could not predict or control war. It was utterly baffled by the problem of how an inert sphere of matter—an egg—turns seemingly by itself into a chick. Yet enthusiasts in the Age of Reason felt Newton held the keys that would unlock all of life's mysteries. Newtonian science swept nearly every other form of analysis from respectability. And Newton himself obtained the power to crush his rivals, dictate what was intellectually acceptable and what was not, and even to steal the credit for the discoveries of his fellow natural philosophers.[228]

When men look desperately for masters of control, they do not peck at the limitations of the new savior. They seize the idea of his power with hungry enthusiasm, for the new sorcerer offers the promise of influencing that which seems impossible to influence.

The same trick of using a few insights into the motions of heavenly bodies to imply mastery over all of the known universe vaulted previous generations of savants to positions of incredible power. Babylonian astronomers, three thousand years before Newton, became power brokers by predicting the seasons and producing a viable calendar.[229] So did the Chinese emperors who made sure that the computations of their own calendar were kept top secret since those calculations were the key to their hold over the state.[230]

In Central America, Mayan and Aztec priests also held massive power by virtue of astronomical virtuosity. These priests were housed in elaborately architected city centers while the peasants who raised the crops they ate resided in hovels in the countryside.[231] Tribute of all kinds flowed to the divine intercessors: food, clothing, and gold. Furthermore, these middlemen for the gods could dictate life and death. The holy men ordered the armies of the Aztec empire to scour distant lands in search of captives. Then the priests sacrificed the prisoners of war by the tens of thousands to satisfy their invisible gods. On occasion, the Aztec clerics would cut open the chests and tear out the hearts of as many as five thousand humans in a single day.[232]

The key to the power of the Aztec priests was similar to that which made Newton the king of science. These Central American men of wisdom made careful astronomical observations and worked out a system for predicting heavenly events. Because the priests could foretell a few celestial occurrences, the citizens around them concluded that priestly power didn't stop there. Surely, the priesthood must have found a way to peer into the invisible machinery that dominates life and death, illness, misfortune, and the luck of the triumphant.

Newton, the Aztec priests, the medieval pope, and the modern doctor all gained power through a simple device—the impression that they held the levers with which man could manipulate an invisible world.

Our cultures, in fact, are our collective fantasies about the worlds we cannot see. They are tapestries of memes. If you were a Sioux a hundred years ago, you believed that there were spirits who manifested themselves in eagles and clouds.[233] If you're a modern westerner, you know all of this is bunkum. If you're a traditional New Guinean, you believe that ancestors hover around your hut ruling the family's affairs like puppetmasters pulling the strings of health, wealth, and happiness. If you're a Christian, you feel that aside from an occasional ghost haunting a house in Amityville, the ancestors have all had the good grace to depart shortly after they died. On the other hand, if you're a Christian, you believe that a man who breathed his last on a cross two thousand years ago was the son of a vast and immortal being hovering somewhere above the visible sky, and that someday this long-departed soul will return to earth and usher in an entirely new order of things. If you're a Buddhist, you know with absolute certainty that this is a figment of the Christian's imagination.

Many of us moderns are convinced that we are above believing in unseen forces that quietly shape our destiny. But are we? Not quite. Our beliefs in invisible powers mold our behavior as surely as the certainty that an ancestor's spirit hovering in the corner of his hut influences the habits of the traditional New Guinean. You've seen coughing and sneezing, but have you ever seen a germ? Only the caretakers of our invisible world have spotted them—the scientists. Yet many of your sanitary decisions may well be based on the fear of these microorganisms, minimonsters of which you've never caught a glimpse. You probably avoid cholesterol, but have you ever had a peek at it? For those who don't use a microscope, it's as ethereal a force as the cloud-riding spirits of the Native Americans.

With poor guidance, we stumble our way through the invisible, sometimes blundering badly. We accept most child-rearing theories, for example, with no evidence. The Sioux are horrified that we hold up a newborn baby and spank it. They believe the baby should be cradled tenderly and given love.[234] Victorians thought that by holding a baby you spoiled it. The same notion persisted in America through the first half of this century. In 1928, J. B. Watson, the leading American psychologist of his time, repeated the concept emphatically in a book that became the child-rearing bible of the next twenty years:

There is a sensible way of treating children. Treat them as though they were young adults. Dress them, bathe them with care and circumspection. Let your behavior always be objective and kindly firm. Never hug and kiss them, never let them sit in your lap. If you must, kiss them once on the forehead when they say good night. Shake hands with them in the morning. Give them a pat on the head if they have made an extraordinarily good job of a difficult task. Try it out. In a week's time you will find how easy it is to be perfectly objective with your child and at the same time kindly. You will be utterly ashamed of the mawkish, sentimental way you have been handling it. . . . In conclusion, won't you then remember when you are tempted to pet your child that mother love is a dangerous instrument. An instrument which may inflict a never-healing wound, a wound which may make infancy unhappy, adolescence a nightmare, an instrument which may wrench your adult son or daughter's vocational future and their chances for marital happiness.[235]

Watson pontificated that mother love was a force that inserted itself into the invisible machinery of a child's psyche, destroying him as surely as an avenging ghost. You and I would not be able to see this psychic damage until it was too late, but Watson could peer directly into the unseen world of the human mind. After all, he was a psychologist.

Research indicates that Watson's advice was just short of criminal. Anthropological studies of the !Kung of the Kalahari show that children whose mothers pet them constantly often turn into far more self-confident adults than the coldly raised progeny of civilized Londoners.[236]

Why did parents follow experts like Watson into what was apparently a pit of error? Why are they today following specialists who say that you must deal with your child's misbehavior only by reasoning with him or that you must encourage a child to vent all his hostile feelings? Because raising our young is another area in which we are wrestling with invisible forces. In fact, we do not know what spoils a baby. We often do not know why he is crying at this very minute. We certainly don't know what effect our picking him up today is likely to have twenty years from now. And when we are pathetically attempting to deal with the invisible, when we have the least evidence of reality, that is when we are most vulnerable to the power of the experts.

EINSTEIN AND THE ESKIMOS

⇒ ⇐

Before the coming of the white man to the north, Eskimos believed that if they cut slabs of ice, organized them in a circle, formed a dome, and lived inside they would please the spirits. Apparently, the plan worked. The contented spirits made sure the Eskimos stayed warm even when the temperature dipped to forty below zero outside. Eventually, Western scientists showed up and tried to explain how the Eskimos had mastered an invisible force called thermodynamics. According to these presumptuous foreigners, the igloo's tunnel-like entrance preheated outside air, the movable snow-block door let in precisely the amount of this air that could be further warmed by the seal-oil lamp inside, and the adjustable hole in the roof allowed just enough of the resulting rising currents out to create the convection that kept the whole thing going. The sturdy igloo builders pooh-poohed thermodynamic nonsense. They knew exactly what invisible powers kept their dwellings warm.[237]

Indians worship an invisible divinity—the cow goddess. As a result, cows eat and Indians starve. We are appalled. Why don't the hungry Indians simply carve up some of the cattle wandering nonchalantly down their streets and wolf down a burger? Anthropologist Marvin Harris has shown that if the Indians slaughtered their cows and threw them between the buns of a Big Mac, far more of them would starve. Harris explains that the Indians survive by using the cows' dung as fuel, their traction to pull plows, and their milk to feed children. Killing the cows would make agriculture impossible, heating unheard of, and milk unavailable. The worship of the sacred cow works because it keeps alive the creatures on which the Indian economy is based.[238]

Pictures of the invisible world can have wild inaccuracies, but every view that flourishes does so because it solves at least one major problem. Balinese religious leaders kept their gods happy by throwing a series of holy feasts that peppered all four seasons. The country's agricultural practices were controlled by the ritual demands of this elaborate holiday calendar. But a few years ago, foreign experts persuaded Bali's farmers to free themselves from the dictates of superstition. Following the enlightened advice, cultivators planted and harvested according to modern schedules. The result was a disaster. Crops began to rot in the fields. Mice and insects got out of hand, eating away much of the harvest that did survive. It turned out that the ornate cycle of holidays with which the Indonesian priests had pleased their gods served a secondary function. It acted as a timing mechanism, mustering the farmers to open and close the sluices of the country's complex irrigation system, and to do their planting on precisely the right days to maximize crop production and minimize rodents and pests. In satisfying a set of arbitrary deities no one had ever seen, the priests had constructed a system more successful than any of Indonesia's up-to-date agricultural planners had yet been able to invent.[239]

Building pictures of the invisible world is the human way of trying to deal with the world we see. Each cosmology-making meme is a problem-solving device, allowing us to master dilemmas that a dog, a cat, or a canary has a great deal more difficulty coping with.

Animals and human beings are both up against a world where most of what determines their fate is invisible to them at the moment. To a monkey in a clearing, food is nowhere in view. Often, neither are the males he's competing with or the females he's competing for. The infants he's contending for the right to father do not, as yet, exist. The predators who could end his life are equally hidden from sight. But he has to deal with all of these to send his genes into the next generation.

To survive, a *human* has to deal with an even more complex invisible world. For a man on his way to work, most of the things that affect him are completely out of sight. Wife, children, boss, competitors at the office, stores that provide his food and clothing, or a natural disaster that could end his existence are all, for the moment, visible only in his imagination. But he has to measure these factors moment by moment to survive. As he sits in a car slowed by traffic, he is aware of his job, its purpose, the paycheck that won't arrive until Friday, the debate he and his wife may have when he gets home. And he has to make predictions.

What pile of paperwork should he dig into if he wants to finish the report whose deadline will come at the end of the day? How much can he afford to spend on a new suit if he plans to take the family to Hawaii for vacation? What should he say to his spouse when he steps back into the house that night to put her in a good mood? What should he avoid saying if he wants to sidestep a battle?

To make predictions like these, scientists construct models of the real world. For example, the nineteenth-century German Bernhard Riemann painstakingly worked out a mathematical picture of an imaginary territory. This curved landscape was warped in an odd way: it arched invisibly into a fourth dimension. Riemann used mathematical equations to feel out the features of this void like a blind man piecing together an "image" of an unfamiliar space by probing with his cane.[240] The diligent German ended up with a mathematical landscape portrait of an "n-dimensional manifold," better known to its friends as "curved space."

Einstein felt he could apply this fanciful worldview to the universe in which you and I live—a cosmos with the three tangible dimensions of height, width, and depth, and one extra dimension: time. In Einstein's view, our universe was indeed curved, forming a hypersphere that bulged outward into a dimension we cannot see.

From Riemann's curved world picture, the frizzy-haired physicist was able to predict a set of hitherto unobserved phenomena. When those predictions proved true, scientists adopted the Riemann model as an accurate portrait of much of the world invisible to them. That fantasy picture of an invisibly curved cosmos has been enabling them to make predictions ever since.[241]

Even animals need predictive powers. To see into the future, simple creatures like the frog have a prewired model of the world. The frog is built with a set of neural trip lines between his eye, the visual processors in his brain, and his tongue. These nerve cells are constructed to follow a set of simple instructions—erratically moving object: flick tongue; motionless object: don't bother. It works, and the frog ends up with food. His nerves embody a model of a planet in which objects flitting by are usually delicious.

But the frog's prewired picture of the world will not change with circumstances. Offer a starving frog an immobilized fly, and he simply will not touch it. His built-in portrait of the universe tells him that only objects that dart around are fit for dinner. If a captive frog goes on long enough ignoring the torpid bits of nourishment offered him, the flaws in his unbending world model could kill him.[242]

More complex animals, on the other hand, build parts of their models on experience. Their pictures of the invisible world are changeable. A dog is capable of quickly developing a model of things it's never seen before. Lock the hound in a room it's never visited, and the beast will immediately check out all the details, building up an image of the place and searching for an exit it's never even glimpsed. Thanks to an ability to imagine walls and doorways it has never seen, the curious canine is able to predict the existence of an escape route.[243]

Albert Einstein used the mathematical model provided by Bernhard Riemann to predict everything from the energy in the atom to the motion of light; but instead of employing math to construct models, our minds most often use metaphors.[244] Our brains are picture-making machines. Every culture has a world*view,* not an algebraic set of cosmic calculations. Uneducated medieval Christians pictured the earth as a flat disk that ended somewhere over the watery horizon. They predicted that small boats that sailed too far from the Atlantic shore would never be seen again. Renaissance intellectuals, on the other hand, revived the old Greek image of the terrestrial surface as a sphere. To Columbus, this meant that he might sail off into the western sunset and emerge with the sunrise on the earth's eastern side. Columbus's optimism was based on a portrait of an underbelly of the planet that Europeans had never beheld, an image of the invisible.[245]

Genes are the form of replicator that dominated the evolutionary marathon for nearly three billion years. But in the latest blink of geological time these strands of nucleotides have been outpaced by the matterless organizers called memes. Among the most potent memes are visions of things unseen. Like genes, memes do not operate in solo, but interlock in the mosaics that form *Weltanshauungs,* worldviews.

A culture's view of its world is generally a vast grid of metaphors starting with the creation of the universe and designed to answer every mystery in life. That diagram of the cosmos is a tool with which we pry open our environment, a tool that creates strange by-products. It offers an illusion of control—the illusion that turns on our immune system and our minds. The worldview also confers power on those who claim to be its guardians—sorcerers, doctors, scientists, and priests. It helps the powerful pull a social organism together. On occasion, as in the case of the Chinese Cultural Revolution, it even impels that social creature's members to lash out and kill.

Yet a culture's picture of the invisible universe, its unifying cluster of memes, accomplishes something more. Though it may be riddled with bizarre errors and ludicrous imagery, a vision of the unseeable produces some small fragment of real mastery. Pictures of the invisible world helped Columbus cross the ocean, the Eskimo tame his winters, and the citizen of Bali regulate his irrigation. Someday they may even help us moderns conquer the medical problems that doctors still insist do not exist.

THE MYSTERIES
OF THE
EVOLUTIONARY
LEARNING MACHINE

THE CONNECTIONIST
EXPLANATION OF THE MASS
MIND'S DREAMS

≫€

The secret behind the problem-solving abilities of worldviews is the same as that behind the success of superorganisms. It lies in the power of networks.

One of the most irritating mathematical dilemmas facing computer scientists is the "traveling salesman problem." Imagine that you're a salesman about to go out on a trip. You've decided to visit ten different cities. How would you figure out which town should come first on your route, which should come second, which should come next, and so on, to give you the shortest mileage? Simple, you'd just sit down with a map and have an answer in an instant. But it's not that easy. Turns out that the potential number of sequences in which you could hit these ten destinations comes to 181,440.

Well, how about flicking on the switch of your desktop PC? Unfortunately, a conventional computer is rather slow at solving traveling salesman problems. Normal computers measure each potential routing one at a time. With close to 200,000 different options to test, that's a time-consuming process. And the traveling salesman dilemma is one that industry encounters all the time. Phone companies laying cable, for example, run into variations on the conundrum that make the simple ten-city example look like child's play. What's to be done?

The stumbling block comes, in part, from the way normal computers are set up. Looking at a problem only one small part at a time, they can't see the big picture. In the mid-1980s, however, a new breed of computer experts began to experiment with machines that can "explore" a broader canvas. These are called "connectionist webs" or "neural nets."

Conventional computers deal with information the way a railroad switching yard deals with boxcars. A train a half-mile long arrives at one end of the yard in Chicago dragging cars destined for Boston, Washington, Philadelphia, and New York. The trains on tracks next to it are also a jumble of cars headed for Boston, Washington, Philadelphia, and the Big Apple. The trick the yard handlers have ahead of them is to take each train apart, couple all the cars headed for New York to one locomotive, and send that locomotive on its way.

At one end of the yard are twenty parallel tracks of trains waiting to be disassembled. At the other are twenty parallel tracks of trains that have been put back together and are preparing to depart. But the critical process takes place on one narrow set of tracks in the center of the yard where all the breaking down and reassembling is done. And each incoming train must wait its turn to get onto that slender stretch of working track.

A conventional computer also stores a sizable batch of information in a kind of holding pen but has to shoot it, one small bit at a time, through a processor where the real work of computation and comparison is accomplished. This is called serial processing.

Neural nets function in a radically different way. They don't use the narrow, railroad-track approach to information processing. Instead, they are shaped like spiderwebs that process information in parallel. The lines of the webs are electrical channels whose conductivity can be raised or lowered. The junctions where the lines meet are switches that can be turned on or off.

Neural nets can solve problems by making rough models of the real world as they learn from data we give them. Here's how you'd solve the salesman's dilemma using a neural net. You'd pick ten junction points on the web to represent ten cities. You'd adjust the conductivity of the lines between them in a manner proportional to the distance between them. (If two cities are one hundred miles apart, for example, you'd set the resistance to one ohm; two hundred miles apart, two ohms; three hundred miles, three ohms.) You'd turn on the current. The switch points representing your ten cities would flicker on and off for a few fractions of a second as the network sought an equilibrium. Then, it would disgorge the answer to your problem.[246]

In essence, the neural network works by using a model of the terrain the salesman is planning to travel. What's the advantage of this electronic model making? California Institute of Technology's John Hopfield showed that neural networks can solve the traveling salesman problem ten thousand times faster than a normal computer.

Neural nets, like the human brain, can infer an invisible world from scraps of visible information. Give the word *bat* to a neural network built by neuroscientist James Anderson at Brown University, and it will respond with a list of the qualities of animals. Give the machine the word *diamond,* and it will spit out geometric shapes. Give the two words together, and the machine will come up with *baseball.*[247] In its limited way, the neural net has inferred a broad picture of a complex and, at the moment, invisible miniworld—the baseball stadium—from two tiny fragments of information.

Humans do the same. At this moment, thanks to the words *bat* and *diamond,* you can picture a runner stealing bases. Look around you. How many runners in baseball uniform do you see before your eyes? Like the neural net, humans can infer a picture of the world from two little words.

The neural network technique does have its drawbacks. In the case of the traveling salesman problem, for example, Cal Tech's Hopfield points out that there is a bit of fuzziness about the net's answers. They are only the best solution 50 percent of the time, but they are close enough for all practical purposes. Ninety percent of the time, the neural nets pick one of the two top answers. That's one of the two best out of 181,440. Normal computers, on the other hand, solve the traveling salesman problem with 100 percent accuracy, but they do it too slowly to be of any earthly value.

Worldviews share the neural network's fuzziness. They are not precise, but they're frequently close enough. They can be wildly inaccurate. It isn't accuracy that counts, however; it's utility. They may be sloppy, but they render solutions to real world problems fast. As neural-net builder Hopfield says, "Biology, by and large, is not interested in finding the best things, just things that are pretty good that can be found quickly."

The creators of electronic spiderwebs have called them neural nets for a reason. These problem-solving skeins of wiring are deliberately modeled on the networks of nerve cells in the brain. In fact, the brain contains webs of circuitry so vast that the machines being built in scientific laboratories are minuscule by comparison.

The worldview you build from childhood is carried in billions of cells whose connecting threads are precisely adjusted to give you your picture of the world. The late Dr. Donald Hebb, one of the foremost theoreticians on the subject of the human brain,[248] called those networks "cell assemblies." He described these as circuits through which stimuli flow like waters through the delta of the Mississippi. If that's true, it would help explain a bit about the nature of understanding. Hebb's model of interconnected mental circuits helps us comprehend why, though we're hit with thousands of random perceptions, very few stick to us. Occasionally, one jumps out as significant because it seems to fit within our pattern of beliefs and can be patched into our existing neural net.

If we believe that life is a battle between Satan and God, some small event can seem proof positive that Satan is out to snare us. If we believe, as the Chinese and the Romans did, that the heavens are filled with messages about our fate, the sight of a shooting star may trigger a sense of imminent calamity. If our belief system says absolutely nothing about a relationship between the stars and life on earth, that same blazing meteorite will seem like a passing curiosity of no lasting significance and may never make it into the brain's circuitry at all.[249] In the case of the person who believes that the heavens portend events on earth, the sight of the shooting star is patched into an outstretched web of neural connections and takes its place in the greater whole. We hunt, over the next days or weeks, for the event it forecasts.

A neural network like this takes a lifetime to build. Without a web of cell assemblies, it would be impossible to recall the myriad events that parade past our eyes and ears, much less to make sense of them. It's easy to see why humans are willing to fight to the death to defend the memes that constitute their belief systems. To allow a faith or ideology to be overthrown would be to abandon a massive neural fabric into which you've invested an entire life, a network that cannot easily be replaced, perhaps that cannot be replaced at all.

When T. H. Huxley's favorite son died, a friend advised him to abandon his "blasted agnosticism" and accept the comfort of Christianity. Huxley answered that he couldn't "alter a set of principles established after so much thought and deliberation merely to assuage his . . .

grief.''[250] He refused to toss away a system of beliefs he'd built up over a lifetime. Early Christian martyrs felt the same way. They preferred dying riddled with arrows or torn apart by beasts to forsaking their worldviews.

With a neural net of enormous size, humans "see" the invisible forces ruling their lives. Neural networks helped the Eskimos invent the igloo, the Balinese create their irrigation system, and Einstein and Bernhard Reimann to paint a four-dimensional universe. Neural nets and the conceptual webs they hold give us our illusion of control over those things that elude our grasp. In fact, they are the imprecise mechanisms that sometimes give us control's reality.

SOCIETY AS A NEURAL NET

※❦

A social group is also a network. If a hunter corners a mother otter, she'll cry out for her mate. If the hunter kills her, her babies will starve. She's only one nexus in a web of relationships.

Marvin Minsky, co-founder of the Artificial Intelligence Laboratory at MIT, thinks of the brain as a society—a society of subassemblies cooperating to learn about the world.[251] The image can easily be reversed. A society is a brain, a learning device that works according to the principles that drive a neural net.

Like switches wired together in a connectionist web, a community of bees is constantly communicating to form a mass brain that can solve problems no single bee could ever tackle. In one experiment, scientists began by placing a dish of sugar water at the exit of a hive. Over the course of time, they moved the water, first a few inches from the hive, then a few feet, then a few feet more—always increasing the distance by a precise increment. The researchers expected that the bees would follow the dish and cluster around it. To their surprise, after a few days, the insects were doing far more than merely tagging along after the moving sugar water. The bees would fly from the hive and cluster on a spot where the dish had *not* been placed—the site where the insects *anticipated* the dish would be put next—and their calculations were right on target.[252] Working as a mass brain,[253] the bees had accomplished something humans are forced to endure in college entrance exams: they had solved the problem of a mathematical series.

The brain of a bee is an insubstantial thing—a slender thread of neural fiber scarcely capable of anything we would call intelligence. But the strength of a neural net does not lie in the limited abilities of any one node in the web. The strength of the connectionist intelligence—its problem-solving ability—is in the web itself: the constant feeling, touching, and communicating between the bees that pool their brains into one. The problem is solved not by a single bee but by the interconnected mass. The social network manages to solve problems in the world around it by the same principle that underlies the neural net. Connections the system finds useful are strengthened; connections that prove useless are weakened.

When a bee flies off, finds a source of nectar, and comes back home, it indicates to its hive-mates where the nectar is located, whether the trip involves battling a stiff wind or breezing merrily through still air, and if the cache of food is packed with nourishment or simply contains a marginal snack. The bee gets her information across by dancing on the wall of the hive, moving in a figure eight whose orientation indicates the direction of the nectar and whose length indicates distance and the difficulty of getting there.[254]

But how does an individual bee gauge the need for nectar in the hive? She does not step inside the habitation to check how full the honeycombs are. Incoming transport bees sense the demand for the cargo they deliver by a simple social cue, the kind of cue that humans have their antennae out for at cocktail parties. If a burdened bee arrives at the hive entrance and workers rush over immediately to unpack her nectar, the incoming flier knows that there is a considerable hunger for more deliveries. The freight carrier is imbued with a sense of energy and rushes off to pick up another load. If the nectar-laden hauler lands on the entrance lip and is left untended, if, in fact, she has to look around forlornly, begging for the attention of one busy worker after another, then she knows the need is not so great. She is hit with a sense of lethargy and bumbles back to work with sluggishness.[255]

When the bee who fails to find an eager welcoming committee goes through a downshift in mood, her response to social cues is more than an individualistic emotional dither. It is the secret to the social network's problem-solving power. For the communal learning machine operates by turning up the pace of elements that are needed and shutting down the speed of those that aren't.

People are also woven together in a superorganismic web, and our mood swings, like those of the bees, give that web part of its problem-solving power. Frederick Erickson, a sociolinguistic microanalyst at the University of Pennsylvania,[256] points out that humans constantly exchange beelike signals. We give our conversational partners cues, nodding our heads, smiling, grunting in the affirmative, gesturing with our bodies, frowning.[257] The average mortal knows from personal experience what the impact of those cues can be. If we run into a gathering of friends, spring a tantalizing bit of information on one of them, and everybody else edges over to hear it, we feel invigorated. Energized by mild euphoria, we may prattle on with additional details about the topic that's just drawn all this attention. If, on the other hand, we spring a piece of gossip that, to us, seems irresistible, and the people near us immediately march away, we become discouraged and are less likely to continue that particular line of conversation. Like the bee arriving with unwelcome food, we aren't motivated to deliver more of a tidbit no one seems to want.

The power of social cues to switch our moods on and off has an enormous impact on the movement of information through the social system. An idea everyone is hungry for will flow rapidly through the human network as the individuals propounding it are encouraged to repeat it in ever-more excited tones. The idea no one cares for is likely to wither away as its adherents grow discouraged and morose.

At its most extreme, those who feel their contribution is unwanted will do more than simply wilt for a few hours. When the revolution of 1917 bore down inexorably on Russia, it eventually became clear to the followers of the czar that all was lost. Protopopov, the czar's iron-fisted interior minister, looked like a man transformed. He appeared shrunken, beaten, and so prematurely aged that he was virtually unrecognizable. When Protopopov was at last forced to resign, he said, "Now there is nothing left to me but to shoot myself." Likewise, as it became clear that the czar's political power had slipped from his grasp and a revolution was in the offing, the ruler of all Russia went through a similar physical transformation. When Count Kokovtsov entered the czar's residence at Tsarskoye Selo for an audience on January 19, 1917, he was appalled by what he found. Russia expert Harrison Salisbury says that Kokovtsov "hardly recognized the Czar, so much had he aged. His face was thin, his cheeks sunken, his eyes almost without color, the whites yellow and the pupils gray and lifeless. They wandered vaguely from object to object."[258]

On the other hand, Kerensky, the leader of the victorious Socialist party, the Mensheviks, was invigorated like the bee who finds her cargo in great demand. Suddenly, Kerensky was catapulted from a position as a factional leader to head of state, leader of the Duma—the parliamentary body that now controlled the country. As he rushed to the Duma's podium and took command, witnesses said that Kerensky seemed to become a physically larger man. "He grew," exclaimed V. V. Shulgin, a monarchist Duma deputy who was present when Kerensky brought order to the confused deliberative chambers. "He grew in the mud of the Revolution."[259]

In neural nets built by computer specialists, nodes that are needed to solve a current problem are strengthened, and those no longer necessary weaken and shut down. The same thing happens with human beings.

Like bees, we are in constant contact. Clint Eastwood rides into town from the desert, carrying his own gun, and shares his thoughts with no one. On the other hand, when you watch a classic Clint Eastwood film, you're generally seated at your television. A half an hour before you saw the movie, you watched the news. That broadcast gave you a sense of where your country stood in the relative hierarchy of nations at that moment. The economic portion gave a feeling of the health of the superorganism of which you are a part. The ads conveyed an impression of the group posture on everything from hairstyles to automobiles. By the time you go to bed at night, you've exchanged information with dozens of people. You've absorbed data that has brought you in touch with every level of your society.

No individual confronts his environment alone. None of us wanders the woods in solitude, killing our food with weapons we alone have invented and made. When you're out of work because of an economic depression, the depression is something over which you have no personal control. Like the bee, the best you can do is rush from one fellow denizen of the society to another, swapping information and hoping for salvation from other human beings. You can curry favor with your former boss, or call all your contacts for another job. Your crisis is a reminder of your dependence on others.

An economic depression is a paroxysm in a human network, a network that produces food 1,500 miles from you, the food that will someday sit on your table. Humans you have never met, tucked in a distant corner of a sprawling economic web, craft your furniture and build your house. You do only the small part a bee does in the corridors of her hive. It is the social system, the superorganism, that performs the task of facing down a hostile nature.

Individuals nourished by this interdependence come up with marvelous things. Shakespeare took a language shaped by the networking of millions of minds over eons of time, pulled historical legends from early England's rise and tales from a Roman Empire that had long since died,[260] then combined them with the zeitgeist of his era and created plays unlike any humanity had ever seen. Sigmund Freud knit together the experiments of the French scientist Jean Charcot, the experiences of his fellow physician of the mind Josef Breuer, the "talking cure" invented by one of Breuer's patients, Greek myths, and the religions of primitive tribes to craft the theory of psychoanalysis.[261]

The brain that makes a decision about whether to throw George Bush out of office or support Bill Clinton's health policies, about whether to follow Mao in a violently destructive Cultural Revolution or demand that Mao himself be overthrown, about whether to turn the individualism of Eric Fromm into a national creed or go hog-wild (as the revolutionary Russians did) for collectivism is not a mass of tissue the size of a cantaloupe. It's millions of those compact masses, meshing through hundreds of forms of contact every single day.

The sense of helplessness that immobilizes an individual who has lost his feeling of control and the flood of endorphins that disconnect the mind and immune system are both manifestations of the superorganismic mechanisms that turn a human society into a learning machine. So are the sense of despair that made a toppled Russian leader feel like putting a gun to his head or an unemployed husband in the early 1930s end his life. All represent the loosening of connections to elements that aren't making a contribution.

When a bullet goes through the heart of a human being, a delicate interaction ceases—a set of relationships between cells. Life ends.[262] The unseen element that we sometimes call the soul has disappeared. The set of interactions that gives a social group its shape, the invisible web of connectedness that knits together a society, the network of structures that make a culture, these are forces whose power transcends the existence of any mere individual. These are the social organism's soul.

Evolution is not just a competition between individuals. It is a competition between networks, between webs, between group souls. The new forms evolving on the face of this planet are not resident only in the features of individual animals or men. They don't merely consist of longer legs or bigger brains. The new forms are impalpable and invisible. They consist of the varieties of cooperation that grow up between solitary creatures—the unseen ties that bind those creatures together into a larger unit. They consist of the entelechy, the shape not of the parts, but of the sum of the parts.

When Japan and the United States battle for economic supremacy, when the Crusaders of Christendom march off to challenge the empire of Islam, or even when rival groups of Red Guards clash, the struggle is not a battle of men but a battle of networks, learning machines bound together by memes, testing their shapes against each other. From a history filled with these contests, the far-flung webs and invisible networks rear up ever higher into a lofty stratosphere of form, hurling the world toward its destination of an ever-more complex future.

THE EXPENDABILITY OF MALES

Men were designed for short, nasty, brutal lives. Women are designed for long, miserable ones.

Dr. Estelle Ramey

In many primitive societies, there are two very separate worlds—the world of women and children, and the world of adult men. A male infant lives in cozy comfort, clinging to his mother, suckling until the age of three or four, playing with other little boys and girls, and being hugged and fussed over by his older sisters. Then, at roughly thirteen, he goes through a brutal expulsion. He prepares for rituals that will separate him from the warm world he has known. These rites will emphasize his rebirth as a new kind of creature—a male adult.

Now the young initiate must carry a weapon and demonstrate his fierceness, winning the respect—and tolerating the ridicule—of boys far older and more experienced in adult ways than he. He must go into the forests hunting for food and enemies. The sections of the village where women laugh and groom the hair of their little boys and the walkways of garden plots where children play are places he must never enter again. He has been expelled into the cold, harsh world of men.

In the fifties, America too established separate worlds of women and men. The American dream became the flight to suburbia. Couples escaped the central cities and bought homes on plots where sheep had grazed a few years before. There, they brought up their families. One result was that father and mother spent their days in environments as different from each other as Mars is from Moline, Illinois. Out in the suburbs was the world of women and children—where doting mamas dragged their kids from dance lessons to Cub Scout meets, stopping off at the shopping center on the way home. The city was another planet, a macrocosm invisible to the kids. It was the unseen world into which the father disappeared every day when he set off for the train, a place whose rhythms were faster, whose ethics were more brutal, whose rewards were more difficult to achieve, and whose punishments could turn a man into an insomniac.

Among Native American tribes of the plains, some boys chose not to make the violent transition to manhood, and remained in the world of women and children. They were called berdache, "women-men." They wore woman's clothing, did woman's work, married men, stuffed their clothes to look pregnant, and even cut themselves to imitate menstruation.[263] Ever since the sixties, young people have dressed in more and more androgynous styles. Could androgyny, like that of the Native American women-men, be the way in which the children of the fifties and sixties, raised in the cozy world of suburbia, have balked against entering the harsh reality into which they saw their fathers disappear five days every week?

No matter what the truth of this conjecture, androgyny may also be a rebellion against a fact of life that is not speculative in the least: the expendability of the male. In nearly every known society, men alone are cannon fodder, laying down their lives to defend the tribe or aggrandize a leader. Males in animal groups and primitive societies may seem rather glorious creatures, accorded the privileges of gods, but, in reality, they are treated by nature like the biological equivalent of paper plates, creatures whose prime feature is their disposability.

When times get tough for the Karamojong in Uganda, they save their scraps of food for their girls and allow the boys to die. In 1979, when Uganda was starving in the grip of civil war, the Karamojong tossed the stiffened bodies of their male children out of the village each night. The only creatures growing fat were the hyenas, who feasted on the discarded corpses.[264]

Male expendability starts in the womb. The egg of the female inches in solitary splendor down the fallopian tube, inviting impregnation. It has no competition. On the other hand, the sperm—the male's contribution to procreation—vigorously swim the lengthy course up the vagina and uterus, beating their long, thin tails in an effort to outrace the millions of their brothers headed for that solitary egg. Only a single spermatozoon—one literally "chosen" by the ovum—manages to finally penetrate the egg's outer membrane and achieve the grand prize of impregnation. The losers die.

But that is merely a preview of the casual manner in which Nature tosses male lives away. Male fetuses are the primary victims of natural abortions, miscarriages, and stillbirths. When times are tough, Nature shows her preference by hiking the rates of spontaneous abortion for males to higher than normal but continuing her tendency to preserve her embryonic daughters.[265] As James V. Neel of the University of Washington says, for males "in utero it's a jungle."

Things don't get any better after birth. In their first few years of life, male babies have a higher death rate than their sisters. Then the nasty habits built into the male genes begin to take their toll. Even in a nice, civilized spot like Alameda, California, where researchers performed a longitudinal study of five thousand adults, males were nearly four times more likely to lose their lives to homicide than females. And they were twice as likely to be accident victims. Their own aggression and bravado did them in.

But cockiness is not the only thing that mows men down. They are twice as likely to be victims of lung cancer, suicide, pulmonary disease, cirrhosis, and heart disease. The immune systems of females work far more efficiently than those of males. How can you encourage the male immune apparatus to function at a higher level? There is a way, but I wouldn't recommend it: castration. The single trick that kicks the male defensive system into higher gear is the elimination of maleness.

Expendability is built into the very genes of males. Kirby Smith of Johns Hopkins University studied four generations of an Amish family whose perfectly average menfolk were missing one small thing. An entire arm of their male gene (the Y gene) was gone. These Amish should have thanked their lucky stars that they were bereft of this particular chunk of masculine materiel. When researcher Smith compared the family with the truncated gene to others in the neighborhood, he came up with an interesting result. Among the normal Amish of the vicinity, men died around the age of seventy. In the genetically impaired family, on the other hand, gentlemen carried on to the ripe old age of 82.3. They lived over twelve years longer than the folks down the road because they were freed from a microscopic piece of poison: an arm of the normal male gene.

One result of these myriad handicaps: in every industrialized country, women live four to ten years longer than men.[266]

But why does Nature treat the lives of males with such abandon? The reasons are simple. If you did away with the vast majority of men on the planet but preserved the women, you would scarcely even dent our species' reproductive capabilities. One man kept around as a stud could easily provide a hundred women with the wherewithal to become pregnant whenever they pleased. Every nine months a one-man, one-hundred-woman collective could produce a hundred babies.

The lives of women, on the other hand, cannot be so casually disposed of. Pare humanity down to one woman for every hundred men, and you'll have one hundred very horny and bellicose guys slicing each other to ribbons or slashing themselves in despair. What's worse, you'll cut the number of possible babies down from one hundred every nine months to one, dooming the human race to extinction. The result? We send our men to war but keep women safe at home. When ships are sinking, it's women and children to the lifeboats first. Let the men founder in the sea. You need each precious woman as a vessel for procreation.

An Arab poem of Mohammed's day, fourteen hundred years ago, is a testament to the indispensability of women. In it, a man is on the road with the ladies of the tribe when a raiding party from a nearby village descends on them. The man's duty is to risk his life to save the women. He fights with the raiders for as long as he can, giving the women an opportunity to get away from the site of the raid and as close as possible to their own tents, where other men will protect them. In his fight, the hero sustains a spear wound in the side. He rides to the top of the hill, blocking the path between the raiders and the fleeing women. Knowing that his life is about to end, he jabs his spear into the ground, wedges its butt under his armpit and leans against the shaft, propping himself up on his camel's back. The enemy raiders, seeing the figure on the top of the hill astride a camel and knowing how fiercely he has fought, are afraid to attack. The sun is at the hero's back. The raiders cannot discern his eyes, but they warily watch his silhouette. Finally, noticing how little he moves, the marauders come to a strange realization: the hero is dead. They have been cowed by a corpse.[267] Like men all over the world, the Arab hero knew the old maxim: nobility consists in sacrificing oneself for the ladies. The male is expendable, the female is not.

Just how disposable males are becomes obvious in the light of statistics revealed by anthropologists William Divale and Marvin Harris in 1976. The pair scrutinized data from 561 primitive social groups. They found that societies constantly engaged in war are very selective about the babies they allow to live. They want boys—male children who can grow up to be warriors—so they weed out the female infants, killing them outright or undernourishing and overworking them. The result is that they end up with 128 male children for every 100 females. So far, it sounds like the males have made out quite well. But when the "treasured" young boys pass the age of fifteen, their fate becomes less rosy. They are sent off to war. And there, they die. On the average, 28 out of every 128 never make it to maturity. Their lives are simply tossed away.[268]

In the modern world, we have taken that primitive principle to unimaginable depths. During the twentieth century, over *ninety million* humans have died in wars.[269] That's more than the entire populations of New England, New York State, California, Texas, and Florida combined. And the overwhelming majority of the casualties were men and boys.

Ironically, nature is most cynical about the expendability of males where the environment provides a relative paradise. Polygamy among primitive human beings proliferates in the tropics.[270] There the men are the most expendable. There, if necessary, women can raise children themselves.[271] The female in a tropical clime can simply walk over to the nearest berry bush (or mongongo nut tree),[272] pluck it, and use it to feed her brood. So men fling themselves against each other in a constant competition to amass the greatest number of wives. The competition reduces the number of men. Polygamy is the privilege of societies where men are ultimately expendable.[273]

Male trouble in paradise is not limited to humans. In tropical climes, male birds wear the gaudiest of all possible plumage: long tails with iridescent feathers, curled plumes bobbing on their heads. They're decked out in the avian equivalent of neon signs flashing a dinner signal to passing predators.[274] The fancy feathers are designed to attract the ladies, but they also say, "Fresh meat, absolutely delicious, stop and have a bite."[275]

Nature, that cruel goddess has said, "These creatures are useless for anything but their seed. Might as well throw them up against the wall and see which ones stick; make them bait for the nearest hungry carnivore and see who survives the chase. Only those with the brightest eyes, the best brains, the quickest wings, will make it. Why should my precious girls waste their time caring for the chicks of incompetents who can't even elude a hyena or a cat? If only one out of a hundred males survive, so much the better. The remaining Lothario can easily inseminate as many girls as he wants. The ninety-nine who don't make it will have proven their inferiority by virtue of their very failure to hang in until mating time.''

In the north, on the other hand, food is scarce, and the season in which you can safely raise your youngsters is short. Nestlings have to be stuffed with delicacies so they can grow as rapidly as possible. When winter comes, if the kids are not ready to fly south they're dead meat. In northern climes, female birds need all the help they can get. The result: male birds are dressed as inconspicuously as their consorts, camouflaged for protection against poultry connoisseurs. After all, the males are valuable. Without them, females can't possibly bring up their young.[276] In northern areas, it also takes more than one *human* to raise an infant. No wonder monogamy tends to be a practice of the north, while polygamy is a custom of the prodigal south.

Males, with their viciousness and violence, sometimes seem like a sex humanity would be better off without. Among mountain gorillas, however, they serve one indispensable function—they're the linchpin of social organization, the key to the superorganism. The gorillas Dian Fossey studied in the Virunga Mountains were organized in small groups of four to ten individuals. Those groups depended on each other for affection, comfort, protection, and breeding. The groups were extremely stable. They stuck together for ten or twenty years. But the central presence that gave them their cohesion was their dominant male, their silverback.

Fossey watched as groups she had known for years lost their silverbacks to old age. When the emaciated and unhealthy elderly male gorilla leaders finally died, the little communities that Fossey had thought would last forever disappeared. Their members dispersed. The females joined other bands, bands dominated by other silverback males. For without the silverback to provide a focal point for organization, resolve quarrels, and offer protection against the incessant raids by wandering bachelors and rival groups, the females could not continue together.

But if you're a man, don't take too much comfort from this fact.

You're still far less necessary than you think. In a vast variety of species, the social order is held together entirely by the female. Elephants provide a good example. The females travel together in a close-knit group, operating under the leadership of a massive Amazon who settles disputes and protects the troop with her fierce charges. The unfortunate male, on the other hand, is excluded during most of the year from polite society. He wanders the forests—a gloomy, solitary brute—until that brief period when he is needed for sexual purposes. Then, for a short time, he and his messy tendency to battle his brethren are tolerated by the ladies, but barely.

There's reason to believe that males are becoming more disposable in modern American society than ever before. The age of information is upon us. Only one hundred years ago, the vast majority of Americans eked out a living by physical labor. That was true even as late as the 1940s. By 1956, it had changed. For the first time, the majority of U.S. workers did no physical labor at all. Most men and women had moved from factories to offices, where they pushed paper, made decisions, held long meetings, read reports, and placed phone calls. Their job was gathering information, weighing it, and spreading it to each other.[277]

In the mid-1800s when the railroads were built, John Henry was a mighty man who drove cold steel through the heart of a mountain, making his hammer ring. Joanna Henry would have had a much harder time of it. In the information age, men no longer have the edge. Women's brains are every bit as quick as men's.

Men were disturbingly expendable in the era of the Arab propped up by his spear. But today they are more disposable than ever. If the women's movement were to decide that the night had come when all the ladies of the world should walk quietly into the bedroom, a butcher knife in hand, and eliminate a snoring burden who insists on drinking beer and smoking cigars while watching Sunday football, the time would be now. Save up enough semen in communal refrigerators, and it would seem the species could move along quite well without males. In fact, an increasing number of women live with other women and turn to artificial insemination when they want to have a child.

Fortunately, most women seem to need men emotionally. They flock to singles bars, put personal ads in newspapers, and sit in corners of the office complaining about the lack of available mates. Bless them.

What's more, economically males have not yet become entirely obsolete. In 1987, the household blessed with a working husband and a working wife earned a median income of $38,346—a comfortable figure at the time. But if you threw out the husband, the income plunged to an impoverished $13,647.[278] Do away with the earnings of the male, and the lone female—left to raise children and bring the bread home on her own—suddenly falls from the upper ranges of the middle class to below its lowest depths. American men are not completely expendable after all, at least not yet. None of this mitigates the fact that in the eyes of Nature, individual men are a dime a dozen. For males, as we will soon see, are the extinguishable junction points in the neural net, the disposable elements that make the social learning machine work.

Could it be that men in this country sense their expendability? Could that be the reason modern males are—perhaps for the first time in American history—working hard to imitate female virtues? Could this account for the premium we have suddenly placed on masculine sensitivity and vulnerability, on the male capacity to nurture children and whip up a quick meal? Could it be that men are trying to hide from the fate they sense that nature has in store for them? The fate of leading short, nasty, brutal, and bloodily ended lives?[279]

HOW MEN ARE SOCIETY'S DICE

><

Why does Nature so cavalierly dispense with the lives of men? Mother Earth is normally a miser, but she turns into a spendthrift under one condition: when she wants to gamble. Her two favorite games of chance are sex and her own version of Monopoly—the competition for territory.

During most of the year every member of an ant colony is considered precious. If a worker ant sends out a smell of distress, others rush over to rescue her.[280] But in midsummer or early fall, the colony invests some of the surplus it has built up during the season of plenty to produce a swarm of sexually mature individuals—new queens and males who fly off by the thousands to try to find a mate. The exercise is a gaudy display of disposability, a ritual in which the colony that normally hoards every scrap throws its most precious product—life—away.

Those males who manage to inseminate a willing female quickly discover that they are marked for termination. Their amorous partner gives them a fatal bite in the stomach or simply leaves them to wander and die.[281] By satisfying their lust with six or seven swains apiece, the thousands of princesses fill their internal sacks with a lifetime supply of sperm. Now a new task awaits these would-be queens: to find a home. A very few succeed in stumbling their way over stumps and rocks until they find a spot where a new nest can be built. This lucky handful manages to establish new colonies. Others try to make their way back to their birthplace—crawling into the nest from which they've come. Most are stopped at the entrance by sentries, attacked, paralyzed, and literally dragged off to the trash heap. Of three thousand hopeful queens, eight may survive.[282]

Nature has wasted lives prodigiously, tossing away ants by the thousands. She does the same among humans, killing hundreds of thousands—sometimes millions—in war. And there too, it is the newly sexually mature whose lives she flings about with abandon. But why? What does nature gain from this? In the case of the ants, her reasoning is simple. Each sexually mature ant's life is a toss of the dice in a casino where the odds are vastly against survival. There are very few places suitable for habitation, and to find them takes tremendous expenditures of time and effort. So the superorganism of the hive produces three thousand pairs of eyes and antennae, each of which will hunt for new real estate as if its life depended on it.

And it does! The innumerable adolescent queens will fan out, taking advantage of random shifts of the wind as they fly, landing to feel their way across the countryside. They will explore every crack in the ground, every crease in a tree. In only two days, they will dedicate 144,000 ant hours to the property hunt. Very few of the old nest's new eyes and ears will live to establish fresh colonies from scratch, but a handful will zero in on a suitable site for homesteading. And the superorganism of the old colony will have achieved its purpose—spreading an offshoot into new territory. You could call this expansionist effort of the ant superorganism a primitive form of imperialism. The annual swarm of adolescent ants is a gamble—and a cruel, costly one at that. The only way to assure a win is to place a tremendous number of bets. The strategy is one of the favorite ploys of the network mind.

Among humans, Nature favors one set of chips in particular for her games of chance—postadolescent males. T. E. Lawrence, in *The Seven Pillars of Wisdom,* rides with an Arab dignitary who, like most Arabs of noble standing Lawrence met, has dedicated almost his entire adult life to raiding—rushing into some cluster of tents a dozen miles from home, killing a few of the men, and stealing the camels, the sheep, and a few old clothes.[283] Together, Lawrence and his friend visit the tomb of the dignitary's son. The brave young man had taken on the warrior hero of a rival tribe—a tribe of cousins—in single combat and won. In retaliation, he had been ambushed by five members of the rival gang. The solitary fighter had not survived the encounter. Though the dignitary has sired a small squadron of male children by his many wives, he is now down to one last son. His other male children have all been gambled and lost.

What would have happened if those gambles had paid off? The victorious sons would have become heroes of the desert, exalted leaders with many wives, much wealth, and the respect and envy of their fellows. For example, Ibn Saud—founder of Saudi Arabia and its king until his death in 1953—started as a desert raider. He had an abnormal tendency to emerge from his battles victorious. The result: through a long series of conquests, Ibn Saud consolidated a territory that, thanks to oil, became wealthy. He indulged in the luxury of naming his new kingdom after himself. The leaders of America, England, and every other Arab state bowed to his power. Presidents of banks and heads of massive corporations hungered for his favor. Superpowers fell over each other in a scramble to sell him weapons.

Ibn Saud's sexual payoff was equally impressive. He sired forty-five ''official'' sons by twenty-two mothers. He fathered at least an equal number of daughters by a bewildering variety of women that included concubines, slave girls, and ''wives of the night.'' As of 1981, roughly five hundred royal children, grandchildren, and great-grandchildren could trace their origins back to Ibn Saud's loins.[284]

Gambling with the lives of men is not limited to the children of Islam. In early societies everywhere from Africa and Asia to Europe and South America, the soldier marching off to war could easily pole-vault from farmer to aristocrat, expanding the territory of his tribe in the process. The Vikings who were ordinary plowmen back home packed their spears and swords in a boat, rowed over the horizon, and became the rulers of Russia, Sicily, and Normandy. Their descendants would someday call themselves czars and kings. The Mongols' property once consisted of a few ponies and the right to graze across Asia's most barren and frigid steppes. Then, in the thirteenth century, they set off, nearly starving on their compact horses, to become nobles in an empire twice the size of the Romans'. Their leader—Genghis Khan—went from an unknown member of a despised tribe to a legend.

The waste of life that created the kingdom of Saudi Arabia, the triumphs of the Vikings, and the conquests of the Mongols is appalling. But behind it lie the same primordial principles that drive the annual swarming of the ants—the superorganism's urge to expand. And Nature's addiction to playing dice with the bones of animals and men.

IS PITCHING A GENETICALLY
ACQUIRED SKILL?

⟫⟪

By gambling with its males, a primitive society learns. It is led by the men who win—those who seem best adapted to the challenges of the moment. These leaders spread their ideas and their ways of doing things—their memes. The not-so-primitive Alexander the Great planted the exalted philosophies of Greece from Persia and Egypt to Afghanistan by slaughtering vast hordes of enemy males on the field of battle. Alexander's massacres are fact. Here's another example from the world of speculation.

Humans are built to eat meat. The craving for it is wired deeply into our system. There is one hormone—cholecystokinin—designed to carry a message from the full stomach to the brain, quieting the appetite. The digestive system refuses to send that hormone on its way until fats and proteins move from the stomach into the intestines. In other words, your body withholds this hormone to keep you hungry until you've swallowed some meat.[285] What's more, as anthropologist and food expert E. N. Anderson puts it, ''We need unusually large amounts of protein and cannot synthesize as many of the amino acids as some mammals can.''[286]

If men are like chimps, we started off as herbivores miserable about our vegetarian diet. Clumsy male chimpanzees who lumber around on all fours occasionally run across a crippled rabbit or a wounded baby gazelle. They do not try to pet and comfort the suffering beast; instead, they smash it with a rock and drag the carcass home for a feast. When the butchers get back to the troop, the entire crew of chimpanzee females and children gather in a frenzy, hankering for some small morsel of the meat. Mothers, babies, and even the unlucky males who weren't in on the kill turn into sniveling sycophants. They crawl toward the hunters, holding out their hands pathetically, begging for all they're worth. The haughty benefactors at first clutch the corpse close to their breasts, then turn magnanimous. To each of those who have humbled themselves, they dole out some small scrap of flesh. The ritual continues until every chimp has had its portion. Such is the power of chimpanzee lust for a decent serving of steak tartare.[287]

Like chimpanzees, we started out as involuntary herbivores. And like the chimps, we must have craved those rich, red, high-protein treats. But we are far luckier than our simian cousins. Chimps occasionally pick up a heavy stick and slam it to the ground, but hurling a stone more than a few yards is beyond their power and aim is out of the question.[288] Our ancestors, on the other hand, eventually managed to stand on two legs instead of four, so they could see the browsing brutes about them from a distance. And their hands lost the clumsiness that still dogs our knuckle-walking relatives.

Let's try to imagine the consequences of that newfound dexterity for our primordial ancestors. One day some prehuman, licking his lips at the sight of a passing antelope and dreaming of a solid meal, gets an idea. Instead of picking up a stick and pounding the ground as chimps do to frighten a predator, he'll throw the branch with all his might at the grass-nibbling beast. The fast-moving length of wood knocks out the antelope and opens a whole new meat locker to a hungry mankind.

As the man who brings home the bacon, the early human with the bright idea experiences a sudden fame. When he comes back to camp, the folks crowd around him, bowing and begging, showing great deference. Soon his regular habit of showing up with slabs of meat gives the man authority,[289] and the stick-throwing hunter becomes a leader. A vast number of ethological and psychological studies have shown that social animals—from birds to humans—imitate their leaders' behavior. Soon, all the young men are throwing sticks and stones, dreaming of becoming the chief to whom all bow and beg, the trendsetter at whom the girls want to fling their bodies.

Those men within the tribe whose genetic endowment has left them with awkward hands and poor coordination (you can include me among them) are left out of the new fashion. Since women shun them, the number of their babies are few. The natural hurlers, on the other hand, are idolized by women and treated with deference by other men. Theirs are the new children of the tribe.

Disturbingly, the stick tossers' ability to bring home protein would not be their only source of appeal to women. Their ability to kill their fellow humans with a well-aimed lob would also help them accumulate mates. Among the Yanomamo Indians of southern Venezuela and northern Brazil, killers acquire the greatest number of wives and father the biggest flocks of children. One skilled Yanomamo slayer had forty-three children by eleven different women.[290] Through genes and learning, the practice of pitching would make swift progress.

This picture may be less fanciful than it seems. The trick of using a thrown object to kill was almost certainly introduced by a male. Among chimps, males are the only ones to kill for meat, though the females are quite anxious to sup on the results. What's more, there is a hierarchy of males even among our relatives the baboons, chimpanzees, and langurs. In the animal world, some become leaders, others mere followers. The dominant males not only attract the most females, they keep other males away from their women. As a consequence, the vast majority of a group's offspring are theirs. Thus any genetic trait that brings a creature to the top of the heap is certain to snap through the crowd in a few short generations.[291]

The notion of the primitive hurler tossing stones on the primordial plain is not a mere whimsy either. Neurophysiologist William H. Calvin hypothesizes that the art of throwing was responsible for the rapid increase in size and complexity of the early human brain.[292] But all this only explains how the knack for tossing might have raced through one isolated human group. How could it have spread to the other clusters of humanity dotting the prehistoric landscape?

Here's where nature's tendency to gamble away the lives of men comes in. And here's where the urge of one superorganism to pit itself against its neighbors begins to show that—despite its moral unacceptability—the violent competition between societies contributes to the evolutionary process. Chimps and gorillas make war, and their early-human relatives—neurally endowed with many of the same instinctive imperatives—almost certainly followed in their path. So the rock and stick tossers were probably eager to try their new skills out on the skulls of their neighbors, and boys being what they are, the neighboring males themselves were likely to be itching for a good fight. Against a tribe with the knack of tossing, however, a gang that was still using its fists and teeth didn't stand a chance. The ancestors of the great baseball players of the future could only come out on top.

When a gang of mountain apes wins a battle against a rival tribe, it kidnaps the nubile females of the defeated troop. The victorious primates take these captive maidens back home and add them to the harem.[293] Primitive humans do the same. So it's easy to imagine what happens when the guys with the great throwing arms out-toss a group that hasn't learned to hurl: The winners massacre an appalling number of rival men with their well-aimed lobs. Then, in the manner of mountain apes, the winning tribe makes off with the losers' mates. They throw a feast, eat a lot of antelope, force themselves on their new brides, and spread great pitchers across the face of the planet. The murderous males further Mother Nature's plans, introducing new improvements in the line of living things. Masculine expendability proves a part of the cosmic scheme for research and development. And so does the itch of one superorganism to fling itself into battle against another.

OLIVER CROMWELL—THE
RODENT INSTINCTS DON A
DISGUISE

≫€

Suppose we were (as we might be) an influence, an idea, a thing intangible, invulnerable, without front or back, drifting about like a gas? Armies were like plants, immobile, firmly-rooted, nourished through long stems to the head. We might be a vapour, blowing where we listed. Our kingdoms lay in each man's mind.

T. E. Lawrence

Let peace be sought through war.

Oliver Cromwell

In the good old days of the stone chopper and the Neanderthal, the driving force between group battles was the longtime champion of replicators, the gene. But roughly fifty thousand years later, the self-duplicating device calling the shots would be that invisible bit of mental stuff called the meme. For the last 2,500 years or so, when superorganisms have developed a craving for each other's flesh, their attempts at cannibalism have usually begun with battles of words, skirmishes of concepts, and wrestling matches over which picture of the invisible world is the "truth."

Once given a "righteous" cause, men all too frequently thrill to the call of battle. General Robert E. Lee said, "It is well that war is so terrible, or we should grow too fond of it." Lofty classics like *Beowulf, The Iliad,* and countless others are about men who looked forward like nothing else to lopping off a few heads. In *The Varieties of Religious Experience,* nineteenth-century psychologist William James quotes a General Scobe-

Ieff saying, "The risk of life fills me with an exaggerated rapture. . . . A meeting of man to man, a duel, a danger into which I can throw myself head-foremost, attracts me, moves me, intoxicates me. I am crazy for it, I love it, I adore it."[294]

Behind this enthusiasm lies an ancient batch of neural weapons that the meme has found lingering in the musty basement of the brain. To see how these bits of prehuman instinct come in handy, let's pay a visit to a man of principle, a champion of ideals, a hero of piety, and faith—in short, a total puppet of the meme. His name is Oliver Cromwell, and to understand him, let's briefly return to the subcerebral layers Paul D. MacLean called the reptilian and mammalian brains.

In the late 1940s, the German researcher F. Steiniger put fifteen brown rats who had never met each other into a cage. At first, the creatures cowered in the corners, frightened and apprehensive. If they accidentally bumped into each other, they bared their teeth and snapped. Gradually, however, it dawned on some of the males that among this batch of strangers were attractive young females. The gentlemen rodents became Don Juans and went a-courting.

The first male and female to win each other's hearts now had something all the others lacked—an ally. The pair took full advantage of the situation, and terrorized their cage-mates. At first, the lovers simply chased their fellow rodents away from food, sending them scurrying to the safety of the far end of the enclosure. Later, the romantic duo hunted down their neighbors one by one. The female was a particularly quick killer. She would sneak up on a victim as it was quietly chewing a bit of chow, spring with a sudden speed, and bite the unfortunate in the side of the neck, often opening a wound in the carotid artery. Some that were attacked died of infection. Others, mauled and worn down by frantic efforts to escape, succumbed to exhaustion. When the happy couple had finished, they were the only survivors.

The rats had cleared the new territory of competitors, transforming the cage into a spacious land of milk and honey for themselves. A new promised land. Now, they could found a tribe that might—if left to its own devices—thrive for generations to come.[295] A tribe that would carry the parental line of genes.

In the Old Testament, Oliver Cromwell's favorite inspirational reading, the Jews had fought to free their Promised Land of Jebusites,

Canaanites, and Philistines. In fact, their God had *ordered* them to do it. Why? Because God was the caretaker of a specific pool of genes. The Jebusites, Canaanites, and Philistines were not carriers of Jehovah's favored genetic strain. Hence, concluded God, they had no place in the land the Lord had set aside for his chosen pool of DNA.

The commands of the Hebrew God were the same as those that primal instincts had delivered to the rats. What sounded like the voice of the Most High was actually the whispering of the animal brain.

Oliver Cromwell's animal brains didn't just murmur their points of view; they roared. Cromwell was born in Huntingdon, England, in 1599. His father was a cloth fuller who had accumulated a good deal of property in the little town. His mother came from a well-to-do family of farmers.

As a child, Oliver was one of those little monsters whose behavior can make us ashamed to call ourselves human. He grabbed pigeons from neighbors' dovecotes, killed and ate them. He stole apples from nearby orchards. His raids on the fruit trees were so frequent that he was dubbed the "Apple Dragon." Vandalism was another of his favorite sports. In particular, young Oliver enjoyed breaking down hedges. Cromwell had ample time for these escapades. He played hooky from school for months at a time. His parents took little pleasure in his hobbies and beat their rampaging boy endlessly, but the pain didn't even slow young Oliver down.[296]

In his late teens, the beast broke out of Cromwell in a slightly different form. Like a late-twentieth-century teenager in a wealthy Connecticut suburb or an inner-city slum, he was obsessed with violence, sex, gambling, and substance abuse. Oliver delighted in attacking decent women in the streets. The historical records say he "ravished" them. In modern parlance, the term might be "rape." Cromwell did not restrict his assaults to ladies. He was known to bludgeon respectable men with his quarterstaff. To top it all off, the lad drank like a fish. The mammal and reptile brains seemed totally in control of Oliver Cromwell's mind. Eventually those animal puppet masters would prove useful in the service of a meme.

The times were rife with roiling memes. The Catholic church had held Europe in a headlock for over a thousand years. Then, in 1517, a discontented priest in Wittenberg, Germany, had nailed a set of complaints to the door of the castle church. The priest's name was Martin Luther, and his protest against ecclesiastical corruption soon swelled into the movement called Protestantism.

At the heart of the new movement was a meme that accumulated human converts at a rapid pace. According to the old ideas of the Church, men could reach God only through the intercession of Catholicism's priests, bishops, and cardinals. But those who embraced the new meme believed that men could find God in a far easier way. Printing had recently made the written word available to those who'd never had access to it before. Now, the faithful could find the wisdom of the Lord by simply flipping open the pages of a Bible.

One country the new meme soon won over was England. Well, sort of. King Henry VIII, always on the prowl for a wife who would give him a son, had a problem with ecclesiastical authorities in Rome: the pope would not issue enough divorces, and Henry needed lots of them. Taking matters in hand, in 1533, he used the new Protestant movement as an excuse to set up his own church. Henry hired his personal substitute for a pope, the Archbishop of Canterbury, and ensured a source of divorces on demand. Under the new English system, God was still available only through a hierarchy of priests. They just happened to be Anglican priests.

Some Englishmen, infected with Martin Luther's meme, insisted on a more direct route to the deity. They read their Bibles and came up with their own ideas of what God might have in mind. Among those Englishmen were the Puritans. Oliver Cromwell became one of them.

But memes do more than infect minds with abstract ideas. They occasionally mobilize those bits of the animal brain that drove the killer rats to clear their cage of rivals. Oliver Cromwell, a man whose animal brain had always been restless, was a godsend for the galloping meme of Puritanism.

The greed of the meme revealed itself innocently enough at first. The English Protestants were eager for land, so they roused themselves with spurious tales of rebellion and massacre in Ireland. According to widespread stories, the Irish had taken up arms and killed anywhere from 20,000 to 200,000 innocent English Protestant settlers. The villains had supposedly left women and children to wander half-naked through the snows, starving and freezing to death. Rumors declared authoritatively that the demonic Irish had held games to see who could stab his sword the farthest into a captive Englishman's flesh. Some said the Irish had even roasted Protestant Englishmen alive, then eaten the steaming meat.[297]

False as these rumors were, they aroused the English to a fury. A group of primarily Puritan London businessmen put forth a simple proposition: They would raise an army at their own expense to defeat the rebellious Irish and put down the outrage of papacy. In exchange, the promoters wanted to keep the acreage they confiscated from the Irish miscreants. One man who enthusiastically invested in the scheme was Oliver Cromwell. In fact, Cromwell bought two thousand pounds' worth of shares—a fortune at the time. The return on his stake, he hoped, would be a substantial chunk of the "liberated" real estate.[298]

The cranial circuit that drove the rats to clear their cage spread its message through Cromwell's brain. He became possessed by a new idea. Why not cleanse Ireland of the Catholic Irish altogether, purging the land of those who satanically worshipped a devilish Roman pope? Ireland, said Cromwell, should be "replanted." The purified soil should be sown with God's own crop—his chosen people. These chosen souls were of course the Puritans.

Among the rats, the notion of "replanting" their cage had been dictated by a gene. The rodents exterminated those whose genetic complement differed from their own. But genes were not what united the Puritans. Puritans bore the blood of all the tribes who had settled England—the Picts, the Jutes, the Saxons, the Angles, and even a few Normans. These men of holy fervor were drawn together by an idea. They were welded into a social body by a meme.

Then, in 1642, civil war broke out in England. On one side were those who wanted the country ruled by Parliament. They tended to be the folks who believed that you could reach God simply by reading the Good Book. Among the Parliamentarians were Cromwell and the Puritans. On the other side were the supporters of the king. They were believers in authority, men who insisted that the sole way to God was through the priests and bishops of the Anglican church. Each group coalesced into a superorganism, and those two superorganisms locked horns and fought.

Before the war, Cromwell had been a country gentleman of little reputation and even fewer accomplishments. Although a member of Parliament, he had shown no flair whatsoever as a politician. On the battlefield, however, he discovered a side of himself he didn't know he had. War exhilarated him. Killing gave him a rush of pleasure. The Cromwellian mind that so far had merely muddled through life came alive in the heat of battle. It was what you might have expected from someone who as a teenager had enjoyed bashing people with a stick. At the age of forty-three, Cromwell had found himself!

When he was in the heat of the fight, the bullets whistling around him, the swords cutting through arms and necks on every side, Cromwell was sure the voice of God was speaking to him. He heard the whispered messages of a meme. Whenever Cromwell won a battle, it was confirmation that he'd interpreted the Lord's voice correctly. And God rewarded his servant every few days by allowing Oliver to carve up his fellowmen.

The voice of his Lord was good to Cromwell. The fighter had a startling tendency to win his bloody encounters and rose rapidly in the army. Eventually, he became its leader. When Cromwell's side defeated the forces of the king, the Puritan commander took over England's reins of power. The old monarch, Charles I, lost his head, which was severed from his body by a Puritan. Oliver Cromwell became the new ruler of Britain.

The superorganism the Puritan meme had pulled together would soon reveal the depths of its hunger. How? By giving Cromwell the opportunity to live out his Irish fantasy. In 1649, Oliver took his squadrons to the shores of the ill-fated isle. He laid siege to one Irish town after another. His men plundered and killed. Cromwell sometimes asked his troops to exercise restraint, but the English Puritans could not contain their fury at the hellish folk who had performed such foul atrocities. When Cromwell's God-fearing men broke through an Irish city's walls, they swung their swords indiscriminately at men, women, and children. The soldiers declared the massacre just. After all, when they'd gone through the corpses, the victors had found that many of the Irish had long, hideous tails—the marks of the devil himself.[299]

Cromwell's dream of "replanting" Ireland came true. His armies removed the Irish by force from two-thirds of their land,[300] and the warriors of holiness handed the territory over to pious Puritans. Like the pair of rats in the cage, the Puritans had cleared Ireland. A country had been attacked and its inhabitants slaughtered—all in the name of a meme. A defeated knot of ideas—that of the Catholics—was beaten into temporary submission. And the Puritan meme had used violent battle and the dark impulses of the animal brain to radically increase its sway.

Cromwell's personal rewards were enormous. He acquired vast amounts of acreage. Eventually, the Lord Protector (as he was now known) became so wealthy that when he had an ailment, he was able to pay a physician the equivalent of what a middle-class family earned in five years. Often, Cromwell scolded his daughter for hanging around with worldly folk. Oliver, you see, was a profoundly religious man; yet, it never occurred to him that there was anything "worldly" about his accumulation of huge estates. These were merely God's rewards for killing. Sometimes a successful meme can pay its helpers well.

In his own way, Cromwell could see that he was the servant of a force battling to impose its form of organization on the world of men and matter. He realized that he was a tool in a struggle between concepts of truly massive scope. Cromwell regarded war, to quote his biographer, Antonia Fraser, as "part of an inevitable, deeply disturbing, deeply exciting process, by which the way of the Lord had to be fought out in order to be discovered." God's will, Oliver felt, would be found in the cannon's mouth.

When Cromwell and his Puritans won the English civil war, the entire state was forced to adopt Puritan views. Alas for the Puritans, this was not to last. Eventually, the English grew tired of the dour new meme and its henchmen. Two years after Cromwell's death in 1658, the monarchy was restored.

But the battle between Cromwell and the king had never merely been a contest between men; it had been a struggle between ideas. Ideas had attracted vast hordes of men to their cause and pitted them against each other for the possession of England. Ideas had used a person who went from killing pigeons to killing men. Ideas had manipulated a convenient tool named Oliver Cromwell.

IDEOLOGY IS
THEFT

⇒⇐

THE INVISIBLE WORLD AS A
WEAPON

➤❦

Humans rally around ideas because they solve some of our problems, because they offer the biological blessings of the illusion of control, and because they are the threads that hold us together in the vast network of a superorganismic mind, weaving scattered individuals into a cooperative entity of awesome power and size.

But webs of ideas also do more: as hungry replicators eager to remold the world, they often turn their ultimate weapon—the superorganism—into a killing machine. And, contrary to the doctrines of some modern critics, they do not engage in this "hegemonic imperialism" only in the malevolent West.

Two hundred years after the Fall of Rome, a merchant named Mohammed lived in the desert town of Mecca, a bleak and isolated community on a caravan route over which passed camels carrying goods to far-off, elegant cities like Damascus.[301] At the age of twelve, when he was an apprentice to his uncle—a trader—Mohammed had made his first trip to cosmopolitan Syria to learn the export-import business. When he reached twenty-five, Mohammed married a well-to-do woman of forty and became a respectable, wealthy burgher, a man whose opinions were listened to. But all that changed when Mohammed reached a mid-life crisis at thirty-nine. He began to have visions. He'd been sitting in a cave in the mountains one day, he said, praying in solitude, when the angel Gabriel had appeared in a blinding light, grabbed him in a bear hug, and forced him to read a message from God. Since then, claimed Mohammed, he'd been functioning as God's spokesman on earth.[302]

Some modern scholars feel that Mohammed's visions may have been the result of epileptic fits.[303] The citizens of Mecca would have found this diagnosis believable. When Mohammed planted himself on street corners and declaimed the new truths that the angel Gabriel had communicated to him, his fellow Meccans were certain that this formerly upstanding, middle-class man had gone mad. They mocked Mohammed or ignored him. One put a slimy camel fetus down his neck as he was praying. Another tried to strangle him. Only a few believed him. Among the believers were a handful of close relatives, one good friend, and a disconcerting number of slaves.

The citizens of Mecca were none too happy with the havoc Mohammed's new notions wreaked on their households. Slaves who'd abandoned the tried and true religions stopped their household chores, running off to pray and wash themselves at all kinds of strange hours; but Mohammed would not keep his visions to himself. When a plot was hatched to murder him, Mohammed fled. He sought refuge in a community where his views might be a bit more welcome, over two hundred miles away in Medina, another town isolated in the desert along the caravan route. In Medina, Mohammed found more willing listeners. During the course of a few years, he was able to build a following large enough to dominate his adopted city's politics.

The fledgling prophet was no man of peace. He consolidated his hold over Medina by ordering opponents assassinated.[304] Then he masterminded a series of assaults on passing Meccan caravans and the armed escorts sent to protect them. When Meccans, fearful of Mohammed's new power, attacked the outskirts of Medina, the "blessed one" led his faithful against the intruders and won. The holy man's military success impressed some of the fierce tribes that wandered in the hills outside of town. They signed up with the new, battle-tested religion. A few years later, the prophet marched his troops two hundred miles to the Jewish town of Khaibar and conquered it. He killed all of Khaibar's nine hundred men, and carried off the women and children as slaves.

At last, Mohammed was ready to take revenge for the indignities his former neighbors in Mecca had heaped on him. In A.D. 630, eight years after he had fled, the prophet led an army of ten thousand followers back to his old hometown. The Meccans were not particularly interested in being treated as the Jews had been the previous year. They gave up with scarcely a fight. Thanks to the heavily armed Islamic squadrons parading through the streets, Mohammed was able to convert Mecca's inhabitants to the beliefs they had formerly scorned as the ravings of a madman.

The sword of Islam was not sheathed once Mohammed's birthplace had been conquered. The city's wealthy traders and illiterate bedouins joined the army that had begun in Medina, and went out to conquer the world for their new belief. They were astonishingly successful. In short order, the legions of Islam overran the ancient empires of Persia, Mesopotamia, and Egypt. During the next hundred years, the Mohammedan hordes spread across northern Africa, taking Algeria, Morocco, and Libya. They invaded India—attacking towns that had defied even the invincible Alexander the Great. They snipped off parts of Spain and nearly conquered France. They even faced the mighty forces of the Chinese army at Talas in central Asia.

Within a few generations of Mohammed's death, these followers of a street-corner ranter, these men from backwater towns and primitive desert tribes, had built an empire of enormous size; but their victories wouldn't stop there. In coming centuries, Mohammedans would repeatedly make the Europeans tremble—eventually attacking even Vienna.[305] They would seize African lands as far away as the Sudan and the Niger. They would convert Afghanistan, win over the Mongols, and spread their rule as far as the Pacific islands of Indonesia and the Philippines. The notions of a man who had claimed to meet an angel in a cave would spawn battles whose bloodshed would soak the earth for the next fourteen hundred years.

THE TRUE ROUTE TO UTOPIA

꧁ ꧂

The appeal of prophets often lies in their ability to paint a picture of an irresistible utopia and to convince us that this better world is almost within our grasp. Marion Keech, the woman who communicated with extraterrestrial Guardians, promised her followers that they would shed all earthly ills and bathe in blessings they could scarcely imagine—after they had been whisked away from our decaying galaxy. William Miller, the founder of Seventh Day Adventism, predicted that God would come to rearrange the world we know and that those who followed Miller would find themselves possessors of a sparkling new paradise. And Karl Marx explained that the elimination of capitalism would trigger the creation of a whole new human nature, one that would flood the greedy dens in which we live with brotherly goodwill.

The supernatural predictions of Keech, Miller, and Marx all failed to materialize. And yet, in a strange way, every one of them bore the seeds of a hard-nosed truth: the power of ideas to draw individual humans into a structured mass *can* make the utopian prophecies of worldviews come true. If the system of belief pulls together a large enough superorganism, the faithful will, indeed, taste a bit of heaven.

Nearly two thousand years ago, a group of believers drew together around Joshua, a young carpenter from the relatively impoverished northern hill town of Nazareth, convinced the parable-spinning woodworker was the Messiah. The believers were Jews, a people who had been living under the oppressive thumb of distant conquerors on and off for nearly 750 years.[306] At the moment, the Jews were dominated by the arrogant, militaristic Romans, who had swallowed up the Hebrew state and made it part of their empire. The prophets of old had predicted that someday a savior would come who would liberate the Jews from their subservience. That champion would lift Israel from its abject subjugation and make it the first among nations, giving it a prosperity and happiness that would never end.

The followers of the populist preacher (later called Jesus by the Greeks) believed he was the long-hoped-for redeemer.[307] Unfortunately, the Roman overlords got wind of the sermonizer's presence, sensed his subversive potential, and executed him. That, however, did not stop his followers. They still believed fervently that he was the Chosen One. Surely, they said, he would triumph over death and return to them.[308] In fact, he himself had predicted his resurrection. Any day now, the faithful told each other, he would come back, bringing with him the new earthly order.[309]

God never quite arrived. Or did he? Over the next three hundred years, the waiting followers of the martyred Galilean became a minority spread from one end of the Roman Empire to the other. Decent Romans spurned and mocked them. Occasionally, they were the subject of vicious persecutions. But like a magnet, the Christian beliefs drew a steady stream of new converts, pulling their devotees into a mass so solid some called it "the body of Christ."[310] By A.D. 310, there were massive, well-organized Christian communities in every major city of the empire.[311] Despite an official policy of repression, Christian views were argued ever more openly in the corridors of power. Finally, they reached the ears of the emperor. Apparently, these Christian notions made an impression.

In A.D. 312, Emperor Constantine was on the eve of battling a rival at the Milvian Bridge outside of Rome. Tradition has it that he looked into the sun and thought he saw the sign of the cross in its blazing face. The Roman ruler took it as an omen. He felt that in the impending bloodbath, Christ would give him victory. Constantine's forces won the day. And in the coming years the emperor made the religion that he felt had brought him his triumph the official creed of the state.[312]

In Constantine's eyes, Christ was a god of war. The nearly illiterate emperor had the cross emblazoned on his soldiers' shields and banners.[313] The real Jesus, a country preacher of peace, would have been horrified, but that scarcely mattered now. Christianity took over the Roman armies, administration, and wealth. In coming years, it would go farther still, absorbing one barbarian tribe after another. And, indeed, the Church's glory would outlive the Roman Empire. As the Christian congregation grew to engulf Europe, its key figures would prosper with it. Christianity's pope would become one of the wealthiest and most powerful men on the Continent—a figure capable of overawing kings and humiliating emperors (as Pope Hildebrand did to Holy Roman Emperor Henry IV). Its cardinals and bishops would bedeck themselves in a splendor so regal that earls and barons would turn green with envy. And Christians by the millions would take upon themselves the privilege of killing, torturing, and raping those who weren't members of their triumphant creed.

Christ may have failed to arrive with a band of angels to transform the ravaged lands of Israel into a paradise, but the belief preached in his name had lifted the humble and given them glory. It had elevated believers from their lowly status as contemptible cultists, placing some in thrones and palaces, and making them the lords and masters of nearly all they surveyed. For the Christian elite, life did indeed become a bit of heaven on earth.

Meanwhile, the growth of the Mohammedan superorganism in the East brought a similar shower of heavenly rewards. As the empire of Islam grew in the seventh, eighth, and ninth centuries, so did the wealth and power of its adherents. Every time the soldiers of the faith conquered a new city, the treasures of the defeated metropolis poured into the coffers back at the center of the empire. Former small-time merchants from dusty desert towns moved into the palaces of the conquered cities as the new overlords. And these nouveau riche religionists showered their wealth and good fortune on their families back home. Descendants of the backward traders of Mecca and Medina, along with great-grandchildren of desert chieftains who had thrown their lot in with the new religion, became as gods on earth. They possessed the finest homes the architects of the day could conceive. They had slaves and harems, dancing girls and the most succulent food. They held the power of life and death over the newly subjugated citizens of the conquered lands.

The writings of Abû 'Ali al-Muhassin al-Tanûkhî, a judge in tenth-century Baghdad, gave an astonishing sense of the Islamic elite's opulent life-style. Al-Tanûkhî recalled, for example, that the Abassid Caliph "Mutawakkil desired that every article whereon his eye should fall on the day of a certain drinking bout should be coloured yellow. Accordingly there was erected a dome of sandalwood covered and furnished with yellow satin, and there were set in front of him melons and yellow oranges and yellow wine in golden vessels; and only those slave-girls were admitted who were yellow with yellow brocade gowns. The dome was erected over a tessellated pond, and orders were given that saffron should be put in the channels which filled it in sufficient quantities to give the water a yellow colour as it flowed through the pond.'' When the servants unexpectedly ran out of saffron midway through the afternoon, the caliph ordered them to fetch yellow fabrics from the public treasury and soak them in the water channel so the leaking dyes would keep the pond flowing with liquid of the correct hue. The afternoon's decorating scheme cost the empire's citizens fifty thousand dinars.[314]

Allah had promised great rewards to the faithful, and he had delivered in spades. But no holy force stood behind the presentation of his gifts. The mechanism responsible was something far more down-to-earth. Mohammed's ideas—like those of Christ—had been a social glue creating a superorganismic bond. And the growth of the superorganism possesses the one power every prophet from Jesus to Marx has dreamed of—the capacity to deliver a small slice of utopia.

WHY MEN EMBRACE
IDEAS—AND WHY IDEAS
EMBRACE MEN

∌€

Humans grab at ideas because ideas knit them together in groups of people who agree with them. They provide the comfort of companionship
and mutual aid. That's one way memes seduce humans into their power.
What are some of the others?

Memes ride the human mind by offering the men and women who
spearhead their cause a richer life. Fidel Castro, who carved out a slice of
the New World to feed the Marxist-Leninist meme, acquired the run of
half-a-dozen homes, a fleet of Mercedes-Benz limousines, Russian Gazik
jeeps, luxuriously wood-paneled helicopters, a fishing villa, boats, gourmet foods, fine scotch, and what may be the ultimate luxury—a personal
pâté chef.[315] Memes frequently make the temptation to riches and power
even sweeter by disguising the pursuit of these prizes as selfless idealism,
ascetic dedication to a cause. Castro, after all, was not a greedy man but a
dedicated ''idealist.'' Memes also seduce us with the illusion of control,
thus tweaking our hormones into higher gear and turning up the vigor of
our immune system. What's more, the insights and technologies they
produce sometimes actually do help us get a handle on the elusive forces
of our fate.

These enticements are some of the reasons humans embrace the
meme. But why does a meme grab onto humans? So it can use a social
bunch as a tool for self-expansion, driving a superorganism like a tank
(more about that in our next chapter). Memes have an ultimate ambition:
taking vast chunks of the world into their possession and restructuring it
according to their form.

It may seem strange to call a meme ambitious, but the mere shape of a successful meme dictates its acquisitive behavior. In fact, the evolutionary race between concepts guarantees that those that develop the cleverest lures are most likely to survive. Take, for example, the religious memes that include the notion of hell. Anyone who doesn't bite the hook enthusiastically is guaranteed a dire fate indeed. Who says so? The meme. The casual nonbeliever is supposedly setting himself up for a hot time in the old town sometime shortly after his death. The meme—be it Christian or Moslem—comes complete with vivid images of an infinite skillet in which the unwise will sauté for infinity.[316] (Buddhism handles the problem differently: fail to follow the precepts of the faith and you may spend your next incarnation as a sextaplegic cockroach.)

These visions of horror work wonders. Terrified humans by the score allow unprovable concepts to take up residence in their skulls. After all, grabbing the meme is the only way to avoid ending up browned and crispy. What about those converts who occasionally doubt that the religious meme is really on the up and up? Or those who are tempted to subject the meme to the hostile light of logic? The successful meme, like any parasite, has barbs with which to prevent the would-be rationalist from shaking it out of his system. Only those will be saved, says the religious meme, who have faith. And what is faith? It is a blind and unquestioning conviction, an absolute willingness to harbor the meme forever, never trying to dislodge it from your gullet. Spit me out, says the meme, and you will tempt a fate worse than death.[317] No, memes do not plot their conquests. They do not have to. In the manner of all true replicators, memes work automatically to resculpt as much of this lowly world as they can possibly grasp, and they have an invaluable ally.

Memes fan out across the planet carried by vigorously scheming hosts. These humans—out for idealism, gain, guts, or glory—spread the meme with a vigor and enthusiasm that would have made Johnny Appleseed's fruit tree planting look lazy by comparison. The Romans conquered the Gauls and turned all Gaul into a Roman province; and Roman memes leaped eagerly into Gallic minds. The Saxons overwhelmed the Britons and turned them into peasants in a new England ruled by a Saxon aristocracy;[318] and Saxon memes spread through the brains of the subjugated tribes. Americans took Hawaii and brought it under Washington's rule, schooling the island's children in American ideals, American memes. The Soviets seized Poland, Czechoslovakia, and East Germany. Then, the commissars put the workers of these newly "liberated" states through political indoctrination sessions every day when work was over. They jailed those who disagreed with the new official dogmas.[319] Eventually, the Soviet invaders shaped dependent nations whose leaders turned to Moscow for all major opinions, and whose goods streamed into the Soviet economy like blood cells flowing from the hand to the heart. The Russians had fed entire new populations to their meme. And the meme, in turn, had given the Russians a reward—a dramatic increase of prestige and power.

At the center of each society is an imperious master—the meme. The gunboats of nineteenth-century America, the tanks of the Soviet Union, and the armies of Islam were merely the arms through which a meme reached out to grab fresh matter. They were the hands with which the meme reshaped raw substance in its own peculiar way.

RIGHTEOUS
INDIGNATION = GREED FOR
REAL ESTATE

Man has much more to fear from the passions of his fellow creatures than from the convulsions of the elements.

Edward Gibbon

There is no nation, it seems, which has not been promised the whole earth.

Elias Canetti

An amoeba making its way through a pond filled with fellow creatures has a habit those other beastlets may find irritating. The amoeba is a sluggish, microscopic blob that wobbles slowly through the fluid it calls home, looking for something to eat. When it encounters some other enthusiastically wriggling microcreature, it gradually enfolds the neighbor in a watery embrace. To do it, the rubbery amoeba doubles up around the hapless creature, pinches itself together, and literally surrounds the drop of water that contains its guest. Then it sucks the drop and its inhabitant deep inside its own body. The ingested droplet now appears to a microscopist as a temporary bubble (technically known as a vacuole) moving within the amoeba's transparent form.

The amoeba floods the bubblelike dungeon in which it has trapped its captive with digestive fluids. Slowly, those liquids pick apart the proteins, amino acids, oxygen, and hydrogen that make up the body of the squirming prisoner. The host absorbs the resulting soup. Then his metabolism busily reassembles the components of his erstwhile boarder, this time fashioning them into sections of amoeba. One entelechy has disappeared. Another has been replenished.

William Buckley, the archconservative, and Arthur Schlesinger, a highly credentialed liberal, were arguing one day on Buckley's TV show about America's budget deficits. They disagreed about nearly everything. The one notion they concurred on enthusiastically was that bureaucratic agencies—such as government departments—mindlessly and inexorably attempt to increase their power, their budget, and their size. Like hungry amoebas, social groups have an automatic desire to grow.

Schlesinger and Buckley had noticed a mechanism that guides the behavior of every kind of social pack, from a cluster of converts in a brand-new religion to the followers of a fresh scientific idea or the patriotic citizens of a state. Jeffrey S. Wicken, of Pennsylvania State University, feels that this hunger is characteristic of all evolving complexes of components—from genes and memes to massive ecological systems. He says, *"For any evolving system,* innovations and strategies that focus resources into the system, while at the same time stabilizing the web of energetic interconnections of that system, will be selected for."[320]

Superorganisms are hungry creatures, attempting to break down the boundaries of their competitors, chew off chunks of their opponents' substance, and digest and redistribute it as part of themselves. *Human* conglomerations have an advantage over those of other species, for in their voracity they are driven by two henchmen: the meme and the animal brain.

Egged on by its co-conspirators, the ravenous voice of a superorganism calls out to charismatic men and women. Disguised as revelation or inspiration, it has spoken to humans as diverse as Mohammed, Saint Paul, Moses, Hitler, Joan of Arc, Mahatma Gandhi, Saddam Hussein, Lenin, and the Ayatollah Khomeini. Its message varies, but under the many disguises is one imperative: Gather a group together and awaken them with my words. Take all those who find themselves in the condition that I describe and weld them into a mighty force that will impose its dominion on a large swatch of the world.

The voice of the superorganism calls out to those on a lower level as well. To them, it dictates sacrifice. The converts have a sublime perception of truth and feel caught up in a frenzied oneness with some superior being whose power leaves them in awe. But the holy vision or lofty secular ideals that create this thrill may be merely the voice of the larger social beast calling for some ultimate contribution—demanding that a seventh-century Mohammedan hurl himself against the defenses of a city far from his ancestral home or that his descendant drive a truck of explosives into an American office building.

Americans, too, have heard the cry of the superorganism. We have been eager to funnel fresh food into the hungry maw of our society. Albert Beveridge, an influential American senator at the turn of the century,[321] had a habit of making statements like the following: "[God] has made us the master organizers of the world to establish system where chaos reigns. . . . He has made us adept in government that we may administer government among savage and senile peoples. . . . He has marked the American people as His chosen Nation to finally lead in the regeneration of the world. This is the divine mission of America. . . . We are trustees of the world's progress, guardians of its righteous peace."[322] Beveridge's words were designed to justify swallowing foreign societies and reassembling them as pieces of the American organism. Two territories that the senator and his colleagues were particularly eager to absorb were Cuba and the Philippines.

If it possessed intellect, an amoeba hunting for fellow creatures whose substance it can absorb would find such rationalizations for digesting its prey quite handy. Just think, his hunger represents an attempt to regenerate a senile world!

An ideology is usually a high-minded mask for a group's itch to take power and resources from other social groups. It's a meme—a cluster of ideas anxious to fatten on the substance of a superorganism's neighbors. Hans Morgenthau, the political theorist, has said that men don't willingly accept the truth about human nature and especially about political nature. The aim of politics, Morgenthau says, is not to make people better or to alleviate their misery: it is to increase the power of one man or group of men against the power of another man or group of men. Morgenthau says our enemies are never as bad as we make them out to be, and we are never as good as we think. We're convinced that we're moral. And we know damn well that our enemy is only out for power and resources, but has no morals at all. Yet, we, too, are out for power and resources. And our enemy, like us, has a moral sense. He uses that moral sense just as we do, says Morgenthau, to narrow the aperture of his consciousness and ignore his appetite for power.[323]

Hidden by the positive attributes of political and religious movements is the rapacious desire to redistribute resources, removing a chunk from their superorganism and adding it to ours. Marxists have a slogan: "Property is theft." They explain that capitalism is an excuse for plunder. It allows the property-owning classes to rob the workers of the fruits of their labor. But under Marxism's sophisticated arguments about the dialectic principle in history lies another form of thievery. For Marxism's implicit message boils down to something like this: The dirty capitalists have cornered all the goods. They hoard the tools of production, and they end up with most of the riches that result from industrialization. Those filthy bastards, let's knock them off their sacks of greenbacks and divvy up the loot.*

Marxists deplore imperialists, but Marxist revolution and imperialist conquest have something in common. They're both the expropriation of someone else's property by violent means.

Marxism and imperialism aren't the only urges for plunder disguised by pious finery. The "just" God of the Old Testament promised the Hebrews someone else's milk and honey. When Mohammed brought the "merciful" words of Allah to Mecca, he handed out a rationalization that helped the Meccans snatch a third of the known world from its previous owners. The Crusaders later used Christianity as an excuse to try seizing many of those lands back from the Mohammedans.[324]

Ideas do more than merely bond a group together. They justify that group's expansion. Like the hungry amoeba, the superorganism is anxious to grow. It is anxious to feast on the flesh of its neighbors. Ideas dress the act of cannibalism in the garb of rectitude.

---

*Some are under the impression that Marx called for compassionate social justice—an equitable redistribution of wealth. However, the founding father of modern communism made it clear that this was not what he had in mind. In his *Communist Manifesto,* he attacked socialists who "want to improve the condition of every member of society . . . [and] wish to attain their ends by peaceful means." Marx vilified these moderates as "fantastic . . . reactionary . . . fanatical and superstitious," dismissing them as creators of "castles in the air" (Karl Marx and Friedrich Engels, *The Communist Manifesto.* London: Penguin, 1967, 116–17).

SHIITES

Moral indignation is jealousy with a halo.

H. G. Wells

Battles don't just take place between societies; they take place between the groups within a social unit. And ideology is what galvanizes these groups for their struggle over turf and power. In 1917, Russia was in serious trouble. Russian citizens had entered the First World War with tremendous enthusiasm. Then they had bogged down on the eastern front for years, losing battle after battle to the Germans. The demands of the war had strained the country's infrastructure past endurance. Transport and supply had broken down. Soldiers fought without bullets, went through the winters without adequate clothing, and made it through each day with scarcely enough to eat. The faltering railroad system ceased bringing adequate supplies of food to the cities. In Petrograd and Moscow, citizens waited at three A.M. in line for a scrap of bread, their faces literally turning blue as the temperature dipped to minus forty.[325]

A frustrated mass of humanity was boiling with anger, anxious to find someone to blame for its misfortunes, eager to uncover a scapegoat. The Bolshevik ideology gave them the perfect target for their hostility. Lenin said that all these problems were the fault of the propertied classes and the bourgeoisie. Russia was at war because the propertied classes demanded it. Russia was losing because the propertied classes were sabotaging the efforts of the soldiers. People were starving because the propertied classes were hoarding all the food.[326]

The answer: destroy the propertied classes and, in the process, take from them everything they had—their homes, their clothing, their power. And that's exactly what happened. The enraged populace had toppled the government of the czar long before Lenin arrived from exile. But now, inspired by his oratory, hordes of common men and women rushed into the elaborate homes and well-stocked stores of the better-off, snatching their food, fur coats, and furniture.

The new ideology did not solve Russia's problems, nor did the theft of clothing from the closets of the well-heeled substantially enrich the masses. The war which the Bolsheviks had promised to end dragged on. The shortages of food became more intense. Iron and coal production ground to a virtual standstill.[327] To make matters worse, Russia was plunged into a bloody civil war. In that internal battle, more than fourteen million Russians were killed. Over five million simply starved to death.[328] Others were shot by neighbors whipped to a murderous frenzy.

The peasants had been promised the greatest fruits of the new order's redistribution of resources and were told that they would receive the land expropriated from the landlords. But, as Yevgeny Yevtushenko wrote in the official Soviet Writers' Union Weekly *Literaturnaya Gazeta,* "The promised gates of paradise proved to be a trap."[329] Years later, the land the peasants had been assured would be theirs was taken firmly from them by the revolutionary authorities under Stalin. Those who objected did not fare well. According to the Russian government's *History of the USSR,* "In some regions 15–20 percent of the peasants were deported; for every kulak [relatively well-to-do peasant] deported three or four middle or poor peasants were arrested." Peasants were packed into unheated cattle cars and sent to distant mountain locations. They lay half-naked on the frigid floors of railway stations along the route, dying of typhoid and simple starvation. In all, nearly fifteen million of them were killed. Aleksandr M. Nekrich, a longtime member of the USSR Academy of Sciences' Institute of History, calls the result "the first socialist genocide."[330]

But under the new ideology, power and resources *had* shifted. They had slid from the old propertied classes to a new elite. The new privileged class was the pyramid of dedicated Bolshevik bureaucrats, who now had an absolute hold over the land, the factories, and the food. An ideology had been the tool a leader—Lenin—had used to unify a group. Ideology had been the weapon with which that aggregation of humanity then seized the resources of another. Ideology had been the force that allowed one superorganism to coalesce from chaos and swallow its neighbors.

Like Marxism, religions often tell the folks on the lowest rung that it's morally imperative for them to grab the goods of the class on top. In the fourteenth century, a group of groaning individuals wandered from city to city in Europe, whipping themselves as they marched. They generally came from poor homes, and their acts of self-punishment were more than the exercises of self-denial that at first they seemed. These flagellants had been told by their leaders that the end was about to come. When it did, rich cardinals and bishops would be tumbled from their lofty castles, tossed out of their comfortable beds, and torn away from their jewelry chests. The poor, those who had mortified themselves in the name of God, would take over. Meanwhile, these humble folk satisfied their sacred ambitions by killing and robbing Jews.[331]

Mohammedanism, too, appealed at first to the poor and the downtrodden. And even Christ's basic motto was "the meek shall inherit the earth." The Savior was not promising some intangible piece of sky, some heavenly wisp of cloud; he was offering real estate.

Ideology is not only the mechanism that allows a superorganism to pounce; it is the indispensable armor with which one group *inside* a society girds its loins for warfare against another. In the early years of Islam, battles developed between two major Moslem sects, each professing its own form of orthodoxy. But under the surface of the ideological struggle was an entirely different kind of a fight: a confrontation between subcultures for dominance of the Islamic world.[332]

The battle between the Shiites and the Sunnis began only a generation or two after Mohammed's death. On the surface, the debate was over who should inherit the power of the Prophet. Who, in short, was the man Allah had chosen as the new spokesman of his truth. On one side, the Shiites believed that the proper leader of Islam was Ali, who had taken Mohammed seriously at a rather difficult time. When the merchant planted himself on corners and attempted to preach the truths he had picked up in his cave-bound conversations with an angel, not a soul outside his own home believed him. His only follower was his wife!

Ali, the Prophet's cousin, had been ten years old when Mohammed began his harangues. The youngster listened attentively to the seemingly lunatic new religion. Then the boy threw in his lot with his older relative.[333] The Prophet probably welcomed the new believer with disproportionate enthusiasm. If he couldn't make a dent among the adults, at least he had won over a child.

When Mecca's town fathers plotted to murder Mohammed while he slept, young Ali acted as a decoy. The killers crept up to the Prophet's bed in the middle of the night, only to discover that the sleeper stretched out upon it was the faithful youth.[334] Meanwhile, the holy man was hot-footing his escape through the desert. Not long after, Ali made the long trek to Medina to rejoin his mentor. To forge even closer bonds to the Chosen One of God, Ali married Mohammed's daughter, Fatima. Young Ali became Mohammed's lieutenant and standard-bearer when the Holy One led his followers in raids on passing caravans from Mecca, sustaining sixteen wounds in one battle alone. Furthermore, the Prophet had treated Ali as an adopted son. Now that Mohammed was dead, the Shiites believed Ali was the man all Mohammedans should obey.[335]

Ranged against the Ali-following Shiites were those who believed in the legitimacy of the Banu Umaiya. In the days when only Mohammed's wife and cousin had believed in his utterances, another had come to uphold the truth of his mission. It was Mohammed's best friend, Abu Bakr, the first male adult to embrace the Koran.[336] From the moment of his conversion, Abu Bakr had stood by the Prophet with an astonishing loyalty. When Mohammed fled from Mecca, the only man he trusted to travel with him was Abu Bakr. When Mohammed had moments of doubt or weakness, it was Abu Bakr who sustained him.

In A.D. 632, Mohammed grew sick and feverish, then weakened and died. His last words were "No one is needed now but that friend [Allah]." But someone else *was* needed: a caliph, a successor; and Abu Bakr became that successor. He organized armies of Arabs and sent them out to conquer. And conquer they did, beginning the process that would quickly dismantle the ancient empires of the Byzantines and Persians, digesting them as segments of an entity whose name the world was only now beginning to hear—the empire of Islam.[337]

Ali waited patiently in the wings as Abu Bakr led the faithful. The young man was certain that his day would come. When Abu Bakr died, Ali and his Shiite followers were confident the mantle of power would descend on him, but it did not. Instead, it passed to members of the Banu

Umaiya tribe. The Banu Umaiya, like Mohammed himself, were Mecca's merchants. They were men of the world, accustomed to dealing with the polished citizens of Damascus, Cairo, and Baghdad. They were organizers, able to understand and administer the affairs of state. All these skills were desperately needed in a superorganism that would go from the possession of one town to the digestion of Persia, Armenia, Syria, and Egypt in only thirty-three years.[338]

But the Shiites were not willing to take the Banu Umaiya's succession lying down.[339] They claimed that the usurpers had stained the holiness of Islam. With their captured wealth and expropriated palaces, their elegant robes and princely ways, the Shiites claimed that the Umayyads had strayed from the spiritual path of Mohammed's truth. The followers of Ali staged a series of battles and assassinations to purge the Prophet's legacy, but they lost.[340]

On one level, the conflict between the Shiites and the Umayyads (whose successors are today's Sunni Moslems)[341] was over religion. It was an argument about which is the true Islam—the Islam of wealth, sophistication, and ultimately corruption, or the Islam of purity, self-denial, and attention to the poor. But on another level, the struggle was something else: It was the grappling of two superorganisms to see which could swallow the other, a wrestling match between two massive subcultural divisions among the followers of Islam. It was a contest between the people of the country and the merchants of the city, a fight between rich and poor. While the Umayyad followers were city dwellers sustained by trade, the adherents of Ali, the Shiites, were uneducated, accustomed to living at the subsistence level, and dedicated to the notion that bloodshed was the only source of a man's nobility. The Shiites identified with the poverty of Ali himself, a man so penniless he'd been forced to turn to relatives to obtain the money he needed to raise his children. These illiterate wilderness dwellers felt that the spartan way of life that had fallen upon them as a necessity was holy but that the customs of the city were not. The Shiite followers of Ali overlooked the fact that though Mohammed had preached generosity toward the impoverished, he himself had been a city dweller, a man of wealth, and—like the Banu Umaiya—a merchant.

The battle was more than a contest between abstract ideas. The Shiites were fighting for the right to install members of their own group as the governors of the recently conquered cities of Cairo and Damascus and as the rulers of Mecca and Medina. They were out to guarantee that when the treasures of battle flowed back from the conquered capitals of distant lands, a Shiite would divvy them up, and Shiites would receive the jewels, tapestries, and slaves.

After the Shiites lost, the Umayyads became the rulers of an empire that would someday extend as far west as Spain and as far east as India. But the battle between subcultures in the Arab world did not end. In fact, it became a semipermanent feature of Islamic society. In later ages, new reformers sprang up, inveighing against the wealth and luxuries that had corrupted the old leaders. The reformers preached poverty and purity. They railed against the corruption of money, fine palaces, and showy clothes, and they promised to sweep clean the holy places of leadership. Gathering followers, the reformers frequently toppled the old rulers. Then a strange thing invariably occurred: The ascetic leaders who had praised the simple ways of the desert moved into the palaces of the materialists they had driven out. Served by slaves and fattened on the fine food and pleasant ways of the city, they, too, grew self-satisfied and wealthy.[342]

These zealots had made a mass of humanity hungry for the goods of those on top and had justified that hunger as saintliness. They had pulled a social beast together with ideology. And each had used the momentum of a ravenous superorganism as a crowbar to pry open the palaces of power.

Today, the descendants of the Shiite bedouins attack a new set of city sophisticates, machine-gunning them in buses, torching their shopping districts, or bombing them in cafés. This time, the vulnerable city folk against whom the Shiites direct their rage are Egyptians, Frenchmen, Germans, and Americans. The alleged motivation of today's army of the faithful is religious idealism, but the real goal is a bigger piece of the action. Once again, ideology has become the device that justifies a superorganism's need to nibble the flesh of its neighbors.

POETRY AND THE LUST FOR

POWER

⇒)€

The hunger of *subcultural* superorganisms disguises itself in strange and mysterious ways. For an example, let's return to modern medicine, which we normally think of as an objective science, above the mere vagaries of political motivations. Like any picture of the invisible world, medicine wears the austere mantle of objective truth. But, in reality, today's medical beliefs were once an ideology one group used to seize power from another.

Homeopathy was developed by a German physician named Samuel Hahnemann in the early 1800s.[343] Hahnemann—who had studied traditional medicine at Leipzig, Vienna, and Erlangen—believed in giving patients highly diluted dosages of what he called "remedies."[344] These were a vast array of substances that actually produced low levels of the symptoms they were designed to cure. Hahnemann was using a principle similar to the one employed today by "clinical ecologists," medical specialists who treat allergies to food, pesticides, plastics, and other substances by giving minuscule quantities of the substance to which the patient is allergic. These microdoses seem to help the body fight off the allergen's negative effects. Treatments of this nature have been demonstrated in recent years to reduce symptoms ranging from depression and uncontrollable rages to backache and skin irritation.[345]

It is not surprising, then, that Hahnemann claimed to be able to treat a tremendous variety of human diseases. His methods were so successful that by 1900 there were homeopathic medical schools and hospitals all across America.[346] What's more, homeopathy seemed to work. In the yellow fever epidemic of 1878, the mortality rate of patients treated by traditional methods was 16 percent. The rate of death for the patients who had been fortunate enough to find a homeopathic doctor, on the other hand, was less than half that.[347]

But merely being able to produce cures isn't enough to establish a medical discipline. A new scientific order is a social organism—a collection of humans welded together by a common belief—and freshly hatched superorganisms are often extremely vulnerable. Older social clusters would prefer to kill such youngsters before they can grow. In nineteenth-century medicine, the entrenched and hostile social beast was another professional clique, a coterie that had labored for generations to establish its authority.

Homeopathy's rival was allopathy. Allopaths, with their dogmatic belief in bleedings, mercury, and opium, waged a pitiless war to discredit the newcomers.[348] They founded the American Medical Association to purge the medical profession of their rivals.[349] And through government manipulation and a public smear campaign, they succeeded, hounding the homeopaths out of the medical schools, out of the medical societies, and out of access to the carefully cultivated image of medical infallibility.[350]

At first glance, this was a battle between two scientific truths, two systems of belief. But under the surface, it was a struggle between superorganisms over the lucrative proceeds of the medical trade.[351] The allopaths won.

As we've mentioned before, today's doctors—the heirs of the allopaths—are only able to deal effectively with about 50 percent of the complaints brought to them. They haughtily dismiss the other 50 percent as representing nonexistent ills. Yet many of the symptoms they overlook may be produced by the very same problems the homeopaths claim to have dealt with successfully. The result is that you and I are saddled with a medical community whose "knowledge" is the result of a battle between subcultures. Because we are in the hands of the winners, a set of cures that could heal us has almost disappeared.

Even a wisp of imagination as fragile as poetry can be used in the squabble between subcultures over who will get the goods. In the first century

A.D., highborn citizens of the Roman Empire were obsessed with ambition. Many drove themselves night and day to win honors in the eyes of
their countrymen and to rise in the hierarchy of the state. They entered
what was called the *cursus honororum*—the racecourse of honors. The system was simple: If you were just starting out, you competed with other
young men to win a government position reserved for neophytes. Once
you landed the prized new post, you were able to work your tail off
and clamber to the next position up the ladder. You showed your worthi-
ness for advancement through a variety of means—diligence, splendid
speeches in the Forum, donations of warships and monuments to the
state, and presentations (at your own expense) of massive public specta-
cles. If the crowds and those in power liked what you had done, you
moved step-by-step upward on the stairway of honors. And, finally, if you
had labored long and hard enough, you might attain the ultimate prize,
becoming one of the two consuls, the highest officers in the land and the
supreme commanders of the army. The *cursus honororum* was a splendid
motivator. It impelled Rome's best and brightest to dedicate nearly all
their energies to the betterment of their society.[352]

But the Roman poet Horace wrote loftily that the struggle for political power was vain and meaningless. True beauty and happiness, according to his poems, were to be found in the quiet moments of a private life
in the countryside, isolated from the tension and bustle of teeming Rome.
How elevated, how inspired, how ethereal Horace's delicate verses and
their sentiments seemed. But were they indeed so far from the low rum-
blings of ambition?

The Roman *cursus honororum* was open strictly to men of noble birth.
Horace was not an aristocrat. In fact, he was the grandson of a slave. Horace's father had been a freedman who'd done well enough to send Horace
to school with the upper crust. After graduation, Horace had continued to
hobnob with the elite. Nonetheless, he wasn't one of them and could not
participate in the traditional race for power. He could, however, spend
time on the farm a wealthy patron had given him. There he could quietly
write his heart out. Curiously enough, Horace's poetry said in subtle ways
that the men who strove for state honors were not very important after
all. They were grasping for something mean and low. They deserved far
less respect than one might think. The men on whom all eyes should be
fixed, the men who truly merited respect (and by implication the men

who merited the prestige and power that follows where respect leads), were the artists who labored contemplatively in their cottages. And who was the king of the cottage artists? Who was the man to whom all the goods should fall? Well, Horace, of course.[353]

Over the generations that followed, Horace's ideal of meditative withdrawal would take an ever-greater hold on the Roman spirit. Men who once had eagerly looked forward to their days on the racecourse of honors became ashamed of their earthly ambition. Instead, they went off to the country to contemplate their souls. The best men of Rome were no longer eager to participate in bettering their state, and Rome's vigor slowly slipped away. The frail ideas of Horace's poetry had masked his personal desire for power. And they had shifted the energies of the nation, reordering the Roman superorganism's values, resculpting its psychic metabolism. The sublime ideals of the poet had produced hard-nosed consequences in the real world.

Poetic exhortation and medical debate are the lambent dance of the most insubstantial of all human phenomena—analysis, emotion, and imagination. Poetry and medicine have produced blessings of sometimes awesome proportions. Yet the elevated and often incomprehensible colloquies in which doctors and poets indulge are sometimes far more than at first they seem. They hide the greed of superorganisms.

WHEN MEMES COLLIDE—THE
PECKING ORDER OF NATIONS

⇒◉←

Nature's way of testing any self-replicating device is competition. For over three and a half billion years,[354] she has set the products of the genetic system in a race to see who can corner the good things of this life. Like a driver strapping himself into his machine at Le Mans, each string of genes has hunkered down into the creature it constructed and driven up to the starting line. The winners of the moment are still here. The losers have retired from the track. *Homo habilis, Australopithecus,* and Peking and Cro-Magnon men—all had their moment in the sun and are gone.

Bodies are usually the *genes'* racing machine, but *memes* have driven a radically different kind of device onto the field. Their contraptions of choice are extended social groups. These superorganismic vehicles are big and complex, but their advantages are awesome: speed, maneuverability, and incalculable horsepower. The Le Mans of superorganisms has a set of very simple rules. To understand how they work, let's take a quick look at some of the strange battles between beings of a smaller size—chickens, monkeys, you, and me.

Just after the First World War, a Norwegian naturalist named Thorlief Schjelderup-Ebbe decided to spend some time on his parents' farm watching the quaint ways of chickens. Schjelderup-Ebbe's study uncovered a subtle form of competition disguised as barnyard peace. When the hens were fed, they approached the trough with extraordinary decorum. Though all were hungry, none ran up and grabbed whatever she could. First, a rather regal-looking hen stepped up to the container of grain and proceeded to dine. The others simply watched. Then another came forward to partake of the meal and eventually stepped aside. Yet another marched up to take her turn.

Schjelderup-Ebbe took careful notes. When he reviewed them, he realized something surprising. The order in which the feathered diners took their turns at the trough was not arbitrary. Far from it. Every day, the same bird went first, the same went second, and so on down the line.

When Schjelderup-Ebbe tossed a strange chicken into the yard, he made another discovery. The normally peaceful birds quarreled like bar-room brawlers. It seemed that the stranger was trying to establish a place in her new society. That meant she'd have to shove some of the other birds below her. And the already-established fowls were not going to tolerate the humiliation of being moved to a subservient position without a struggle. They fought against downward mobility for all they were worth.

Schjelderup-Ebbe wasn't content to simply watch the feathers fly. He kept note of exactly who was attacking whom, and he counted every vicious peck. When the naturalist toted up the numbers, lo and behold, another strange phenomenon emerged. Some birds had received hardly any jabs at all. No one dared touch them. Others were pecked beyond endurance. They were easy targets, and nearly everyone wanted to get in a lick.

The birds no one laid a beak on were distinguished by more than just their invulnerability. They also happened to be the creatures who stepped up first to enjoy a meal. And the birds who ended up with many a hole in their feathered coats had their own unique dining distinction. They were always among the last in line.

Schjelderup-Ebbe had discovered that in the world of chickens there is a social hierarchy, a division into aristocrats and commoners—a lower, middle, and upper class. The alert researcher called the phenomenon a "pecking-order."[355] It wasn't long before naturalists were discovering similar social orders in a bewildering variety of species.[356]

Research on pecking orders (known technically as dominance hierarchies) has gone on now for roughly seventy years and has yielded some startling revelations. Position in the pecking order determines far more than just how many feathers you lose. It readjusts your life-style, your chances of survival, your sex life, and your physiology.

The pecking order can determine whether you live or die. According to the father of sociobiology, E. O. Wilson, the highest wood pigeons in a pecking order go to bed at night with a full stomach. The birds on the bottom frequently don't. The lower-class pigeons perch on the evening roost with a meager meal working its way through their gut—just enough fuel to carry them through the night. If the weather grows unusually cold or if they don't find food the next day, some of those birds on the underside of the heap may not survive. On the other hand, the comfortable birds on the top of the tree nearly always manage in even the most troubled times. Their right to all the best food and lodging makes sure of that fact.[357]

Pecking order privileges include sex. Three male and three female rats were left alone in a cage to battle out their social system and sleep with whomever they liked. Two males lost out in the struggle for top position. They apparently also ended up with an empty dating calendar. When the females gave birth and their pups were examined, it turned out that all of them had been sired by the one dominant male. The experiment was run another twenty-one times, with different rodent participants on each occasion. The final result was that the dominant rats managed to father a walloping 92 percent of the young.[358]

Where you end up in the pecking order can even change your physical makeup. A dominant male monkey has a higher sperm count, more visible testicles, and a far more regal posture. The monkeys who don't make it to the top skulk around stoop shouldered and less sexually potent. But if some meddling researcher kidnaps the ruling simian and leaves his lordly spot vacant, the round-shouldered subordinates will grab for the empty throne. The monkey who ends up on top will undergo a change. His testicles will drop, his sperm count will climb, and his hunched posture will melt away, replaced by an authoritative upright strut. The new king of the castle goes through a biological transformation simply because he's moved up on the hierarchical ladder. For a monkey's physiology, position in the pecking order is everything.[359]

Humans not only undergo the same changes when they're under society's heel, but their blood pressure goes up and stays there. The result is an increase in the odds of heart attack and stroke, and a loss of mental swiftness.[360]

Blood and sperm are not the only bodily substances to shift their concentration in response to pecking order changes. In monkeys and humans, when groups fight, the *winners* snag a hormonal prize: their testos-

terone level rises. Testosterone—the male hormone—inspires confidence and aggression. A fresh jolt of it in the blood invigorates the victors. For the losers, it's a different story: Testosterone level plummets.[361]
The body shifts into resignation. Low baboons on the totem pole carry
additional consequences in their bloodstream. Their circulation is flooded
with glucocorticoids—stress hormones that constitute a slow internal
poison. The baboons on top do not suffer this chemical corrosion because
their bloodstreams are relatively glucocorticoid-free. Once again, position in the pecking order reshapes physiology.[362]

After a while, top or bottom position in the pecking order gets to be
a habit. Numerous studies show that a creature who has won a fight is
more likely to win the next one. An animal who has lost barely shuffles
through his next contest. The odds are high he'll lose again.[363]

This may explain a phenomenon that crops up in Julius Caesar's battle narratives. Caesar frequently confronted tribesmen bred from birth to
fight, men who prided themselves on their ferocity. But when the Roman
legions won a decisive victory, the proud barbarian warriors sometimes
bowed their heads and marched meekly into slavery. The barbarian
women—who had once been equally defiant—held up their children to
the Romans and begged to be spared. Then they gave themselves up, cav-
ing in to a subservient fate. The humiliation of defeat changed these fierce
fighters into beaten men. A quick slide from the pecking order's top to its
bottom seemed to radically alter their personalities and even their phy-
siques, apparently by tilting the captives' internal chemical balance.

Biochemical resignation explains why barnyards are not a perpetual
battlefield. Chickens seldom get into major brawls. True, when a stranger
steps onto the scene, the intrusion triggers a riot. But when the dust dies
down, the feathered ladies settle into a stable order. The dominant female
once again luxuriates in her prerogatives, and the lowliest pullet endures
her ignominious lot in life. The pecking order's hormonal shifts help insure this peace. Those who win are flush with internal chemicals of pride,
and those who lose are numbed by glandular drugs that lull them into submission.

The irony is that even the chickens on the bottom gain an advantage
from the endogenous chemical brew that leaves them too lethargic to
fight over their fate. If pecking order positions were up for constant grabs,
each creature in the yard would have to spend her time in attacks and self-

defense. The nonstop battle would waste every bird's energy and time. Well-fed chickens would grow scrawny watching out for ambush instead of scratching in the dirt for food. And some would do worse than merely lose weight: they'd actually die from their wounds.[364] A long-term truce has its benefits, even if it leaves you under everybody's heel. At least it lets you live your life in quiet, freeing up your time to poke around for grubs and worms. In the last analysis, the cocky aristocrats, the status-conscious crones in the middle, and even the picked-on runts have a solid reason to go with the status quo.

The struggle for position in a pecking order is not restricted to individuals. It also hits social groups.[365] There is a form of pecking order Schjelderup-Ebbe never studied—the pecking order of superorganisms.

On the outskirts of an Indian village lived two tribes of langurs—lanky, curious monkeys. It was easy to see which group had made it to the top of the pecking order. One langur clan had staked out a territory in the center of town. This group lived the life of Riley. Its members hung around the bazaar, waiting for a small boy to toss aside a half-eaten fruit or for a passerby to drop a crust of bread. They picked through the garbage for gourmet treats—wilted vegetables or melon rinds. When the rains came, these pampered langurs stretched out under the overhangs of roofs. When it was sunny, they turned those roofs into a playground. Getting along from day to day was a breeze.

The second crew of langurs lived in the hills just outside of town. Life for them was anything but easy. They were forced to comb the ground for edible shoots. They stripped the greenery off the trees for dinner, and they dug the occasional insect out of a rotting log for a protein treat. When it rained, they huddled miserably under the dripping leaves. In the pecking order of local groups, their superorganism was on the bottom.

Every once in a while, the monkeys from the hills became fed up with their gritty lot in life. They came down to the village where the pickings were easy. The monkeys of the bazaar, however, were not particularly interested in opening their neighborhood to these intruders from the wrong side of the tracks. The privileged lady langurs who lived in town got together and chased their underprivileged cousins back to the woods from which they'd come.

But no social group's position in the pecking order is written in stone. One day, the male leader of the bazaar langurs was fooling around with a sidekick in the middle of the town road. The hill langurs stood at the edge of their territory, grudgingly watching the fun. Suddenly, a car came hurtling down the tarmac. The lordly male of the elite group looked up startled—but too late. The car sideswiped him and sped away. The leader of the bazaar troop was dead.

Sensing a sudden shift in the balance of power, the chief of the hill tribe sauntered over to the aristocratic ladies of the ever-so-snobbish bazaar set—the same supercilious aristocrats who had chased him and his companions away an endless number of times. The hill langur leader donned the simian symbols of authority—an upright, arrogant posture and a self-confident stride. He swaggered straight to the queen of the bazaar clan and mounted her. Knowing that with their old leader dead she and her companions had just slid down the pecking order, the formerly snooty matron gritted her teeth but gave in. Her face was screwed up with conflict. But even as he copulated with her, the new master of the high-class neighborhood looked around him with an expression that seemed to radiate a casual contempt.

The troop of langurs from the hills, the group that had always been treated by its bazaar-dwelling cousins as second class, had taken a giant stride up the ladder of status. The luxurious real estate of the town market now belonged to them. The troop from the neighborhood at the center of the town, on the other hand, had taken a dive toward the bottom.[366]

Human superorganisms also have their pecking orders. The Soviet Union and the United States struggled for generations over who was number one. Tanzania and Chad are painfully aware of their positions on the bottom of the heap. They belong to a bloc whose pecking order position is indicated by its very name: the *third* world.

There's good reason for a group to want to climb as high in the pecking order as it can. The superorganism at the summit has the best territory, the best food, the best of everything. That's why some ant species go to war. The ant colonies that win increase their territory and build insect empires. The larger the size of an ant society's territory, the better each ant citizen is fed and the bigger each worker is able to grow. When it comes to sex, the winning colony scores an extra bonus. During the mating season, it is able to produce more winged, sexually active queens and males. As a result, even its chances to start fresh offshoots is greater than those of its less successful neighbors.[367]

There are other good reasons for wanting your group to reach the top of the pecking order. Remember Jane Goodall's tribe of chimps? After many years, the clan split in half. One gang stuck with the old home territory, while the other wandered off to start life in a fresh new place. Between them, they established a pecking order of groups. The tribe that stayed home had the largest membership and the choicest land. It was clearly on top. Eventually, this favored clan began to pick on the troop that had pulled up stakes. At first, the privileged group merely advertised its superiority by harassing the gang on the bottom. Later, the mood of the dominant clan grew ugly, and the top tribe wiped out its rivals. Being on the bottom of this pecking order turned out to be fatal.

No wonder human groups so often try to move up by manipulating ideas or by making war. The Helvetians in the days of Julius Caesar were one of those superorganisms driven by the lure of pecking order glory. As Caesar tells the story in his *Conquest of Gaul*,[368] the Helvetians lived in a state of considerable size parked roughly where Switzerland is today; but the Helvetians wanted more land, more slaves, and, above all, more power. In fact, they wanted to rule over every tribe in sight.

The Helvetians did not just grumble about their pecking order aspirations, they did something about it. They laid out a methodical plan of conquest. They planted and harvested for two full years to lay in a store of supplies. They bought up all the oxen and wagons for hundreds of miles around. Then, in the third year of the grand plan, they packed all their earthly possessions, got together their wives and families, burned their twelve major towns and their four hundred villages, and set off on the glorious campaign. Caesar swears that over a quarter of a million Helvetians marched out to seek their fortune. There were so many of them that it took twenty-one days just to get the entire group from one side of the river Rhone to the other by boat and raft.

As the Helvetians marched, they pillaged and plundered, enslaving inhabitants of the territories on their route. They were doing just fine until they met the armies of another superorganism with an equally blind ambition to climb the pecking order. That rival social cluster was the clump of humanity known as Rome.

Caesar put superior engineering to work. He built an instant bridge.

In one brief day, he crossed the river that had held up the Helvetians for nearly a month. His men were better disciplined and infinitely better organized. They outmaneuvered and outfought Helvetia's massive force of warriors. When it was all over, the Helvetians begged for peace. Caesar was relatively kind: he sent the rebels back to the territory from which they'd come and ordered them to rebuild the homes they'd burned. The Helvetians had gambled mightily—and lost. Of the 368,000 who had left to conquer Europe, only 110,000 remained. Like Helvetia, the Roman superorganism had once sent its citizens out in military swarms to descend upon their neighbors. The Helvetians had failed to achieve their dreams of pecking order triumph. Rome, on the other hand, had not. She took over the entire European barnyard.

A strange thing happens to the *memes* of the superorganism that mounts the pecking order's peak. They spread as rapidly as the germs of plague, exultantly leaping from mind to conquered mind. Today, most of the populations of Europe, South America, and North America speak languages rich in Roman words. They do their public business in buildings adorned with the flourishes of Roman architecture. They read and write the Roman alphabet. And they look to the days of Rome's glories as those in which their own civilizations were forged. For the meme of Rome gambled on driving its superorganism up the hierarchical ladder of nations. It placed its bets on making its host the first chicken at the trough. And the meme of Rome hit the jackpot.

SUPERIOR CHICKENS MAKE
FRIENDS

~≫⋇≪~

There is one more advantage to moving up the superorganismic pecking order: friends. A group high on the ladder of nations has them. A group at the bottom doesn't. That simple fact is vital to the spread of memes.

In 814 B.C., Phoenician merchants from the coast of what is Lebanon today sailed halfway across the Mediterranean and set up a trading colony on the northern shores of Africa. They called the new encampment Qart-Hadasht—"new town." Europeans mispronounced the Semitic name, and, in the mouths of these westerners, it came out as "Carthage."[369]

The sea-going traders of Carthage did very well for themselves. They built ships to carry rare goods from one country to another, explored the coasts, and looked for barbarian towns whose craftsmen made strange objects. Carthaginian trading vessels showed up as far away as the Baltic, the Cameroons, and even an unheard-of island we know today as Britain.[370] After all, you never knew when some backwoods trinket might fetch a steep price back in the centers of sophisticated civilization—Babylon, Nineveh, Memphis, or Thebes.[371] Occasionally, the shipboard entrepreneurs' hunt for new commodities turned up raw materials of considerable value. For example, in Spain the Carthaginians discovered they could buy tin that had been trekked across seas and mountains all the way from Cornwall.[372]

Business for the Carthaginians was brisk. Back home, they dug an extra harbor and enlarged the size of their merchant fleet. To protect their seafaring commercialists from pirates, they constructed a sizable navy. Carthaginian warships were miracles of technology—swift, narrow galleys driven by sail and oar that could accelerate like a jackrabbit and ram an enemy vessel hard enough to snap it in half.

The Carthaginians soon started colonies, planting settlements straight across Africa's northern lip and bringing distant Spain into the Carthaginian sphere of influence. In the barnyard of Europe's central sea, the Carthaginian superorganism was soon at the top of the pecking order. Though the population of Carthage was small, its power was vast. Why? On those rare occasions when the Carthaginians went to war, their armies were accompanied by mobs of foreign troops. Their astonishingly flexible cavalry came from one north African tribe—the Numidians. Their sling-ers—slingshot-wielding equivalents of today's sharpshooters—came from yet another North African nation: the Balearics. And their infantry was supplied by the peoples of Libya.[373] Since the Carthaginians were on top of the heap, everyone wanted to share in their good fortune.

Then a tribe that had been in diapers when Qart-Hadasht's first buildings went up decided to challenge Carthage's supremacy. At first, the idea seemed like a joke. The Carthaginians were the most skillful sail-ors in the world. The upstarts, on the other hand, had no idea of how to rig a sail or work an oar. In fact, they didn't even know how to build a ship.

All that soon changed. The shore-bound challengers were Romans. What they didn't know, they were more than willing to learn. In 260 B.C. the enterprising citizens of the Italian city-state managed to find the wreck of a Carthaginian warship that had run aground. Roman military engineers pored over the battered vessel, examining every detail. They took it apart and noted each trick of the boat's construction, then built a copy of their own. When the Roman technicians tested their warship, it worked as well as the original. So the Romans rapidly hammered together an entire fleet, turning out 220 ships in only three months.[374] These traditional landlub-bers were now the proud possessors of a navy.

The soldiers of the seven-hilled city set out on the seas for conquest. Rome attacked Carthage's central base for Mediterranean trade: Sicily. The Carthaginian merchants could not hold out against Latin ferocity, and the island became a Roman possession. As part of the peace settlement, Rome also demanded—and got—every island that dotted the sea be-tween Sicily and the African coast. Then Rome broke the treaty under which she had been granted this windfall, and seized Carthaginian-controlled Sardinia and Corsica as well.[375] With these enemy outposts in the middle of their sea-lanes, the Carthaginians were in trouble. Their merchants could no longer sail in safety from one market to another. Every Carthaginian trading vessel was in danger of Roman naval attack.

But a Carthaginian general named Hamilcar Barca had an idea. He wanted to put the Romans on the defensive.[376] And he was determined to open new sources of supply to replace those the Romans were cutting off. Hamilcar Barca's plan was to outflank Roman might by turning one of the old Carthaginian trading partners into a massive province. That territory was Spain. Its conquest would allow the Carthaginians to monopolize Spanish goods, carrying them by land, if necessary, back to Carthage's customers in the East. What's more, a massive Carthaginian domain to the west of the Roman sphere of influence might make the toga-wearers feel less secure about their Mediterranean rampage.[377]

To put his plan in motion, Hamilcar Barca kissed his wife and most of his children good-bye, took one nine-year-old son as a companion, mobilized his soldiers, and marched well over a thousand miles to the Spanish coast. There, his troops demonstrated just who was on top of the local pecking order. During nine years, they won a steady stream of battles against the native troops. Spanish tribes eventually flocked to the Carthaginian side. A small mob of Iberian, Celtiberian, Tartessian, and Gallic chiefs offered to ally their men with the Carthaginian military machine.[378]

Barca not only enlarged his army but he enriched the Carthaginian treasury depleted by Roman trade obstructions, pumping shiploads of silver, timber, and tin from the hills of Spain into the Carthaginian commercial pipeline. Watching Hamilcar Barca's every move was his son Hannibal.

Despite Hamilcar Barca's brilliant maneuvers, things in Carthage were going from bad to worse. Roman leaders cooked up phony stories of Carthaginian evildoings, spread the propaganda to the Roman population, then declared one "just" war after another to avenge the manufactured slights.[379] Carthage, which thrived on trade not war, was being hammered into bankruptcy.

Eventually, Hamilcar Barca died, but his son Hannibal was determined to carry on. The twenty-nine-year-old came up with a daring scheme to save the Carthaginian homeland.[380] He would stamp the

Roman infection out at its source. Leaving his younger brother behind to hold Spain, Hannibal would gather up his army of allies and spring a surprise attack on Rome itself. He'd do it by approaching the city from a route no one could possibly anticipate. He'd take his forces—elephants and all—through the treacherous mountain passes far to Rome's north.

The strategy seemed foolproof. For years, the Romans had been certain that no army of any size could cross the mountain slopes alive. Since invasion from the north was clearly ludicrous, the generals of Rome left their northern flank virtually undefended.

Crossing the Alps with nearly seventy thousand men, horses, and elephants was no easy matter.[381] Pack animals and battle steeds plunged off the sides of thigh-wide mountain trails to their death. Men bogged down in the snow, exhausted and starving, then gave up on their lives. And the primitive folk who made the mountains home inflicted damage with surprise attacks. But in the end, Hannibal and his followers accomplished the impossible: they clambered down from the treacherous mountain slopes to the "impregnable" northern Italian plains.

When Hannibal's troops descended from the Alpine forests, the Romans were astonished and totally unprepared. Hannibal pulled off stunning victories against the armies Rome sent north to meet him. At the battle of Cannae, for example, the Carthaginians wiped out a Roman army that outnumbered them two to one.[382] Between 50,000 and 70,000 Romans died, including 80 senators and 29 military tribunes.[383]

Hannibal shrewdly took advantage of the old pecking order rule: friends flock to the beast on the top; they abandon the beast on the bottom. Whenever the Carthaginian triumphed, he reduced his Roman prisoners to slavery. Since the Romans had been bolstered by armies of allies—troops from the conquered tribes of the Italian boot—the savvy commander sent his non-Roman prisoners back to their native towns unharmed. All he asked was that they carry a simple message: Carthage had nothing against the non-Roman Italians.[384] The gesture worked. As Hannibal demonstrated his strength, one Italian tribe after another deserted Rome and threw its lot in with Carthage. After all, it looked like the city of the seven hills was slipping down the pecking order fast.

The Greek historian Polybius says that, at his peak, Hannibal managed to hold together an army of extraordinary diversity: "He had Libyans, Iberians, Ligurians, Celts, Phoenicians, Italians, [and] Greeks who had naturally nothing in common with each other.''[385] It was the Carthaginians' seemingly unstoppable winning streak that kept them all together.

Hannibal outfoxed the Roman forces every time they met. He stormed across the Italian peninsula, taking towns and overwhelming garrisons almost at will. Soon panic seized the Roman populace. They imagined the Semite from North Africa showing up outside the walls of their city any day or night. But Hannibal was in no position to attack the center of Roman strength. Marching over the mountain passes, he had been unable to carry the siege equipment that would allow him to defeat the city's heavy fortifications. Hannibal waited for that vital hardware to be sent by sea. He charged back and forth across Italy during thirteen long years, hoping for the ships that would deliver these vital armaments. Unfortunately, they never came. The city council of Carthage sent out ship after ship to aid Hannibal, but not one boat ever made it. The Romans may have been doing poorly on land, but they still controlled the Mediterranean.[386]

Eventually, the Romans ordered an up-and-coming young citizen from a distinguished military family, Scipio Africanus, to confront Hannibal's brother—Hasdrubal—back in Spain.[387] (You may recall Scipio from an earlier episode in this book. He was the general who marched into the Roman Senate toward the end of his life and tore up his account books, outraged that he'd been accused of corruption.) Scipio was as good at defeating Hannibal's brother as Hannibal was at trouncing Romans. When Hasdrubal lost, he moved down the pecking order. And as the defeated Carthaginian's feathers grew unkempt, his Spanish allies one by one deserted him.[388] In fact, they threw their lot in with the new master of the Spanish barnyard—the Romans. The rule that pecking order victories bring you friends had boosted Carthage's strength for years. Now it was beginning to work against her.

Hearing of the Carthaginian defeats in Spain, Hannibal's Italian allies, too, trickled away. Carthage was sinking on the hierarchical ladder, and none of the tribes wanted to go down with her. Finally, Scipio secured all Spain for Rome and turned his attention toward the Carthaginian mother city.[389] In 203 B.C. Hannibal, who had held Roman citizens in terror for over a decade, was forced to flee for home to take over his native city's defense. He went with scarcely an ally left.

When the conflict was over, Carthage was beaten.[390] Her trading empire was gone, and her colonies were in Roman hands. The North African and Spanish troops, which had long been the mainstay of her military power, turned their backs on her. And Hannibal, who had built an army of allies to corner Rome, became a fugitive.[391] The town that had ruled the Mediterranean roost was friendless and alone.

The final result was that the Carthaginian meme died out, replaced by that of Rome. Carthage's language disappeared. Her religion was forgotten. Aside from Hannibal, her great men became historical obscurities. Even one of her most glorious trading colonies in Spain, Gadir, lost her Carthaginian name. Henceforth, the city would be known as Cadiz.[392]

The pecking order phenomenon is not restricted to ancient times. Humans in the modern era are still motivated by its primordial rule: friends flock to the bird on top; they shun and even abuse the bird on the bottom. This simple principle has cropped up in the recent history of America.

When the Soviets launched Sputnik in 1957, the achievement made two statements: It broadcast Russia's growing military power. The rocket was an adaptation of an intercontinental ballistic missile, something the United States didn't have at the time. Suddenly, the Soviets were in a position to annihilate North American cities with nuclear weapons. The launch of Sputnik also announced that America was no longer the undisputed master of world technology. Militarily and scientifically, the Soviet Union had taken a massive leap up the hierarchical ladder, and America had been nudged a small step down.

The result in the third world was electrifying. Minor nations rapidly fled from the country they felt was slipping and stampeded to the side of the superorganism shinnying up the pole. Third world newspapers broadcast their hatred of America and gloated over her humiliation. "Russians Rip American Face" screamed a headline in Bangkok's *Sathiraphab*. A professor in Beirut, Lebanon, said his students were so jubilant, "You would have thought they launched it themselves."[393] In third world minds, the Soviet triumph was helping the downtrodden nations vicariously trounce the beast on top—the United States.

Sputnik was not the only event of that era to telegraph a Soviet rise. In 1949, a group of Marxist-Leninists were cheered on by the Soviets as they took control of the most populous country on the planet, China. In

1956, Moscow ordered its tanks into Hungary, and the United States didn't dare come to that country's defense. And, in 1960, an island nation just a speedboat ride from Miami declared its right to throw its lot in with the Soviet bloc.[394] As the Russians rose in the pecking order, their pool of friends increased, and as we sank, the number of our faithful companions declined.

The cumulative result was a dramatic increase in the number of Soviet allies and a drastic shrinkage in ours. During the early fifties, our friends in the United Nations were so numerous that any issue submitted to a vote went resoundingly our way. By the mid-sixties, frequently almost every nation voted against us.

For a superorganism, slippage in the pecking order can produce catastrophe. The Yanomamo, the "fierce people" of the jungles around South America's Orinoco River, pride themselves on constant warfare. Fighting is their way of life, and to make war you need allies. The tribe on the top of the pecking order gets the most allies, and the tribe on the bottom gets the least.

Like the chicken at the bottom of the pecking order, everyone takes advantage of a defeated tribe's helplessness. Rival groups raid the low-ranking clan over and over again. Enemies lay in wait by its gardens. Knowing this, the shunned tribe is forced to hide out in the jungle, where it can't get decent food. Succulent yams lie in the soil of their garden plot, but these harassed people cannot dig them up. They know they'll be setting themselves up for ambush.

Stronger tribes attack the group on the bottom to steal its women. Knowing this, the men of the friendless clan remain perpetually on alert. Slowly, the tribesmen without allies grow more starved and exhausted. Sometimes, in the end, they no longer have the strength to survive.[395] Deprived of friends, the tribe at the pecking order's foot may finally disappear, its most desirable members and goods absorbed by its rivals. The superorganism that once was strong simply ceases to be. It's lonely at the bottom. No wonder most creatures prefer to be on top.

WORLDVIEWS AS THE WELDING TORCH OF THE HIERARCHICAL CHAIN

Philosophers are men hired by the well-to-do to prove that everything is alright.

Brooks Adams, brother of Henry Adams,
to Judge Oliver Wendell Holmes

Poetry, science, ideology, and religion—the blindmen's canes with which we feel out the invisible world, the glue that binds us together as a collective creature whose cells are individual souls—help stir the social beast to a pecking-order-inspired cannibalism. And once the battle is over, the meal is complete, and the rival society is no more, poetry, ideology, and religion may serve a new purpose: They often become the torch that welds the citizens of the swallowed loser into a new pecking order. They help turn captive chunks of the vanquished group into parts of the newly enlarged superorganism.

For example, Hinduism has seemed to its admirers in the West a profoundly spiritual view of the world. It rejects materialism, lays aside earthly desires, tells its adherents to go with the flow, to accept the world as it is, to build up a positive karma, and to strive for Nirvana in a selfless world. What could possibly be more benign? Under the surface, however, the Hindu religion is not what it seems. In fact, it was the device with which one conquering group managed to validate its theft of power, prestige, and goods from a rival superorganism.

In approximately 1,500 B.C., a cluster of Aryans drove their herds of cattle from Iran to northern India through the Hindu Kush mountains.

These were men whose lives centered around two things: their cows and their fighting. So inextricably were the two woven together that the Aryan word *gavishti* had two meanings: the first, "to search for cows"; the second, "to fight." On the Indian side of the mountains, these violence-prone Iranian cattle herders found a people far more sophisticated than they were. The Iranian intruders could neither read nor write. The people native to India, however, excelled at both. The Iranians had never seen a building more complex than a temporary hut. The Indians had lived for over a thousand years in elaborate cities. But apparently the Iranians had something that the Indian inhabitants lacked: an eagerness to fight. During the next few hundred years, the Iranians relentlessly attacked the indigenous Indian population and brutally beat the unfortunate locals into submission. It was a pecking order triumph par excellence. The Iranian invaders reduced the Indians to the shameful role of a conquered people and declared themselves the lords of the land.[396]

But where does a lofty and otherworldly religion fit into all of this? Hinduism was the picture of the invisible world crafted over the following centuries by the priests of the Iranian's descendants. At Hinduism's heart was a simple notion: There were several classes of human beings, as distinct from one another as worms are distinct from lions. First, there were the "twice born"—men favored by the gods with all their holy blessings. Then there were the Shudras and the outcastes, loathsome people so beneath the contempt of the heavenly deities that the gods refused to accept their prayers. The deities had ordained it thus. They had declared in their infinite power that the twice born were to ride forever on the shoulders of the dirtier and humbler classes of men. For the twice born were close to divinity. The lower castes were not. And who were these exalted twice-born mortals? The descendants of the Iranians.

This pious self-aggrandizement of a conquering barbarian tribe led to the Indian caste system. The top three castes were exclusively reserved for the "twice-born" Iranians. One of these privileged orders (Kshatriyas) contained the Iranian warriors and aristocrats. The second (Brahmans) included the Iranian priests (those wonderful folks who had come up with the system to begin with). And the third caste housed the Iranian landholders and merchants (Vaishyas). Down at the bottom of society, squirming like insects beneath the Iranian heel, were the original Indian natives, the occupied peoples. They became the loathed Shudras and outcastes. The defeated Indian Shudras were promptly put to work. They were sent into the fields to raise the crops upon which the wealth of the Iranian nobles, priests, and merchants would soon be based.

The Iranian overlords were fair skinned. The natives who had been placed in a state of permanent humiliation were dark in hue. That complexion difference was embedded permanently in the name of the social structure. The newly initiated hierarchical layers were called varnas—castes—the Iranian word for *color*. [397]

Under the Hindu system, the descendants of the Iranians were born with all the privileges that Hitler's Nazis would someday dream of. Take, for example, the prerogatives of the Iranian priests, the Brahmans. Anyone from a lower caste who jostled a Brahman on the street committed a sin. If he bumped the Brahman with his arm, the arm was cut off. If he touched the Brahman with his foot, the foot was surgically removed. If he sat in a Brahman's chair, he had a ten-inch red hot rod rammed up his nether parts. If he complained to the Brahman about this treatment, he had the same smoking piece of metal shoved down his throat. [398]

Anyone of lower class who sipped water from a pool at which a Brahman was contemplating a drink polluted it. A Brahman could make whatever accusation he wanted against a person of lower caste, and the accused would be punished. But the lower caste citizen could make no complaint against the Brahman. Once a Brahman had married at least one woman of his own superior caste, he could go into the street during festivals, roam about searching for some nice-looking girl from the lower strata, marry her if it pleased him, and discard her when he tired of her charms. But no man of lower social position could marry a Brahman girl. The members of the Iranian race were treated as *Übermenschen*—literally, "overmen."

Why does the Hindu religion tell its adherents to go with the flow, to abhor the things of this world, to set aside earthly desires, to hope only for an improvement of their lot after this life is over? Because Hinduism was designed to keep the conquered Shudras in their place. It told those trapped in the lower castes to be content with their humiliation and shun the appalling actions that might spring from desire and discontent. It instructed them never to overthrow their Iranian masters.

A brilliant invention made the society work: specialization. An entire class of human beings was consigned to agriculture for a lifetime. Another class was given the lifelong specialty of warfare. Presumably, perpetual practice made each better at his craft. And, of course, there were the specialists in religion who spent their time suppressing vanity and desire—the two things that could have torn the pecking order apart. For desire would have made those pinned in the lower orders hanker to move above their humble station.

The armies of professional religionists, the priests and monks, seemed at first glance to serve no useful economic purpose. But, in fact, they were the keepers of an indispensable meme. They declared that if you had the patience to tolerate imprisonment in a lower caste during this life, you would be rewarded later by rebirth in the next caste up the ladder.[399] If you held still long enough, you could actually become an Iranian!

In fact, the Iranian overlords expressed themselves as if they knew that they were fashioning a new superorganism from the swallowed pieces of the society they'd overwhelmed. Their ancient scripture, the Rig-Veda, written in the days when the conquerors were piecing together their new social system, put the argument for superorganism quite bluntly. The Brahmanic priests, said the Veda, were man's mouth. The warriors were his arms. His thighs were the Vaishya (the tradesmen and landowners). And the Shudras (the farmers) were his feet![400] The Brahmans were not the only folk to turn plunder into a permanent condition, grafting the conquered onto the underbelly of their own superbeast. The aristocrats in many civilizations are the fossils of earlier conquering hordes. Their position at society's apex is the residue of robbery.

In England, the titled classes, the folks who hold their noses in the air, are the descendants of Saxon, Viking, and Norman soldiers who pillaged, slaughtered, and raped in successive waves from roughly A.D. 470 to 1066.[401] In Japan, the aristocracy, which has sat securely in place for nearly 1,800 years, is the remnant of a population of nomadic Mongoloid horsemen who came across the sea from Korea in the first century A.D., brutalizing the local population into submission with long iron swords.[402] Yet we still gush with excitement over princes and princesses who show off at elaborate parties and are featured in fashionable magazines, not realizing that the virtue which distinguishes their families from ours was the greater willingness of their ancestors to use violence.

In Japan and England, just as in India, religion, philosophy, poetry, and ideology have all been used to nail the conquered in their lowly place. For a view of the invisible world holds a very visible world together—the world of society. A web of memes justifies the subjugation of those on the bottom, upholds the power of those on top, and sometimes maintains the specialized roles that allow a static society to work. Even more, it clothes the power of those who rule with a sublime halo, disguising over-lords as the chosen of God—or, in the case of Marxism, the inevitable heirs to the forces of history. It sanctifies the pecking order.

WHO ARE
THE NEXT
BARBARIANS?

THE BARBARIAN PRINCIPLE

⇒⟩ ⟨⇐

A thousand men who fear not for their lives are more to be dreaded than ten thousand who fear for their fortunes.

Denis Diderot

A position at the top of the pecking order is not permanent. Far from it. Animals who make it to the peak know that simple fact. They see that yesterday's adolescents have become today's restless adults and watch warily as these youthful challengers size up the odds of knocking their elders off the top of the heap.

Dominant beasts remain vigilant. But a strange thing happens to *nations* at the pecking order's apogee. The dominant superorganism sometimes goes to sleep. It falls complacently into a fatal trap, assuming that its high position is God given, that its fortunate lot in life will last forever, that its lofty status is carved in stone. It forgets that any pecking order is a temporary thing and no longer remembers just how miserable life can be on the bottom. The results are often an unpleasant surprise.

We all know that Rome was picked apart by peoples any respectable Roman could see were beneath his contempt. The barbarians didn't shave. They wore dirty clothes. They were almost always drunk. Their living standard was one step above that of a mule. Their technology was laughable. They usually couldn't read and write, and they certainly had no "culture." What could these smelly primitives do? They could fight.[403]

Rome was not the first superpower toppled by the third world rejects of its day. Egypt, land of the Pharaohs and home of the Sphinx, had been the most imposing power of its time. When the other great dominion of the day—Sumer—was still only a disunited gaggle of embryonic city-states, Egypt united as a six-hundred-mile-long kingdom under its first pharaoh, Menes. That was five thousand years ago.[404] Militarily, Egypt towered over its neighbors. Its public buildings were carved with solemn scenes of Egyptian warriors leading huge masses of conquered peoples off to slavery and beheading the unruly types who refused to go along with captivity. No one could challenge the mighty empire. Not if he wanted to live to talk about it.

Like ants in the colony with the biggest territory, Egyptians lived the good life. Consumer comforts were a dime a dozen. In one splurge downtown, you could stock up on cosmetics, perfumes, and an occasional statuette of a cat to decorate the living room.[405] Bread, beer, and kitchenware were available at the local market in dizzying abundance.[406] A nationwide irrigation system kept the fresh food coming. And a state-run system for storing agricultural surplus made sure that even in a bad season, everyone had plenty to eat.

The government was stable. The system of pharaohs had worked like a charm for 1,300 years (our constitutional democracy has only made it a skimpy two hundred plus). The country's architects were building pyramids while most of Egypt's neighbors were still struggling to put up pup tents in the desert. And things seemed to be getting better all the time.

Numerous tribes roamed like animals in the wastelands far beyond Egypt's boundaries. An Egyptian would have sneered at the notion that these wanderers could ever constitute a serious threat. Then a churlish mob no one had ever heard of swept down from the north. They were cultural nobodies, contemptible yahoos. Their life-style was fourth rate. Aside from their lack of couth, they had only three distinguishing characteristics—they were excellent horsemen, they reveled in violence, and they had a knack for inventing military hardware.[407]

They were called the Hyksos, and they totally overwhelmed the smug Egyptians, chopping up the fine-tuned Egyptian army like chicken liver. Then they rode their chariots into Egypt's splendid cities and took over. They had the gall to plunk one of their own on the pharaoh's holy throne. Adding insult to injury, these gutter crawlers proceeded to rule Egypt for the next 107 years.

Nearly a thousand years later, another superpower would be over-whelmed by barbarians. This time, however, the tale would end in irony. For this superpower itself had begun as a barbarian horde.

The empire of Babylonia had racked up some astonishing accomplishments. Four hundred years before the Hyksos taught the Egyptians the drawbacks of overconfidence, Babylonia's King Hammurabi had concocted a political breakthrough that revolutionized government—the idea of writing down laws. While the rest of the world was mired in illiteracy, the Babylonians were using a script so simple that every corner merchant could keep lists of his inventory and send angry letters to tardy suppliers.[408] Even humble miners were able to write graffiti on the walls of the pits in which they worked.[409]

The Babylonian military machine was a wonder to behold. With its unbeatable strategies, it hacked out a massive realm. By 600 B.C., Babylonian territory stretched in a thousand-mile arc from the Persian Gulf to the shores of the Mediterranean.[410] The Babylonians were so confident in their power that when they ran into resistance from the Hebrews, they tried something no weaker nation would have dared. They transported almost the entire population of the Hebrew kingdom of Judah to Babylon[411] and resettled the Jews in the bustling imperial capital, hoping that a few years of making a living in the Babylonian metropolis would turn this Bible-reading throng into nice, middle-class idol worshippers.[412] (It didn't work.)

Once Babylonia had trounced every tribal people in sight, its major concerns became the other superpowers of the day—the Assyrians and Medes. Babylonia had good reason to worry. Each of these rivals was a mammoth empire famed for military prowess. For example, even the captive Hebrews had commented on the Assyrians' ability to overwhelm peaceful towns by surprise attack, coming down "like a wolf on the fold."[413] What's worse, not long ago the Assyrians had reduced the Babylonians to a secondary power, and Babylon's politicians did not want to see that fate occur again.

As for the threat from the Medes, the Babylonian queen Nitocris was so concerned that she underwrote a strategic defense initiative on the scale of Ronald Reagan's Star Wars anti-ballistic missile scheme. She literally had a new course dug for the Euphrates River, forcing the water to loop around in a confusing tangle that would presumably keep the Medes from mounting a surprise naval attack against her capital. In addition, she fortified the river's edge with huge embankments, and—to slow the Euphrates' current—dug a lake some forty-seven miles in circumference, reinforcing its banks with stone.[414]

With empires like those of the Medes and Assyrians threatening them, who needed to worry about rabble from the hills? But the rabble did, indeed, turn out to be the Babylonians' major problem. A tribe from the rocky Zagros Mountains of southwestern Iran decided it would like to rule the lush valleys where cities flourished and the wealthy wore fine apparel. That tribe was the Persians.

The Persians were unlettered and uncouth. But they loved a good fight.[415] It wasn't long before this hitherto-unknown mob overwhelmed the Assyrians and the Medes, Babylon's two rival superpowers.[416] Then the Persians turned on the isolated Babylonians and won.[417]

The irony came a few decades later. By now, the victorious Persian rulers had turned from barbarians to urbane city dwellers. True, they still traveled up into the hills to eat and drink with the old folks for a few weeks,[418] but then they went back to their estates, their servants, their armies of bureaucrats, and their imported luxuries. They took over the superpowers of the day, one by one, finally subduing Egypt in 525 B.C.[419] The Persian superorganism was now the master of the international pecking order.

Of course, there were still dangers, but the Persians knew exactly where to look for them, or so they thought. Like the Babylonians before them, the Persians were blind to the barbarians and expected trouble only from nations celebrated for military might. They forgot that the real danger often comes from a people everyone has totally dismissed. So the great Persian leader Darius didn't bother with the scarcely civilized yokels who squabbled interminably on a bunch of islands and rocky coasts to the west and who called themselves the Greeks.

The Western upstarts provoked a fight. When some of the cities under Persian rule revolted, the insignificant foreigners sent a fleet to help them out. Then these outlanders proceeded to burn down Sardis, capital of the western segment of Persia's empire.[420]

The Persians, determined to teach the impertinent nobodies a lesson, ordered a naval detachment to administer punishment. Like the group of helicopters once dispatched to rescue American hostages in Iran, the invincible Persian fleet ran into technical problems. It was wrecked in a storm. In 490 B.C., the Persians tried again. This time, they sailed off to the homeland of the upstarts and clobbered one of their pitifully backward towns into the turf. But the nobodies turned the tables: they sent the invading Persian troops running and destroyed seven of the empire's momentarily victorious ships.[421]

The Persians had had it. They were determined to make these half-baked rednecks from a land scarcely marked on the map rue the day they tangled with Persia. Emperor Xerxes gathered an armada of mind-boggling size, its ships numbering well over a thousand.[422] What's more, according to Herodotus, the Persians put together an army of 1,700,000 men,[423] including troops from every territory of the empire—Arabia, Bactria, Media, Assyria, Ethiopia, and Libya.[424] Even the distant Indians contributed their heaviest transport vehicle—the war elephant.[425]

The resulting military force was so vast that it stretched farther than the eye could see. Provisioning it with food and equipment took four years and the resources of an entire continent.[426] Whenever the Persian host marched to a fresh campsite, it literally ate every available bit of food and drank every potable drop of water in expanses dozens of square miles in size. Says Herodotus, "There was not a nation in Asia" the Persians didn't take with them, and "Save for the great rivers there was not a stream . . . [their army] drank from that was not drunk dry."[427] The folks they were marching to conquer couldn't come anywhere near this logistic sophistication.

But the barbarians the Persians considered beneath contempt won the war. In the years before the first major Persian-Greek war, when he was informed that the burning of Sardis had been pulled off by a landing party of Athenians, the exasperated Persian emperor Darius had been forced to asked, "Who are the *Athenians?*"[428] Now, presumably, he knew.

One hundred and fifty years later, a Greek whom even *his* fellow Greeks called a barbarian would conquer the entire Persian empire. His name was Alexander the Great.

The whole thing was as unlikely as the Vietnamese turning around and conquering the U.S. But it happened. In fact, in history it happens over and over again. It happened in 1870 when the French were forced to fight a country that just a few years earlier had been a disorganized clutter of ragtag ministates ruled by comic opera princes. The land of Napoléon was rated by every armchair general as the mightiest military force on the Continent, but France lost. Its army was chopped up like ground round. Its glorious capital, Paris, faced the humiliation of a foreign army marching down its streets.[429] The upstart nation that had brought France to its knees was Germany.

An equally surprising fate occurred to England when it trained its guns on the superpowers of its day in two world wars. After the smoke had cleared, two backward nations of Johnny-come-latelies ended up dominating the world. These countries, whose inhabitants had usually been regarded as just one small step above the primitive, were the United States and Russia.

The lesson: Never forget the pecking order's surprises. Today's superpower is tomorrow's conquered state. Yesterday's overlooked mob is often the ruler of tomorrow. Never underestimate the third world. Never be complacent about barbarians.[430]

ARE THERE KILLER CULTURES?

Man's greatest good fortune is to chase and defeat his enemy, seize his total possessions, leave his married women weeping and wailing, ride his gelding [and] use the bodies of his women as a nightshirt and support.

Genghis Khan

He butchered three of them with an ax and decapitated them. In other words, instead of using a gun to kill them he took a hatchet to chop their heads off. He struggled face to face with one of them, and throwing down his ax managed to break his neck and devour his flesh in front of his comrades. . . . I . . . award him the Medal of the Republic.
General Mustafa T'las, Syria's minister of defense, praising a hero of the 1973 war with Israel before the Syrian National Assembly

Appeasing of governments which revel in slaughter is an invitation to worldwide catastrophe.

Fang Lizhi

Some readers will be outraged by my presumption. How dare I regard any group as barbaric. What appalling ethnocentrism! There are no barbarians. There are simply cultures we haven't taken the time to understand. Cultures to whom we haven't given sufficient aid. Cultures in need of development. Beneath the skin, all men and women are the same. They have the same needs, the same emotions, and the same ideals. If you simply took those folks you speak of so contemptuously out for a cup of coffee, you would discover that they are just like you and me.

But there *are* barbarians—people whose cultures glorify the act of murder and elevate violence to a holy deed. These cultures portray the extinction of other human beings as a validation of manliness, a heroic gesture in the name of truth, or simply a good way to get ahead in the world.

Certain Islamic societies tend to be high on this list. On November 28, 1943, Franklin Roosevelt met secretly with Joseph Stalin and Winston Churchill in Iran. When Roosevelt returned home, he sent a telegram to the Shah thanking the Iranian ruler for his hospitality. The president explained that he'd noticed the hills in Iran were bare. American agronomists had learned to prevent soil erosion and enrich the landscape by planting trees on slopes like these. Roosevelt suggested an experimental tree-planting program. The Iranian leader thanked FDR. But privately the young potentate was highly insulted: According to Moslem standards, the gift demeaned his virility. Stalin was far more understanding of Mohammedan culture. He offered the Shah tanks and planes.[431]

Hafez al-Assad, current leader of Syria, worked hard to solidify his position as the country's undisputed ruler. He didn't do it by selling Syria's citizens on the values of his political platform. Instead, he slaughtered twenty thousand Moslem fundamentalists who opposed him.[432]

According to the *New York Times,* in 1980 Yasir Arafat, the Palestinian leader, had a Lebanese imam (a holy man roughly equivalent to a pastor) shot in the head for refusing to preach the propaganda of the PLO. Then Arafat visited the imam's Lebanese home, took his ten-year-old son aside, explained to the little boy that his father had been murdered by the Israelis, handed the lad a gun, and said, "When you grow up, use this to take revenge." Arafat wanted the boy to be a killer.[433]

Holiness, righteousness, and even day-to-day propriety in Islamic cultures are based on the example of Mohammed.[434] Though Islamic literature praises Mohammed as a man of peace, he was also a military leader. In A.D. 624, the Prophet announced the concept of the jihad—the holy war. He said in the blessed book, the Koran, "I will instil terror into the hearts of the unbelievers: smite ye above their necks and smite all their finger-tips off them. . . . And slay them wherever ye catch them. . . .'' In the next nine years, the man of peace ordered a minimum of twenty-seven military campaigns. He personally led nine of them.[435]

It is not surprising that Moslem jurists would later declare that there are two worlds: the world of Islam, Dar al-Islam, and the non-Islamic world, Dar al-Harb. These two territorial spheres, explained the Moslem scholars, are in a state of perpetual war.[436] According to some Koranic interpreters, any leader who fails to ''make wide slaughter'' in the land of the infidel is committing a sin. A statesman is allowed the temporary expedient of peace only if his forces are not yet strong enough to win.[437] This may explain why Elias Canetti, in his Nobel Prize–winning book *Crowds and Power,* calls Islam a killer religion, literally ''a Religion of War.''[438]

In reality, Islam, like most other religions, has both its positive and its negative sides. It imposes a host of admirable responsibilities on its adherents: for example, *zakat,* the presentation of regular, substantial contributions to the poor. Allah also demands that his followers ''give glad tidings to those who believe and work righteousness,''[439] ''cover not Truth with falsehood nor conceal the Truth when ye know [what it is],''[440] and ''treat with kindness your parents and kindred and orphans and those in need.''[441] However, Allah issues many a darker order as well. And the percentage of modern Islamic adherents who have focused on Allah's calls to combat is dismaying.

Today, the descendants of the Persians who fought the Greeks in 480 B.C. are devout Moslems. In the thirties, one of them labored diligently to become an Islamic scholar. He pored over the Koran for years. As he demonstrated his superior knowledge of Allah's pronouncements, he rose in the ranks of Iranian holy men. Finally he achieved the second-highest title—ayatollah (roughly equivalent to a Catholic cardinal).[442] His name was Ruhollah Khomeini, and he wrote books and pamphlets and even taped and distributed his speeches to inspire the citizens of Iran with sacred virtue. The ayatollah's words roused Iranians to overthrow the shah and usher in a government based on strict Islamic doctrine.

What did the ayatollah's pronouncements say? Among other things, that infidels are like dogs. Their existence is an affront to Allah. Here's how the ayatollah himself put it:

"Moslems have no alternative . . . to an armed holy war against profane governments. . . . Holy war means the conquest of all non-Moslem territories. . . . It will . . . be the duty of every ablebodied adult male to volunteer for this war of conquest, the final aim of which is to put Koranic law in power from one end of the earth to the other."[443]

The leaders of the USSR and of England and the president of the United States are . . . infidels.[444]

Every part of the body of a non-Moslem individual is impure, even the hair on his head and his body hair, his nails, and all the secretions of his body. Any man or woman who denies the existence of God, or believes in His partners [the Christian Trinity], or else does not believe in His Prophet Mohammed, is impure (in the same way as are excrement, urine, dog, and wine)[sic]."[445]

Concluded the ayatollah, "Islam does not allow peace between . . . a Moslem and an infidel."[446]

Though many of us imagine that the promotion of harmony is a prime objective of every major world faith, the ayatollah disagreed. "The leaders of our religion were all soldiers, commanders and warriors," he wrote; " . . . they killed and they were killed." The concept of a peaceful prophet was so alien to the ayatollah that he was convinced Christ's message had been deliberately distorted by Westerners. Said Khomeini, "This idea of turning the other cheek has been wrongly attributed to Jesus (peace be unto him); it is those barbaric imperialists that have attributed it to him. Jesus was a prophet, and no prophet can be so illogical."[447]

Khomeini's dicta may seem irrelevant now that he has long been dead, but his words have gained in influence since his demise. Early in the nineties, Iraq's humiliation in the Gulf War undermined the credibility of the secular Moslem regimes, leaving a power vacuum into which fundamentalism leaped.[448] There are currently roughly 100 million Islamic fundamentalists (rechristened "Islamic revivalists" by some scholars[449]). Activists among them, employing the slogan "Africa for Islam," are making diligent—and often violent—efforts to seize power in numerous sub-

Saharan states.[450] They have gained sufficient favor with South Africa's ANC that Nelson Mandela, in a 1992 visit to Teheran, told the Iranians that Africa must be reshaped along the lines of the Iranian revolution.[451] (Ironically, when South African leader Bishop Desmond Tutu gave a speech to a Palestinian crowd in 1989 lauding Palestinian interests, he failed to realize that the Arabic banners carried by his listeners read, "On Saturday We Will Kill the Jews, on Sunday We Will Kill the Christians!")[452]

Khomeini-style fundamentalists have become vigorous political forces in areas like China's Sinkiang region (where as of 1994 Beijing officials were seriously concerned that the area's inhabitants, influenced by propaganda from Iran, would attempt to break away and found a fundamentalist Islamic republic).[453] Islamic fundamentalists have been involved in the Indian state of Kashmir's vicious civil war.[454] They've been active in Malaysia,[455] Thailand (where Moslem guerrilla forces were fighting in 1993),[456] and the Sudan[457] (where an Iranian-backed fundamentalist regime is engaged in a campaign to subjugate, exterminate, or—according to the United Nations International Labor Organization—literally enslave the black Christians and animists in the southern region of the country).[458] Followers of Khomeini have been moving aggressively in Algeria,[459] Jordan,[460] Tunisia,[461] Lebanon, Kuwait,[462] Pakistan,[463] Afghanistan, Azerbaijan, Turkmenistan (where by 1992 posters and portraits of the ayatollah had become a particularly strong sales item in local stores),[464] France,[465] and, according to Greek Defense Minister Ioannis Varitsiotes and the University of Belgrade's Dragoljub R. Zivojinovic, Czechoslovakia, Albania, and Yugoslavia.[466] In many of these cases, fundamentalists are sweeping elections, manipulating generals, funding insurrections, sponsoring terrorism, or actually taking control.[467]

Islamic fundamentalists have poured money into America's black communities in an effort that has brought more than a million African Americans over to the one true faith.[468] While most of these converts remain peaceful, Al-Fuqra, a predominantly African-American Islamic group under the leadership of the Pakistani Sheikh Mubarak Ali Jilani Hashemi, has declared a jihad in North America and, according to law enforcement agencies, has been involved in bombings, murders, and other forms of bloodshed in Colorado, Arizona, Pennsylvania, and Canada. It has been reported that Al-Fuqra also had a hand in the 1993 effort to blow up New York's United Nations building, the city's FBI headquarters, and its Holland and Lincoln Tunnels.[469]

When the Iranians declared a death sentence on the British author Salman Rushdie, black American imams everywhere from Brooklyn to Los Angeles enthusiastically supported the move. (So did the Moslem head of UCLA's Middle Eastern Studies Department.) Even a loyal African-American Gulf War veteran, won over to Allah in 1991, stated after his change in faith that "soon it [Islam] will take over all of America, then the world."[470]

The U.S. African-American community is only a beachhead. Islamic forces have been attempting to gain control of U.S. media outlets in the hope of using them as propaganda tools for the Moslem point of view. The Saudis and America's Christian fundamentalists battled in the early 1990s for the right to purchase America's second largest wire service, UPI. Ultimately, the Arabs won. In addition, Amal Adam, the former head of Saudi Arabia's equivalent of the CIA, was the primary backer of a British-based firm called Capcom, whose chief officers were the heads of TCI (Telecommunications Incorporated), America's largest player in the cable television game. In 1993, TCI made headlines when it came within a hair's breadth of merging with Bell/Atlantic. Had the effort succeeded, it would have formed what financial analysts universally heralded as one of the giants of the coming interactive media revolution, giving the Saudis additional leverage for American media manipulation.

The ground is ripe for worldwide Islamic fundamentalist expansion. Mohammedanism is currently the fastest-growing religion on the planet.[471] There are a billion Moslems—as many as Jews and Christians combined—and that number is increasing daily. According to Cairo University's Professor Ali Dessouki,[472] fifty countries are now Islamic. What's more, there are massive Mohammedan populations everywhere from Nigeria to Mongolia, the former Soviet Central Asian republics,[473] Southeast Asia, and the Philippines. The countries with the world's largest Islamic bodies of citizenry are not even parts of the Arab world—they are Indonesia and China.[474] To top it off, Islamic public opinion, if the Arabs,[475] Iranians, and Pakistanis are an accurate barometer, is virulently anti-American.

Today's Islam extremism is the perfect example of a meme grown ravenous. Saddam Hussein, in his 1990 drive for expansion, claimed to be following Allah's message. The late General Zia, former head of Pakistan, who masterminded the fundamentalist-led Afghan resistance efforts using U.S. funds, kept a map in his office with all of Iran, Pakistan, Afghanistan, and Soviet Central Asia marked in green. It was the symbol of his ultimate ambition—unified Moslem rule extending through every green-marked territory.[476] In 1990, one enthusiastic Turkish official, Minister of State Ereument Konukman, noted the substantial Turkish populations in the former Soviet Union and China and looked forward to uniting them "under the colors of the Turkish flag."[477]

A fundamentalist clergyman in Lebanon says, "Don't believe that we want an Islamic republic in Lebanon. . . . What Hezbollah wants is a *world* Islamic republic."[478] Cairo constitutional lawyer Dr. A. K. Aboul-magd adds, "I even venture sometimes to say that Islam was not meant to serve the early days of Islam, when life was primitive and when social institutions were still stable and working. It was . . . meant to be put in a freezer and to be taken out when it will be really needed. And I believe that the time has come. . . . The mission of Islam lies not in the past, but in the future."[479] Dr. Abd El Sabour Shahin of Cairo goes a step further and warns that Western civilization makes a big mistake when it "thinks it will endlessly remain dominant."[480] Even secular Moslem intellectuals teaching in the top universities of the United States and Europe have joined the expansionist bandwagon, calling for a leader who will pull world Islam together into an unstoppable force.[481]

"Islam will . . . take over the world," said an Egyptian in Cairo in the late eighties to a crew from Britain's Granada TV. No isolated, gray-haired zealot, he was one of a new breed of young university graduates, members of the middle class, and professionals, often among the highest achievers in their region. These religious devotees do not have a happy fate in store for those of us in the West. Explained the young Egyptian, "Islam is a tree that feeds on blood and grows on severed limbs."[482]

In the early and mid-nineties, a spate of books and articles appeared proclaiming that, despite such rhetoric, Islam poses no geopolitical danger. Abul Aziz Said, of the School of International Service at American University, said point-blank that "Islamic fundamentalism is not the enemy of the West." "Islamic fundamentalism," he declared, "is a defensive social and political movement, a reaction to westernization and

modernization." It is, he insisted, "an attempt to restore an old civilization, not create a new empire." Yet, later in his article, Said said that ancient imperial triumphs were at the heart of the "world influence"
fundamentalists were legitimately attempting to "regain." And the veil
slipped a bit from his true feelings when, zeroing in on his conclusion, he
declared that "imitative responses of Muslims to the challenge of the
West . . . evince . . . identification with the 'enemy.' "[483]

John L. Esposito, former president of the Middle East Studies Association, criticized "the creation of an imagined monolithic Islam" and
contended that those apprehensive about fundamentalism "fail to account
for the diversity of Muslim practice."[484] Palestinian-born Columbia University scholar Edward Said echoed the assertion that diversity renders the
notion of an Islamic threat, in Said's word, "phony."[485] However, diversity within a cultural community does not necessarily halt its expansionist
drive. The European West spread its often brutal control over every continent while so divided and "diverse" that it was engaged in an almost
nonstop series of internecine wars. And early Islam conquered a territory
almost equally vast while its leaders squabbled and fought and its religious
sects were rent by schism.

Esposito, like many other writers on the topic, justifies the ferocity
of anti-Western Islamic sentiments by reminding us that "many in the
Arab and Muslim world view the history of Islam and of the Muslim
world's dealings with the West as one of victimization and oppression at
the hands of an expansive imperial power." There's no question he is
right.[486] However, the Islamic world held the upper hand in the struggle
between the Occident and the Levant for over 1,100 years. The West
managed to turn the tables briefly when the Crusaders established a short-
lived Middle Eastern toehold. But the Crusader states were not planted
on undisputed Moslem land. The heartland of the Islamic empire, the section bordering the Mediterranean rim, was a deeply Christian area, a vital
spiritual and economic core of a "Western" imperium, which, for over
six hundred years before Mohammed's birth, had included the non-Arab
provinces of Turkey (known then as Asia, Galatia, Bithynia, Pontus, and
Cappadocia—where Saint Paul established many of the first churches),
Syria (whose city of Damascus was one of the earliest major Christian centers), Israel (homeland to the Jews since roughly 1200 B.C. and, despite
Roman efforts to expel the native population, still dotted with Hebrew
villages when the Moslems arrived, sword in hand), Egypt (populated at

the time by rabidly Christian descendants of the pyramid builders, along with significant numbers of Greeks and Jews), Libya (the former Cyrenaica), Tunisia (Carthage and its environs, where Saint Augustine was born and eventually became bishop of Hippo), and Northern Algeria and Morocco (then called Mauretania). These were the countries that had produced the Bible, the Christian monastic movement (born in Egypt), Saint Jerome's conversion (in what is now Turkey), Saint John of Damascus, the famed early church historian Bishop Eusebius of Caesarea, Origen, Saint Athanasius, the Aryan heresy, and a significant number of fathers of the Roman Catholic faith and the Eastern Orthodox creed.

The knights of the cross did not retain their reconquered kingdoms long. They took Jerusalem in 1099 and were expelled by 1187. Nonetheless, according to historian Amin Maalouf, the author of *The Crusades through Arab Eyes,* modern Arabs tend to see today's world events as a continuation of the Crusades.

For six hundred years after the fall of the Crusader states, Islamic forces returned to the attack, capturing Greece and chunks of Eastern Europe, raiding towns in Sicily and the Italian coasts for goods and slaves, preying on Mediterranean shipping, chaining Europeans like Miguel Cervantes to the oars of their galleys, and until 1826 forcing the Christian citizens of Yugoslavia and Albania to give up their children to Moslem overlords (who brought up the males on the Koran, then turned them into soldiers known as Janissaries).

It wasn't until 1798 that Napoléon began to shift the balance between East and West again when he briefly invaded Egypt, from which he was ignominiously expelled by the British and the Turks. But the heavy-handed fertile crescent "imperialism" so resented by the Arabs didn't begin until after the First World War, and it lasted less than forty years. Southern Spain remained under the *Moslem* yoke for 781 years and Greece for 381, and pieces of longtime Christian terrain like Saint Augustine's North African homeland and the religious and secular capital that eventually eclipsed Rome in power and splendor—Byzantium—are still in Moslem hands today. Syria, on the other hand, was only under Western control for 21 years, Egypt for 67, and Iraq a mere 15. If one accepts Esposito's reasoning, Westerners—who were bludgeoned by "an expansive imperial" Islam for well over a millennium—have more right to fear an Islamic revival than Moslems have to hate the West.

More to the point, Phebe Marr, of the National Defense University's Institute for Strategic Studies, contends that militant extremist groups dedicated to violence and an absolute rejection of the West are small. In addition, she claims, "The radicals do not have a broad base of popular support. . . . Even in Lebanon, however, where such groups flourish, a poll of university students taken in 1987 indicated that more than 90% disapproved of . . . assassinations, hostage taking, and sabotage of government installations." On the other hand, Marr admits that "there may be only a thin line between the open, mainstream movements and their clandestine [violent] counterparts." She concludes that "the Islamic revival is not only here to stay but is likely to be a leading domestic political force shaping the Mediterranean region during the coming decades. Despite political vicissitudes, the various movements loosely collected under the rubric of 'Islamic Fundamentalism' have shown a staying power that indicates they have achieved both breadth and depth in their indigenous societies."[487]

Like Marr, Abbas Hamdani, professor of Middle Eastern history at the University of Wisconsin, asserts that "to propose a monolithic view of Islam and then equate it with fundamentalism would be wrong. . . . Except for mass followings in Algeria and Tunisia, fundamentalists represent a small segment, although a popular, vocal, and highly motivated one, of the total population. [Hamdani overlooks the Sudan and Afghanistan, both of which, at this writing, were in fundamentalist hands.] Even in Iran, which appears to be totally convulsed in fundamentalism, it is a small minority that has monopolized power."[488] As the case of Iran demonstrates, it only takes a minority to seize control of a country, especially if that minority is enthusiastic about using violence. In Germany's July 1932 elections, 63 percent of the voters cast their ballots against the Nazis. By the November elections, the anti-Nazi vote was even larger. Yet Adolf Hitler was able to achieve dictatorial power only four months later on March 23, 1933, in part because his storm troopers—like the militant gangs controlled by the fundamentalists—were willing to murder their opponents.

Khomeini's works advocate vigorously converting or murdering all those who do not embrace Allah's holy meme. Then they urge a holy war on the nations of the West. The ayatollah wrote, "Any nonreligious [i.e., non-Islamic] power, whatever form or shape, is necessarily an atheistic power, the tool of Satan; it is part of our duty to stand in its path and to struggle against its effects. Such Satanic power can engender nothing but corruption on earth, the supreme evil which must be pitilessly fought and rooted out. To achieve that end, we have no recourse other than to overthrow all governments that do not rest on pure Islamic principles, and are thus corrupt and corrupting, and to tear down the traitorous, rotten, unjust, and tyrannical administrative systems that serve them. . . . If Islamic civilization had governed the West, we would no longer have to put up with these barbaric goings-on unworthy even of wild animals. . . . [Western governments are] using inhuman laws and inhuman political methods. . . . Misdeeds must be punished by the law of retaliation: cut off the hands of the thief; kill the murderer instead of putting him in prison; flog the adulterous woman or man. Your concerns, your 'humanitarian' scruples, are more childish than reasonable." Khomeini had a prescription for such problems: "All of humanity must strike these troublemakers [the governments of the West] with an iron hand. . . . Islam has obliterated many tribes because they were sources of corruption [i.e., sources of non-Islamic influence]." Judging from the ayatollah's rhetoric, the next tribes he would have liked to see obliterated were those in Europe and America.[489]

Allah is rapidly providing Khomeini's followers with a sword to carry out their master's wishes. He has offered Islam the fire in which the Koran says those who follow false faiths are destined to burn: nuclear weaponry. He has also provided the long-range missiles needed to use it.[490] According to the late imam's logic, there may be only one just and righteous thing to do: employ this technology to wipe out recalcitrant heathens like you and me.

The modern growth of Islam is the coalescence of a superorganism drawn together by the magnetic attraction of a meme. But this meme has an advantage: The social body it is trying to pull together has existed as a unified social beast in the past. The old reflexes of solidarity are still there, waiting to be aroused. The meme of the new Islam is not laboring to generate a small and fragile embryo. It is simply attempting to awaken a sleeping giant.

VIOLENCE IN SOUTH AMERICA
AND AFRICA

The Islamic world is not the only place where violence is elevated to a virtue. Some Americans blame themselves for the bloodshed in Latin America, but violence was already endemic to these regions when this country had not yet become involved. In 1827, many years before the first platoon of American marines set foot on Latin America's tropical soil, the German philosopher Hegel described the area's culture. In his lectures on world history, Hegel explained that Latin America was a deeply homicidal place. Revolutions were an almost daily occurrence. Armed violence was the standard method of changing a government or settling a difference of opinion.[491] The situation was so bad that the Latin American revolutionary Simón Bolívar, the man who had freed Venezuela, Colombia, Ecuador, Peru, and Bolivia from colonial rule, gave up in disgust during 1830 and left South America for Europe.

Bolívar said, "After fighting for 20 years, I have reached a few definite conclusions." Chief among them was that Latin America's addiction to butchery would inevitably cause the Central and South American states to "fall into the hands of an unruly multitude and then into the rule of petty tyrants."[492]

American greed had nothing to do with the thirst for murder that dispirited Bolívar and Hegel. There were no massive multinational U.S. corporations in existence during the 1820s. Imperialist ambitions wouldn't infect American presidents for another seventy years.[493] Due to reasons Hegel never described, some Latin American cultures seemed inherently brutal.[494]

Yet many African societies make the perpetual civil war of Latin America seem like languorous peace. For example, in the 1970s, President Francisco Macias Nguema of Equatorial Guinea felt genocide was the best way to consolidate his power. He established the domination of his tribal group, the Fang, by killing fifty thousand of his countrymen. A third of the nation's population fled into exile. Many who stayed behind died in forced labor camps.

Sekou Toure, president of Guinea (a country far to the north of Equatorial Guinea), once said, "Get rid of the vermin . . . there is no room for half measures." The vermin he wanted to exterminate were primarily members of the Fula tribe, especially those whose last names were Barry or Diallo.

From January, 1971 (when Idi Amin, a black Moslem military leader, staged his coup) to 1981, between 100,000 and 300,000 black Ugandans were executed. The majority of those who died were members of the Acholi and Langi tribes who had supported Amin's predecessor, President Milton Obote. The heads of black Ugandans were beaten in with hammers, their legs were chopped off, and they were forced to eat the flesh of their fellow prisoners before they were put to death. To picture the nature of the event, imagine that when Bill Clinton won the presidency from George Bush in 1982, Clinton's followers had cemented their power by torturing and murdering registered Republicans. In an ominous echo of Adolf Hitler's genocidal rhetoric, a former Ugandan civil servant explained that President Amin was simply pursuing a "final solution" to the "Acholi and Langi problem."

But Idi Amin was not just one isolated madman in an otherwise peaceful country. When the bloodthirsty leader was deposed and Milton Obote returned to power, the newly restored president continued Amin's policy of slaughter. Obote killed 100,000 more black Ugandans; he simply did it with less press coverage.

Africanist Ken C. Kotecha, in his book *African Politics,* calls the atrocities afflicting the continent "a policy of annihilation."[495] That policy was rampant during the nineties in countries from Liberia to Somalia, the Sudan, and Mozambique. Why did the political use of the mass grave sweep across Africa to begin with?

When the former African colonies were granted their freedom, each new nation began life with a constitution that guaranteed civil rights to its minorities. Some of those minorities were white or Asian, but most were black. In the first few years of independence, the constitution of virtually every free country was altered. Altered to remove the guarantees of liberty to minorities. Altered to put all power in the hands of one man at the top. Altered to give unopposed authority to the party the "chief executive" ruled, and to the friends and kinsmen among whom he divvied up the country's wealth and power. Needless to say, the minority populations who had been stripped of all rights often objected. Their protests were usually violent, and they were dealt with harshly.

Burundi was typical. In 1962, the country won independence. In 1972, members of the Hutu tribe attempted to rebel against a government dominated by their rivals, the Tutsi. During the Hutu uprising, two thousand people were killed. President Micombero declared martial law, mobilized his army and his revolutionary youth brigades, then went on the attack against Hutu of any kind, whether they'd taken part in the rebellion or not. Micombero's squads rounded up local administrators, chauffeurs, clerks, skilled workers and, in the words of a Tutsi witness, "almost all the Hutu intellectuals above the secondary school level," herded them into jail cells, and either shot them or beat them to death with clubs and rifle butts. Four thousand were killed in the town of Bujumbura alone.

At the Université Officielle, the president's armed forces and youth squads showed up in classrooms, read off the names of Hutu students, and took them away to their deaths.

A third of the school's students were slaughtered. Thousands of Protestant pastors, teachers, and school directors were murdered. When the killing was over, 150,000 had died.[496] In 1993, when a Hutu was elected president, the Tutsis went on the rampage again. This time, 800,000 fled to neighboring countries in an effort to avoid what Foreign Minister Paul Munyembari called "genocide."[497] By 1994 that genocide had spilled over to Rwanda, where more than 500,000 Hutus and Tutsis were killed.

The violence of Africa, of the Islamic world, and of Latin America is not a massive deviation from the human norm. It is a simple outbreak of something we all share. The citizens of these lands are in the grip of forces no human can escape—the animal brain and the battle between superorganisms. There are no righteous societies; there are simply different degrees of depravity.

To establish the American republic, our forefathers exterminated Native American tribes. Native Americans were no better. James Mooney, a pioneering ethnologist whom numerous Native Americans of the late-nineteenth century regarded as a friend, said, "The career of every Indian has been the warpath. His proudest title has been that of warrior. His conversation by day and his dreams by night have been of bloody deeds upon the enemies of his tribe. His highest boast was in the number of his scalp trophies, and his chief delight at home was in the war dance and the scalp dance. The thirst for blood and massacre seemed inborn in every man, woman, and child of every tribe."[498]

Violent addictions still run like ramrods through our statements of ideals. We glorify the bloody war that gave us our independence. We sing a national anthem that invokes images of bombs bursting in air during battle. One of our favorite tunes for schoolchildren is "The Battle Hymn of the Republic," a paean to bloodlust. In it, our God is pictured as a violence junkie who gets his kicks by "trampling out the vintage where the grapes of wrath are stored."

The Islamic world today does not see us as respecting human life. It views Westerners—and Americans in particular—as the ultimate destructive force, the civilization that indulged in two world wars and capped that carnage with the creation of the atomic bomb. In the minds of Moslems, only believers in Islam are true champions of peace and justice. To Moslems, *we* are the people whose hands are perpetually stained with blood.

Like most of us, Moslems see only their better side. And like us, they imagine that their darker impulses do not really exist. Instead, they feel that the urge to destroy and conquer belongs only to their enemy. That's how the Moslem world justifies our imminent conquest and how the Moslem superorganism excuses its hunger.

What is the difference, then, between Americans, Africans, Latin Americans, and Moslems? Why do I claim that they, not we, are the barbarians? It's a question of degree. No American leader has ever followed the path of Syria's Assad and embarked on a mass extermination of political opponents to secure his position in office. No Yankee presidential candidate has emulated Equatorial Guinea's Francisco Nguema and wiped out fifty thousand members of a rival ethnic group in some electoral ward that was rooting for his opponent.

There is a little bit of the barbarian in all of us, but some are far more barbarous than others. There are cultures that idealize carnage. Others—we hope ours among them—put a premium on human life. Some cultures feel that debate is superior to battle, that discourse is preferable to the sword. These cultures stress conciliation, not violence, as a means of conflict resolution. They measure political manhood by the ability to produce *voluntary* consent. Their memes generate democracy and pluralism.

Some of us in the West have a tendency to justify the proviolence stance of third world countries. We turn our backs on African genocide or Syrian political killings as the good Germans turned away from the murder of six million Jews under the Nazis, or we find excuses for it. When we do, we become implicit accomplices in murder. Many of those who romanticize homicidal peoples have gone a step further. They have striven to replace the melting pot's leashed hostilities with "multicultural" enclaves roused to permanent anger by the dogmatic language of ethnic "struggle."

It is important that the societies which cherish pluralism survive. It is critical that they spread their values. It is vital that they not mistakenly imagine all other societies to be equal and their own to be inferior. It is imperative that they not allow their position in the pecking order of nations to slip and that they not cave in to the onrush of barbarians.

THE IMPORTANCE OF
HUGGING

≫◄

Why do some cultures seem abnormally prone to revel in violence? One possible answer comes from the patriarch of American psychology, William James, who said that civilized life makes it possible "for large numbers of people to pass from the cradle to grave without ever having had a pang of genuine fear." James implies that without the omnipresent sense that at any moment they may lose their lives, the beneficiaries of civilization feel far less of the savage animosity, the fierce hatreds, and the deep desires to mutilate and kill that terror inspires.[499] James's notion is intriguing, but let's not forget that we, too, have our hatreds and our violent moments.

Another explanation may be found in a survey of forty-nine primitive cultures conducted by James W. Prescott, founder of the National Institute of Child Health and Human Development's Developmental Biology Program. Some of the cultures Prescott studied took great pleasure in "killing, torturing or mutilating the enemy." Others did not. What was the difference? Says Prescott, "Physical affection—touching, holding, and carrying." The societies that hugged their kids were relatively peaceful. The cultures that treated their children coldly produced brutal adults. Or, to put it more technically, a low score on the Infant Physical Affection scale correlated with a high rate of "adult physical violence."[500]

You can see elements of Prescott's Infant Physical Affection factor at work in Islamic society. Islamic mothers tend to be warm and nurturing, but Islamic fathers treat their children harshly, acting cold, distant, and wrathful. Their justification is an old religious proverb: "Father's anger is part of God's anger."[501] When he reaches puberty, an Arab boy is expelled from the loving world of his mother and sisters into the realm of men.[502] There, hand-holding between males is still allowed, but physical affection between men and women is frowned upon. A vengeful masculinity stands in its place. The result: violent adults. For an indirect glimpse at how the principle works, let's meander into the world of the bedouin.

Bedouin culture is the mother of all Islam. The bedouin are desert wanderers who, until recently, traveled with tent and camel through the Middle East and across northern Africa, driving their flocks of sheep and goats, and organizing caravans. The city children of Mecca, where Mohammed was born, were given out to bedouin nurses to be suckled. Mohammed himself was nursed by one of these bedouin "foster-mothers," and spent his childhood years among the shepherds of the desert.[503] The bedouin also made up the bulk of the armies with which Mohammed's followers went out to conquer the world.

The old bedouin ways have by no means disappeared. In 1978, an American graduate student of anthropology went to study "interpersonal relationships" among the bedouin of the western Egyptian desert. Her name was Lila Abu-Lughod, and she had a unique advantage in penetrating the most intimate aspects of bedouin life: Abu-Lughod's father was an Arab. In fact, he accompanied his daughter to Egypt and introduced her to the head of the family she would study, for had Lila appeared outside the nomads' tent, pads in hand, explaining that she was a scientific researcher, her quest would have been over before it began. The bedouin would have noted that she was a woman alone, which could only mean one of two things: Either her family cared nothing about her, in which case any man who ran across her could do with her as he willed, or she had committed a deed so immoral that her family had thrown her out, in which case any man who ran across her could also do anything with her that suited his fancy.[504]

With her father to make the introductions, however, Abu-Lughod was accepted as a good Arab girl and was taken into the household as a stepdaughter, living among the bedouin women as one of them. In the process, she saw details of Arab society from which Westerners are ordinarily shut out.

Abu-Lughod returned with some extremely revealing observations, including the manner in which bedouin society outlaws close, warm relationships between men and women. Romantic love is "immoral." Wives are expected to act aloof and uncaring about their husbands. A wife refers to the gentleman with whom she occasionally shares a bed simply as "that one" or "the old man." When a husband brings in a new bride, the previous spouse is supposed to show no jealousy, no emotion, no sense of hurt.

Husbands and wives are not to be seen together in public. Kissing or hugging openly is considered disgusting, indecent, almost inhuman. A couple who indulge in such a moment of warmth would be subject to contempt, fury, and hatred. Men spend very little time with their wives and scarcely ever mention them.[505]

In relationships between the sexes, a display of caring is despicable. Anger is what wins respect. The new wife of Rashid, one of the young men in the village Lughod was observing, ran away. Rashid was distraught, but among the bedouin, a man is not allowed to reveal his emotional wounds—especially if they are inflicted by a woman. Rashid's pained reaction was considered weak and scornful. Even his relatives scolded him. Later Rashid began to rage. Now everyone approved. This was the manly thing to do! Then the abandoned husband demonstrated a response that the other members of the tribe could be proud of. He began to search for someone to blame. Rashid interrogated women and children to see if one of them had annoyed the runaway wife so badly that she had been impelled to depart. Finally, he concluded that the girl had fled because of sorcery. The one behind the evil deed: his senior wife. The furious Rashid cursed his first wife and punished her by refusing to talk or visit with her. With this act of retaliation, everyone was happy.[506]

Author Leon Uris, who reviewed considerable anthropological research to compile his vision of Arab family life in *The Haj*,[507] believes that Arab village dwellers exhibit this same coldness toward their kids. Children, Uris claims, are seldom shown warmth, but they *are* frequently punished harshly. Hisham Sharabi, Omar al-Mukhtar Professor of Arab Culture at Georgetown University, goes a step further and asserts that Arab children are "repressed" to an intolerable degree.[508]

Like the bedouin, Middle Eastern city dwellers put a premium on violence, anger, and revenge. One young Palestinian discovered that his unmarried sister had become pregnant, irredeemably sullying the family honor. The virtuous young man wiped away the shame by killing the girl and cutting open her belly with a knife. According to French sociologist Juliette Minces, who has lived and researched extensively in the Middle East, incidents like this are extraordinarily common.[509] No wonder the noted Arab social scientist Halim Barakat has blamed the plight of the Levant on the structure of its families.[510]

The Arab peasants who stumbled across the famous Nag Hamadi Gnostic Gospels in an Egyptian cave in 1945 were just a few weeks away from a far more "important" deed at the time: they were planning to avenge the death of their father. A few weeks after their accidental contribution to archaeology, Muhammad 'Ali and his brothers tracked down their father's killer, murdered him, cut off his arms and legs, then ripped out his heart and ate it. The cheerleader urging them on was none other than their mother. And it's quite likely that the man on whom these faithful sons were venting their rage had done away with their father out of obedience to the same ancient laws of vengeance.[511] The British orientalist Sir Charles Lyall sums up the Arab lust for violence with one blunt aphorism: "Who uses not roughness, him shall men wrong."[512] Could the denial of warmth lie behind this Arab brutality?

It wouldn't be the first time that a lack of physical affection has gone hand in hand with a love of inflicting pain. In sixteenth- and seventeenth-century England—the England of Shakespeare and Elizabeth I—displaying love to your kids was considered utterly inappropriate. Young humans, cursed by the original sin of Adam, still carried the devil within them. His Satanic majesty could be chased away only with a good thrashing. "Spare the rod and spoil the child" was a deadly serious maxim.[513]

The youngsters of England in those days displayed a brutality the bedouin would have understood. They tethered chickens in the yard, then pelted them with stones until the tortured creatures finally died. They burned cats alive and pitted animals against each other, encouraging the beasts to tear each other limb from limb. And all of this was considered good, healthy fun. Said one approving poet about cock-throwing—tying a bird to a stake or burying it up to its neck in the ground, then letting schoolchildren stone it to death—" 'Tis the bravest game."[514]

When the sixteenth- and seventeenth-century British reached adulthood, they didn't outgrow their love of violence. Englishmen set dogs on bulls for sport. The dog would clamp its teeth on the bull's nose, tear off its ears, and shred its skin. In the end, either the dog would slash the bull's throat, ripping its jugular and killing it slowly but painfully, or the bull would gore the dog and trample it to paste. One way or the other, the crowd would be amused.[515]

The British didn't restrict their delight in pain to animals. They whipped and hung their criminals in public; and huge audiences showed up with picnic baskets to watch. But a few hundred years later, the British memes evolved into another form, and parents changed their mind about how children should be raised. They offered a bit more affection, and soon the scenes of brutality in English streets came to an end.[516]

In much of Arab society, the unmerciful approach of fathers to their children continues, and public warmth between men and women is still considered an evil. Perhaps this is why a disproportionate number of Arab adults, stripped of intimacy and thrust into a life in which vulnerable emotion is a sin, have joined extremist movements dedicated to wreaking havoc on the world.

THE PUZZLE OF COMPLACENCY

≫⊱ ⊰≪

In a world where some cultures elevate violence to a virtue, the dream of peace can be fatal. It can make us forget that our enemies are real and can blind us to the dark imperatives of the superorganismic pecking order.

For thousands of years, China was an empire of unbelievable size and stability. Its technology and wealth were the envy of its neighbors. In 221 B.C., the Chinese laid out a standard length for the axles of carts. The result: a wagon could roll over tens of thousands of miles of highway, and its wheels would fit precisely in the ruts left by previous travelers.[517] The Chinese had paper money and uniform standards of weights and measures while Europe was still blundering through the Dark Ages.[518] Chinese weaponry and military strategy were light years ahead of anything else around. While Roman emperors were still relying on mechanical catapults, Chinese generals were deploying gunpowder mortars.[519] As early as the fourth century B.C., Chinese princes were already sending armies of half a million men into battle, and those legions were equipped with hardware the Europeans of their time could not even imagine. They had trigger-operated crossbows, chain-mail armor, and swords and spears of a miracle metal—steel.[520]

But the Chinese were periodically blinded by their own power. Slipping into the cheerful conviction that they could simply wish warfare away, they overlooked the potential of barbarians. One of the first to make that mistake was Chinese Emperor Wu Ti. In A.D. 280, Wu Ti took a good look at the colossus over which he ruled and discovered that it was in economic trouble: trade was in a shambles. The people were poor and burdened with unbearable taxation. When Wu Ti examined the problem more carefully, he quickly found its root. China was being dragged into the pit by a burden that had grown like a cancer: her military budget. Taxes were sopped up by the needs of a massive army, and most of the country's coins had literally been melted down to make weapons, forcing merchants to abandon money and rely on primitive barter. There was so little cash available that even government bureaucrats had to be paid in grain and silks.

But there was good news on the horizon: Wu Ti's armies had just overcome the two great powers that had for years posed the empire's major military threat—the muscular kingdoms of Wei and Wu. Now the moment had arrived when China could discard her military burden, lighten the load on her people, and set her economy free.

Reducing the military budget was a good idea, but the Chinese took it too far. In the year 280, Emperor Wu Ti made a staggering announcement, one that must have gladdened the hearts of Chinese everywhere. He decreed a general disarmament. The anvils of the swordmakers and the armorers grew silent. Generals were commanded to decommission their troops. Soldiers were ordered to go back to civilian life. The government hoped that its former infantrymen would settle down as farmers and become tax-paying citizens, helping to replenish the drained coffers of the administration. It sounds like a prescription for utopia, doesn't it? But the blissful state of permanent peace never quite materialized. The Chinese had disregarded the Lucifer Principle.

Hovering outside the country's borders was a federation of nomadic tribes that lacked China's civilized refinements and gift for high-tech creativity. But its leaders had studied every nuance of Chinese art, administration and weaponry. And they possessed one significant edge. They had no compunctions about killing. In fact, it was their favorite sport. This tribal constellation was called the Hsiung-nu. We know them better as the Huns.

At first glance, no one could possibly have thought that the Huns were a serious threat. Their army consisted of a mere fifty thousand men. The recently disbanded Chinese legions had been as large as a million. But in A.D. 309, the relatively small Hun military machine descended on the Chinese capital, Loyang. The Chinese defended themselves doggedly, but they were at a severe disadvantage. Having retooled for peace, they were no longer equipped for war.

After two years of fighting, the Huns marched into the defeated city and took the Chinese emperor—the descendant of the mighty sun, the linchpin who held together the heavens and the earth—captive, and in 313, they killed him. Within three years, the alien forces had completed their mop-up operation and seized the entire western sector of the critical Chinese north. Chinese princes, generals, and wealthy landowners fled for their lives. The day belonged to the Huns. An empire bigger than all the European states combined had fallen, all because it ignored the danger of the barbarians.[521]

You'd think the Chinese would have learned from that mistake, but they didn't The Hunnish rule of China lasted a generation. Then, in 329, it came to an end. For the next two hundred years,[522] the Chinese were ruled first by one barbarian group, then another. It took a long time before China was finally able to restore her ancient glory, but once she had regained power, she slipped a second time into the blissful complacency that shutters the eyes of those on top of the pecking order. And her second careless stumble carries even more lessons about the dangers ahead of us.

In the eleventh century, once again convinced that she could use her great strength to usher in an era of peace, China turned to diplomacy and did so brilliantly. She discovered that it cost far less to pacify her enemies with tribute than it did to maintain an elephantine army, so she paid her enemies off. To keep these hulking powers from her throat, she worked insidiously behind the scenes to stir up trouble. Not trouble that would threaten her own security, but that would create squabbles among her enemies. After all, the more they quarreled with each other, the less they'd bother the Chinese.

The whole scheme worked like a charm. It worked so well that both the Chinese and their enemies were able to dismantle their military complexes and pour the savings into the domestic economy. That diverted treasure produced a burst of prosperity.

As usual, the Chinese and their superpower enemies had blithely dismissed the significance of the unwashed rabble beyond their frontiers. In 1114, that rabble—the Tungusic Juchen—kicked themselves loose from the superpower sphere of influence and prepared for war. Those preparations took a full eleven years. But when they were over, it was time for the major powers to watch out. First, the Juchen attacked China's biggest enemy—the Kitan. The Chinese were delighted. The primitive Juchen had just removed their greatest international problem. But the emperor and his subjects rejoiced a bit too soon. The Juchen wheeled around, hungry for an even bigger conquest.

In 1126, the backward people who only sixteen years earlier had been the lowly puppets of a superpower fought their way into an unprepared Chinese capital and decided to stay. The Middle Kingdom had been undone by both diplomacy and disarmament because its inhabitants had forgotten about the barbarians.[523]

Behind the threat of barbarians is a simple fact. Social superorganisms itch to move up on the hierarchical ladder, and many of those who want to ascend would like to do so at our expense. The legitimate wish for peace often blinds us to this fact. But there is another impulse that also distracts us from the danger of barbarians: the itch to battle our fellow citizens.

The Roman emperor Constantine converted to Christianity in A.D. 324.[524] Eighteen years later, he moved the capital of the empire to the old Greek colony of Byzantium.[525] Constantine explained that he was acting on direct orders from his new God,[526] Jesus, who presumably preferred to leave the city of pagan deities and Christian persecutions behind. At that moment, Byzantium—in what is today Turkey—became the administrative hub of Europe.[527]

By A.D. 600, Byzantium was the mistress of a territory that included the lands of modern Egypt, Palestine, Syria, Greece, Yugoslavia, and parts of Libya, Algeria, and Spain.[528] She was rich beyond measure and had signed a highly favorable peace treaty with her biggest superpower rival—Persia.[529]

But things were not as cheerful as they seemed at first glance. The Byzantines had the nasty habit of blaming everything in sight on each other and of fighting—to the death—over every trivial detail on which they disagreed. The most famous of these conflicts were the battles between the Greens and the Blues—the conservatives and liberals of the day. The two groups were followers of different teams in the sporting matches at the local hippodrome, but they were also rivals for the soul of the city. One was led by aristocratic landowners, the other by merchants and industrialists. One supported orthodox religion; the other was drawn to unconventional spiritual notions.[530] The Greens and Blues killed each other over petty issues like which words truly belonged in a prayer. They had pitched battles in the streets over economic policy, toppling statues and burning down public buildings as they went. They murdered each other during riots at sporting events. And they were not content with one or two accidental homicides during a heated melee. On one occasion, the Greens hid stones and daggers in baskets of fruit, showed up at a solemn festival, and massacred three thousand of their Blue opponents.[531]

Eventually the Greens and the Blues led a revolution in which they indulged their love for murdering their neighbors. They killed the sons of the emperor before his eyes, then slew him as well. The revolutionary factions installed their own imperial candidate, Phocas, on the throne. This reformer indulged in a political bloodbath of devastating proportions.[532] Determined to return the state to "moral" and religious purity, he rubbed out anyone whose views were the least bit unconventional.

With the Persian superpower out of the way, the Byzantine citizens thought that only their internal opponents mattered. They forgot that other enemies might exist somewhere outside the comfortable borders of their own land.

While the Byzantines attacked each other in the streets over fine points of theology, a horde of bedouins and backwater merchants rode out of the unfamiliar deserts to the southeast, roused to a frenzy by a new religion that made battle a holy deed. These armies of Islam snatched nearly everything the bickering Byzantines had. The Mohammedans took Syria, Palestine, Armenia, Mesopotamia, and Egypt in a mere twelve years.[533] Then they gradually pried North Africa, Spain, Greece, Yugoslavia, and the rest from Byzantine hands. In 1453, even Byzantium itself would fall, to become the new capital of an Islamic empire. The city's name—Constantinople—would disappear from the map, replaced by the Turkic label "Istanbul."

Like the citizens of Constantine's capital, we would rather fight each other than acknowledge a simple reality: that there are people in the outside world who relish the opportunity to destroy us. Our 1980s antinuclear movements were *not* directed at makers of atomic bombs in Pakistan, Iran, Iraq, or North Korea, just at those in Washington. Our protesters against American meddling in El Salvador or Nicaragua didn't care about the deaths of millions in Cambodia, the murder of twenty thousand in Syria, or genocide in black Africa. They focused their attention solely on American misdeeds.

Our internal fights often prevent us from even seeing external dangers. Under Ronald Reagan in 1983, the United States sent its marines to Lebanon as a peacekeeping force in the hope of slowing the civil violence that was tearing Beirut apart. One day, a member of the ayatolla-inspired Hezbollah drove a truck filled with explosives into the building where the marines were encamped, killing over two hundred men, most of them unarmed and sleeping.

American politicians and reporters did not condemn the groups that had declared all Western infidels impure as excrement, fit only to be smashed with an "iron hand." Governmental agencies did not search for the murderers and attempt to stop them from attacking Americans in the future. Instead, U.S. journalists and politicos kicked off an investigation to find an American villain—one of our own citizens who had "failed to provide the base with adequate security." The tendency to bicker internally had totally obliterated our ability to look carefully at outside threats. And we will never be able to overcome threats we refuse to see.

No one stays on top of the pecking order forever. This is a difficult lesson to learn. Debate is a necessity, but if it becomes irrational, violent, and blind to the menaces beyond our borders, it can doom us as surely as it did the Byzantines.

POVERTY WITH PRESTIGE IS BETTER THAN AFFLUENT DISGRACE

Gifts make slaves.

Claude Lévi-Strauss

We Americans have attempted to use every one of the old Chinese techniques to establish peace. Since the 1890s, we have repeatedly toyed with disarmament. (In the 1920s, we actually achieved disarmament in a limited form. The dramatic international scaling back of military forces helped set the stage for Hitler.)[534] We have extolled the virtues of diplomacy, and when it comes to our own barbarians, we have used the third weapon in the Chinese arsenal of pacification—tribute. We justify our payoffs to backward nations with a new philosophy, one that probably never occurred to the bureaucratic sages of the Chinese empire. We explain that our gifts are development funds, designed to bring peace by uprooting the very causes of discontent and war. We call our new form of tribute "foreign aid."

In many cultures, however, giving things to people is a way of humiliating them. It is a sneaky technique for drawing attention to the recipient's lowliness on the hierarchical ladder. Take, for example, the "big men" of Melanesia and New Guinea. In the days before traditional practices were supplanted by Western ways, a young New Guinean would work like a maniac to raise himself in the eyes of his peers. He would strain feverishly to boost his yield of pigs, yams, and coconuts. He would recruit his wives, children, and relatives to join in the frantic race for agricultural productivity. If all went well, he would take the profits and plow them into building a men's clubhouse. When the neighbors—pleased with the clubhouse food and entertainment—were sufficiently impressed, the struggling entrepreneur would ask them to join his growing army of pig, yam, and coconut growers.

The grand climax of the young man's effort would come when he challenged a local "big man"—a high-placed figure revered for his powerful following. The contender would do it by inviting his older rival to attend a feast. At the grand dinner, the upstart would banquet the elder with a deluge of pork dishes, coconut pies, and sago almond puddings. The young man's followers and those of the guest would count every dish of food that hit the table. If the mountain of delectables the rookie offered was large enough, the big man knew he was in serious trouble.

The elder would go home and spend the next year spurring his followers to new heights of productivity. Then he would invite the young challenger to a feast at his place. He, too, would heap the table with pies, roasts, and puddings. And, once again, the crowd would keep a breathless count, for if the older dignitary failed to lay on as rich a feast as the young man had the previous year, it would all be over. The venerable gentleman would be shamed. As he plunged down the ladder of prestige, his followers would desert him, and the callow whippersnapper who had mounted the challenge would leap dramatically upward in the pecking order. Now he would be the big man. In New Guinea, the man who could not give as much as he received earned only one reward—disgrace.[535]

The New Guineans were not alone in regarding the giveaway as a technique for inflicting humiliation. The Kwakiutl people of the Pacific Northwest were famed for their potlatches. In the potlatch, a Kwakiutl chief would invite a rival and his tribe over for a visit, then shower the guests with gifts. The greater the pile of presents, the more the guest would lose face, plummeting down the pecking order. Among the Kwakiutl, to give away goods is divine, but to accept them is less than human.[536]

Even our recent ancestors were aware of generosity's subversive power. Medieval European aristocrats threw an annual feast and invited the peasants in to stuff themselves. The ritual drove home the fact that the noble was on top and the peasants on the bottom. The Anglo-Saxon word for someone on the crest of a social heap—lord—was a testament to the put-down power of the handout. The word's literal meaning: "loaf giver."[537]

The role of the giveaway as a hierarchical weapon goes back to our cousins the chimpanzees. In an earlier chapter, there is a description of what happens when one of these meat gourmets is lucky enough to kill a young gazelle or a baby baboon: females, children, and even his rivals crawl toward the hunter, lowering their eyes and stretching out their hands with palms upturned. They whimper, squirm, and cry.[538] Such is the power of generosity to elevate the giver and cast down those who receive. No wonder those on whom we lavish aid are not particularly fond of us.

Compassionate gestures have a purpose we seldom admit: they confirm our feeling of superiority, gratifying us with the certainty that those who receive our "help" are, indeed, below us. This makes the recipients loathe us. They'd gladly exchange the food and blankets we send for the opportunity to look down upon their "benefactors."[539]

The fathers of our foreign policy feel that by alleviating hunger, poverty, and disease, we can pull the pins out from under the urge to shed blood and make the third world love us. The philosophy hasn't worked. The abasement of the charity recipient is only one reason. Another: our official definitions of want bear little relationship to the reality of the human psyche. We assume that humans desire food, clothing, and shelter, but we forget that people crave something far more vital: status and prestige.[540] They yearn to move up in the pecking order!

Our relief agencies ship food and medicine to the poor of South America, but when allowed to buy what they prefer, women of South America's underclass purchase something they consider more vital than penicillin or protein-rich nutrient: they spend their precious funds on lipstick. Lipstick brings the admiring glances of men and the envy of women. To the shanty-dwelling women of South America, that pecking order bonanza is worth more than a well-balanced meal.

We should know better than to think that the citizens of underdeveloped countries are motivated by the simple desire to escape poverty. We have the evidence right here in the United States. In Harlem, a hotbed of deprivation, the driving desire of teenagers is not for something of practical merit; it's for status symbols. According to Claude Brown, author of *Manchild in the Promised Land,* adolescent boys above Manhattan's 125th Street feel compelled to wear a new pair of designer jeans twice a week, to "show fly" (to dress up), and to wear high-priced, status brands like Fila and Adidas. One teenager told Brown, "It's embarrassing not to have a pair." In Harlem, prestige frequently means more than food, shelter, and clothing. Far more.[541]

Why shouldn't it? Exactly the same instinct works its will on the wealthy folk downtown. These resplendent souls will waste substantial sums of cash to purchase flimsy plastic luggage simply because it bears the logo of Vuitton.

Claude Brown has an explanation for this: teenage Harlem's preoccupation with prestige is the fault of a society afflicted by materialism. Brown fails to realize that virtually every tribe or nation ever studied has been obsessed by some sort of status symbol. Even naked, spear-carrying Pacific islanders wore "penis cones," whose decorations showed off their rank. All human cultures—including the "classless" societies engineered by Marxism in its prime—have been in the grip of the pecking order.

So powerful is the pecking order impulse that pride has frequently meant more than survival to human beings. Pilots in the First World War refused to wear parachutes because safety devices were not "manly." The fliers chose going down in flames over slipping a notch in the pecking order.

So have many others. In A.D. 70, the Romans attacked Jerusalem. One group of Hebrews stubbornly marched to a desert fortress called Masada. For years, they held out against the Roman legions. Then, when the Jews could resist the attackers no longer, the entire band committed suicide.[542] These fighters preferred death to a fate as featherless chickens on the bottom of the imperial pecking order.

No wonder the emotion that follows a disastrous hierarchical downfall is called "mortification." The root of "mortification" is the Latin *mortis,* death. Disgrace is, to many humans, as dire a fate as physical extinction.

Humiliation and the insidious force of the giveaway can trigger superorganismic cataclysm. Take the example of Iran. From the late-nineteenth century until the Second World War, the Iranians felt the disgrace of living in the shadow of the superpowers. Iran was addicted to superpower help, ashamed of its dependency and resentful of the resulting influence the major nations achieved over its affairs. In 1879, for example, the shah asked the Russians to raise and train a police force in the north. The resulting "Cossack Brigade" had Russian officers and Iranian noncommissioned officers.[543] In 1907, the shah grew impatient with the newly convened parliament—the Majlis. His solution was to call in the Russian-led police brigades, bombard the parliamentary building, and reestablish autocratic rule.

Supporters of a constitutional democracy turned for help to Britain. Ten thousand of them took refuge on the grounds of the English embassy. That same year, the British and the Russians put their heads together and carved Iran into three spheres of influence—a Russian sector in the north, a smaller British zone in the south, and a neutral territory in the middle.

Playing politics with the great powers made the Iranians feel like a small dog running between the legs of giants, a dog in danger of being trampled. One Iranian-born writer said bluntly that Britain and Russia were quarreling over Iran's dead body.[544] Then, during the Second World War, a new white knight, the United States, arrived to save the Iranians from the dishonor of superpower domination.[545]

Americans opened new oil fields,[546] and trained and equipped the Iranian military.[547] American corporations started subsidiary operations in Teheran. The American government gave the Iranians money, helped place the country on the road to development, and propped up the Iranian ruler, the shah, counseling him on every nuance of policy from internal security to the management of his image in the Iranian newspapers.[548] Then American executives and advisers moved into luxury villas in walled-off Iranian suburbs, hired Iranian servants, and lived like kings.

Some American actions were Machiavellian. Others were generous. Both were destined to incur resentment. It wasn't long before the Iranians felt their old sense of humiliation and realized that they were still in the pecking order's lower depths. Even the shah felt the Americans despised him. According to his occasional confidant, the Soviet ambassador, the shah picked quarrels with the United States on minor issues to release his frustration—the frustration of a chicken who feels how low on the pecking order he has slid.[549]

In the fifties, one Iranian leader, Muhammad Mossadeq, championed a move that would shame the Americans and restore Iranian pride. As premier, he planned to snatch the oil fields from the British and Americans, and to make them Iranian national property.

When Mossadeq spoke of Islamic pride, he literally brought tears to the eyes of his fellow Iranians.[550] Islamic extremists were willing to kill in

Mossadeq's name, and kill they did. Among others, they assassinated the incorruptible Prime Minister Ali Razmara. Fear of Mossadeq's fanatical supporters was so great that no imam could be found to say prayers at Razmara's funeral. When one holy man was offered three thousand pounds to perform the services, he answered that "he valued his life at a higher rate than this."[551] Terrified by Mossadeq's growing power, the shah fled the country.[552]

In 1951, the fiery premier began his nationalization of oil. The result was a disaster. At least, it would have been a disaster if all the Iranians cared about was food, shelter, and clothing. Britain closed down the refineries, and vast numbers of British Iranian Petroleum Company employees were thrown out of work. Tribal chiefs accustomed to living off of oil royalties went empty-handed. Mossadeq's administration was starved for lack of cash, and government employees went from week to week without pay. The Iranian economy became a basket case.[553] But the Iranians did not complain. Why? The feeling of power was worth the price. Pecking-order pleasure centers reveled in bringing down those on high.

The Iranian euphoria was not to last. Both the United States and Britain were worried about the loss of this valuable piece of real estate. With British encouragement, the CIA arranged Mossadeq's overthrow.[554] The shah returned, more beholden to America than ever, and the oil fields went back to the foreigners. The Iranians, however, never forgot their moment of pecking order triumph.

Iran did very well under American tutelage. Poverty plunged, education and health care spread through the land, women gained new freedoms, and the standard of living skyrocketed.[555] American policymakers were proud of their accomplishments. By the measure of food, clothing, and shelter, the U.S. had helped Iran accomplish miracles. But both our State Department and the shah had forgotten that pride, dignity, and dominance—the needs of the pecking order impulse—can be far more pressing than the demands of the body.

In 1972, after thirty-one years in power, the shah at last felt that his people had attained happiness. He decided to celebrate by throwing himself a party. The monarch invited sixty-eight kings and heads of state,[556] housed his guests in air-conditioned, silk-lined tents complete with living room, bedroom, and kitchen, and fed them mountains of caviar and food prepared by chefs flown in from the world's most expensive restaurants.[557] He presided over a military parade in which his troops dressed in uniforms from an ancient time. The soldierly garb belonged to the era of Cyrus and Xerxes—history's great Iranian rulers.[558] Twenty-five hundred years ago, these Persians had built an empire, and the shah dreamed of doing the same.

Though the country owed much of its progress to the Americans, a rabble-rousing clergyman said the Yankees had placed the Iranians in chains and robbed them of their self-respect. The cleric understood the needs of the pecking order far better than the shah. Despite the increases in the standard of living, the Iranian people were seething with frustration, and, contrary to Western assumptions, it wasn't because of political oppression. Savak, the secret police, was brutal, but the country offered far more freedom than any of its immediate neighbors.

In 1978, when the man of God called for it, Iranians rioted. They ran through the streets by the hundreds of thousands. At first, the shah's police and soldiers felt helpless to end the demonstrations. Finally, they joined the demonstrators. The shah who had given his people everything that American policy defines as happiness was driven from the country, and the clergyman who understood the pecking order's hungers returned from exile to become Iran's new ruler. He was a man we've met before: Ayatollah Ruhollah Khomeini.

The ayatollah pulled off a pecking order trick of astonishing proportions. He preached a view in which Iran was suddenly at the very peak of a new kind of hierarchy. In his rhetoric, Iranians were transformed from mere followers of America to *leaders*—leaders of an Islamic revolution that would soon sweep the world. The Iranians, said the ayatollah, were morally superior to the infidel dogs, the inhuman Western devils, who preferred their stereos to the words of Mohammed. The Iranians followed the words of Allah; the Americans did not. The Iranians championed the cause of Mohammed; the Americans did not. The Iranians followed the basic rules of decency—they kept their women in black, forbade kissing and hugging in public, and outlawed the nudity of bathing suits—while the Americans flung their shamelessness in the face of all mankind. The Iranians were the righteous men who would scourge the continents in the name of God; the Americans, on the other hand, were the contemptible followers of the Great Satan.[559]

The ayatollah had turned the pecking order upside down. The

Americans, the children of the devil, were at the bottom. And the Iranians—the blessed of Allah—were on the top.

What does this indicate about our foreign policy? Poverty is a relative term. The poor are simply those on the lowest rungs of the hierarchical ladder. Move everyone up—including the poor—and the impoverished will still be on the underside. The poor in the shah's Teheran had things Iran's ancient emperors never dreamed of. Cyrus the Great might have offered half his kingdom for one slum-dwelling Iranian's transistor radio or for the antibiotics given to a single poverty-stricken child. What had made old Cyrus a god on earth was not his food, clothing, and shelter but his position at the top of the pecking order.

The nations of the third world accept our handouts gladly and even ask for more, but they often hate us for our "generosity." They resent us as bitterly as the New Guinean big man humbled by a flood of earthly goods or the Kwakiutl chief shamed by his rival's largesse.

Even if we eliminate starvation and disease, only one thing will allow third world nations to overcome the emotional laceration of their pecking order fate: an upward move. Such is the nature of the hierarchical ladder, however, that whenever a creature moves up, someone else must be shoved down. Many would like the one stomped toward the bottom to be us.

WHY PROSPERITY WILL NOT BRING PEACE

Evils which are patiently endured when they seem inevitable become intolerable once the idea of escape from them is suggested.

Alexis de Tocqueville

There's yet another flaw behind our belief that by eliminating hunger and elevating the income of the third world, peace will descend upon the earth, and that by eradicating starvation and poverty at home, we will cause muggings and murders to melt away. History indicates a rising standard of living and a bigger plate of food may be the very catalysts that unleash a storm of violence!

Thanks to rising oil prices, per capita income in Libya went from $40 in 1951 to $8,170 in 1979.[560] The increase was dizzying. Libya's citizens had once lived lives as desert nomads, barely finding enough grass in the sandy wastes to feed their sheep. Now, they had trucks, houses, radios from Sony, and high-tech watches from Seiko.[561] By 1979, in fact, the average Libyan pulled in a higher income than the typical Italian or Englishman. The increase in standard of living, however, did not create a more peaceful society. In fact, Libya's murder rate went up dramatically.[562] Furthermore, under the leadership of Muammar el-Qaddafi and his revolutionary dreams of glory, Libya began exporting terror worldwide.

Good times can be just as damaging to peace here in the United States. When the economy goes up, murders do not go down; they rise! The most startling fact revealed by a study of the relationship between American murder rates and the economy from 1929 to 1949 was that homicides actually plunged dramatically during the Great Depression.[563] In addition, military historian Robert L. O'Connell contends that since the beginning of recorded time, periods of optimism and bursts of new weapons development have gone hand in hand.[564]

War and the dreams of conquest are fueled, it seems, less by poverty than by the heady whiff of new riches. In the twelfth century, the Mongols were a nomadic people, living on the plains of eastern Asia. Their economy was based on their horses. They drank mare's milk and made war on horseback.[565] The Mongols supplemented their diet on long trips by pricking their horses' necks, drawing blood, mixing it with a bit of the millet meal they carried in a pouch at their side, and eating the ruddy paste while galloping across the countryside.

In the late twelfth and early thirteenth centuries, the Mongols were gifted with an economic boom—good weather increased their supply of fodder,[566] allowing a spectacular growth in the number of new colts. The Mongols, interpreting the boost in pasturage as a sign that God had granted them the entire world, took off on a rampage. Within less than seventy years, they had conquered a territory nearly twice the size of the Roman Empire. It included China, Russia, Persia, Syria, Iraq, and chunks of eastern Europe.[567] That empire was not won with gentle persuasion. One Mongol descendant, Tamerlane, was remembered for building 120 towers of severed heads in the opulent city of Baghdad. Needless to say, the skulls had all been parted from their supporting vertebrae by Mongol swords.[568]

But why does carnage so frequently follow a boost in well-being?[569] One clue may come from the following puzzle: Murder rates rise after a war. You'd think they'd go up the most in the losing nations, whose citizens are frustrated and gnashing their teeth over their misfortune, but they don't. Murders increase the most in the country that *won*.[570] The same phenomenon has been observed in animals. When two groups of rhesus monkeys were squeezed into a territory smaller than what they were accustomed to, one pack aggressively asserted its right to lord it over the other and hogged the available real estate. As the winning gang beat its rivals into submission, a strange thing occurred. The members of the losing troop "fought less among one another. But within the dominant group, which was in the process of acquiring new space, aggressive interactions increased."[571] Why?

The answer takes us back to the influence of the pecking order on testosterone. In studies of combative monkeys and of competing college wrestling teams, a simple fact has emerged.[572] Testosterone levels go up in the winners and down in the losers.

Testosterone makes winners restless, confident, and aggressive.[573] The steroids taken by athletes, for example, are a synthetic variation of natural testosterone. These drugs can induce a boldness that borders on insanity. Psychiatrists Harrison G. Pope, Jr., and David L. Katz, of McLean Hospital and the Harvard Medical School respectively, interviewed forty-one steroid-using bodybuilders and found, among other things, that one of these athletes was convinced he could jump from a third-floor window without harm. Another had bought two expensive sports cars, then cockily driven them at forty miles per hour into a tree while a friend videotaped the feat.[574]

Testosterone is the very elixir of feistiness. Inject a young rooster with this remarkable hormone and the bird struts away to look for a fight. What's more, he usually wins the battle, and moves up in social stature.[575] One reason may be indicated by the work of Turkish researcher Una Tan, who showed that testosterone can actually increase physical prowess.[576]

According to Edward O. Wilson, even "hens given small doses of testosterone become more aggressive and move up in rank within the dominance hierarchies of the flock."[577] In fact, injecting a few hens with testosterone propionate is a good way to trigger a barnyard revolt.[578]

A social group that has just had a stroke of good fortune is filled with men and women who've won in a big way. The leaders are on a testosterone high! No wonder the Mongols, flush with fresh prosperity, thundered across the Asian plains in search of battle; they were bursting with the hormones of aggression. But what is the logic behind this hormonal commotion? Surely, a stroke of good luck should make people more contented with life, not less. It should trigger a burst of gratitude, not a restless desire to gallop off and grab even more. But biology refuses to knuckle under to this common sense.

The Arizona spadefoot toad lives in one of the driest deserts in the world. Its survival is a miracle. To live, the toad needs water, without which its cells would shrivel and die. And it needs whole puddles of the stuff in which to reproduce. Yet months go by in the Southwestern desert without rain. Sometimes those months stretch into years. How does the spadefoot toad hold on to life?

The beast follows a simple strategy. When times are tough, it saves its energy. As the desert grows dry and the sun becomes hot, the toad burrows under the sand and goes into hibernation, slowing its metabolic rate to a crawl, and conserving every drop of water and fuel stored in its flesh. There, the toad lies motionless, month after dreary month. If the amphibian were to emerge from this lethargic state too soon, its poorly timed outburst of optimism could be deadly. It might burn the water and nutrients it needs to see it through the coming months. Digging back under the sand could not save it, for the reserve supplies packed into its body would be gone. So in its days of impoverishment, when the desert floor is parched, the wise toad stays quietly underground.

A sudden burst of prosperity brings the toad back to the surface. When a rare downpour soaks the Arizona land, the spadefoot toad is jolted awake by a hormonal surge. He shakes off his sluggishness, is seized with enthusiasm, and scrambles into the open air, searching madly for a puddle. When he finds one, he croaks for all he's worth, hoping to entice the ladies of his species to gather 'round. Within a short time, the puddle is a hotbed of social action. Males and females fling themselves into a sexual orgy. Within twenty-four hours, the spadefoot toad's paradisal puddle is filled with the results: a squirming horde of tadpoles.[579] Poverty makes the spadefoot toad passive and inert, but the coming of prosperity whets a spirited desire to get even *more* out of life.

The spadefoot toad is following a basic biological law. That same principle makes the rapid rise in good fortune among humans a dangerous thing indeed. Nature shuts down the expenditure of energy when resources disappear, but she unleashes energy when fresh resources arrive. She makes those who are deprived sit still and endure their fate, but when good fortune lifts the curtain of hopelessness, biology gives the lucky souls who've landed on an upward track a burst of manic zeal.

The connection between the human supercharger testosterone and the hyperactive states of creatures like the toad is not merely metaphoric. Experiments indicate that testosterone is the hormone that jolts hibernating creatures out of their torpid metabolic state when environmental resources finally return.[580] Testosterone has an equally impressive impact on creatures who do not sleep the tough seasons away. During the winter, when times are tough, the canary is a silent bird. But when spring comes, his body produces a testosterone surge that results in a sudden desire to sing. Testosterone produces this musical enthusiasm by, among other things, triggering a growth of neurons in the brain.[581]

You can see a similar biological conservation device at work in yourself. You sit down to a meal. A half-hour or less after you've started eating, you begin to feel warm.[582] The food you're chewing hasn't reached your bloodstream yet—in fact, it will take hours before it is digested.[583] So where does the sudden spurt of fuel that warms you come from? The body has held energy in reserve, just as it does in the case of the spadefoot toad. Those stored calories are designed to tide you over in case you skip lunch or find yourself in the middle of a famine. Once the first bite of a new meal passes your lips, however, your metabolic regulators conclude that there's new food at hand and release some of the hoarded nutrients into your bloodstream.

At least three times a day, your body uses the logic of the toad, holding its reserves until it senses the arrival of fresh resources. The body uses the same strategy when you go on a diet. It senses that the food supply is disappearing, assumes that you may be forced to make it through the next few months with almost nothing to eat, and slows your metabolism to build a stockpile.[584] Since it holds onto the energy tucked away in your fatty tissue like a miser clutching his money, you have trouble losing weight.

These food strategies are controlled by a part of the brain known as the hypothalamus.[585] The hypothalamus also regulates anger and the urge to attack. Cats whose anterior hypothalamus is electrically stimulated will maul a rat. What's more, like Mongols setting off joyfully on a raid, these cats will learn to navigate a maze simply for the pleasure of pouncing on a rodent placed at the labyrinth's end.[586]

Like the spadefoot toad, human cultures in periods of hopelessness go into dormant passivity. The untouchables in India seldom attempted to overthrow the system that held them in miserable subjugation. They were resigned to their lot.

But give a social group a jolt of resources, and suddenly it is infused with energy, optimism, and restlessness. Servants may feel ready to grab the knife with which they have been cutting the meat for the master and put it to the master's throat. Nations that have been wallowing passively in the slough of despond look for an opponent to bash in the hope of gaining fresh territory. The Arabs, for example, stayed dormant until oil wealth hit them in the early seventies; then their terrorists assaulted the West.[587] Their murderous exuberance was the product of a chemical cocktail, a biological potion dosed with testosterone.

The lesson is simple. Defeat makes superorganisms sleepy. So does poverty. But a military win or a shower of new wealth rouses social energies, inspiring the pecking order instincts to lift their contentious heads. And when a society is aroused, watch out.

Helping those less fortunate than ourselves is a moral necessity, but don't expect it to bring stability. And certainly don't look for gratitude, or peace.

THE SECRET MEANING OF "FREEDOM," "PEACE," AND "JUSTICE"

> *And let us bathe our hands in . . . blood up to the elbows, and besmear our swords. Then we walk forth, even to the market place, And waving our red weapons o'er our heads, Let's all cry "Peace, freedom and liberty!"*
>
> Shakespeare, Julius Caesar

But what about freedom, justice, and equality? Isn't the goal to put all nations on an equal footing? Isn't that what peace should be about? An equality of nations will never exist in our lifetime. Why? Because peace, freedom, and justice are deceptive concepts. Hidden beneath their surface are the instincts of the pecking order.

The barnyard chickens studied by naturalist Schjelderup-Ebbe had their periods of peace, but they never had equality. No matter how quiet things were, there was always a dominant bird, and there was always some unfortunate chicken trampled to the bottom of the social ladder. This state of things is not restricted to fowl down on the farm. Chimpanzees, baboons, and apes—the animal relatives with whom we share the greatest number of social instincts—are all prisoners of deep-rooted hierarchical drives. Apparently, so are we. When we preach the ideals of freedom, peace, and justice, our intentions are less than honest.

One man's freedom is all too often another man's oppression. That's true whether you're a comfortable citizen of a civilized society or a barbarian restless to bully your way up the ladder. For example, Julius Caesar's arch opponent, Vercingetorix, used the promise of freedom to unify the Gallic tribes, leading them in many of their battles against the Romans.

The Gauls were not tiny clusters of primitives in fur loincloths. Their leaders often read and wrote Greek,[588] and argued the philosophy of the cosmos.[589] The Gallic populace was divided into substantial nations, each of which had hundreds of thousands of citizens and was led by its own politicians; but the Gallic realms seldom saw eye to eye. Vercingetorix pulled these states together into an alliance that could face the Roman might by using the magic word *freedom*. He proclaimed passionately that the Gauls must unite to fight for liberty from Roman oppression.

What Vercingetorix failed to explain was his precise idea of how that liberation would work. Freedom, as he saw it, would consist of unifying all the Gallic nations so they could operate with one mind. Whose mind? Vercingetorix's, of course. To ensure solidarity, Vercingetorix tortured and killed those who disagreed with him. On occasion, he was more humane. He cut off the ears and put out the eyes of those who didn't share his views, then sent the mutilated back home as a warning to anyone else who might be tempted to entertain independent ideas.[590] The freedom Vercingetorix offered the Gauls, in short, was the exchange of one tyranny for another.

Peace is another word abused by those with hidden pecking order goals. It usually means, "Since I'm on top, let's keep the status quo"; or, "Now that I've managed to climb on your back, would you please be kind enough to sit still."

Justice is the term used by those on the bottom of the heap who are itching to move up. When these folks refer to "the struggle for justice," they generally mean, "Let's keep fighting until I come out on top." Once the devotees of justice have seated themselves on the uppermost rung of the ladder, they too almost invariably become staunch defenders of "peace."

Stripped of their moral disguises, the slogans of freedom, peace, and justice are often weapons that those attempting to achieve hierarchical superiority use to stuff the rest of us into the lower ranks of the pecking order. This can be true when the slogans are uttered by individuals. And it can be true when these words are used to motivate superorganisms.

During the early days of the Iranian revolution in 1978, Americans were horrified to learn about the atrocities of the Shah's rule. We discovered that Muhammad Reza Pahlavi, our staunch ally since 1941, had employed a secret police to deal with his opponents. The Iranian covert police force, Savak, had carted Pahlavi's critics off to prison and treated them in the most appalling ways. Savak officers had stripped young women and burned their nipples with cigarettes.[591] They had kept one elderly, arthritic politician in a cell filled with water up to his waist.[592] They had beaten a female college professor for reading political poetry at a public meeting, broken up gatherings of opposition groups with goon squads, bombed the homes and offices of civil rights lawyers, and shot (according to their own admission) 174 urban guerrillas after secret trials.[593] To top it all off, they had been guided by American advisers.

In 1978, when the shah was forced to flee his country, the ayatollah[594] said the revolution was a passionate grab for justice. It was, Khomeini declared, a necessary movement to free Iran from tyranny.[595] But when Khomeini flew from exile in Paris to take over the reins of the upheaval he had fomented, it became obvious just how deceptive the words *freedom* and *justice* can be.

Over the course of the next year, Khomeini installed men like himself in power. They were Moslem clergymen who had spent their lives studying the Koran and teaching its ways. The power of these spiritual shepherds—roughly equivalent to our priests, ministers, and rabbis—was soon vast. In the name of a revolution declared to bring freedom, the Iranian holy men shut down ten newspapers. Freedom of speech, they agreed, could lead people to criticize Islam, a sin that was not allowed. Real freedom meant the right to worship as you were told.[596]

When Khomeini took power, his revolutionary followers set up kangaroo courts and dragged one victim after another before the bench—executing a total of ten thousand "criminals and counter-revolutionaries."[597] The "revolutionary tribunals" met at midnight, carried on their proceedings in secrecy, refused to let the accused defend themselves, and often did away with victims because of some private grudge held by one of the self-appointed "revolutionary judges."

Iranian prime minister Mehdi Bazargan felt this was an outrage, but Khomeini defended the operation of the tribunals as true Islamic "justice." According to Iranian historian and journalist Shaul Bakash, "Khomeini took the view that the insistence on open trials, defense lawyers and proper procedures was a reflection of 'the western sickness among us,' that those on trial were criminals, and 'criminals should not be tried, they should be killed.' '' Khomeini's notion of justice, then, was our idea of despotism.[598]

The "crimes" for which people were exterminated under the ayatollah's "revolutionary courts" ranged from membership in the wrong political party to "corruption on earth." On July 3, 1980, two middle-aged women and two men were dressed in white robes in the town of Kerman, led to a plot of open ground, buried up to their necks, then stoned to death. The women were charged with prostitution, the men with "sexual crimes." The judge who had passed sentence hurled the first stone. The country's chief prosecutor was wildly enthusiastic about the executions. Stoning appears in the holy book, and the exultant high official said, "We approve of anything in the Koran."[599]

Not everyone agreed that the state should be ruled by a gang of ecclesiastics, but the clerics vigorously enforced their dominance, organizing personal armies, encouraging the creation of revolutionary brigades armed with machine guns, and controlling the Hezbollah, gangs of Moslem zealots armed with clubs. If secular-minded groups tried to throw a rally, the Hezbollah would charge into the crowd, clubs in hand, break a few heads, and send the ralliers running. If the leaders of the opposition proved adamant, the holy men had them seized by revolutionary brigade members, arrested, tried by handpicked "revolutionary tribunals" under the clergymen's direct control, and imprisoned or executed. Often the leaders shot or jailed had originally fought alongside the clerics to overthrow the hated shah. The newly declared "criminals" had been under the impression that they were struggling for their freedom.

Moslem clergymen, led by their ayatollah, "Islamized" society— feeding their citizenry to a time-honored meme. Opposition to that meme was one of the worst crimes of all. When President Bani Sadr criticized the clergymen's rapidly growing, ultrareligious political party, the holy men sent club-wielding gangs of Hezbollah to beat up Sadr's supporters. They arrested the president's staff members, shut down the newspaper in which he published his opinions, and rallied mobs under his window to chant "Death to Bani Sadr." (Incidentally, when the crowds grew tired of repeating Bani Sadr's name at the top of their lungs, the leaders signaled them to switch to another chorus: "Death to the United States.") Finally, the clerics impeached Bani Sadr, who fled for his life.

Other critics were less fortunate. Firing squads dispatched a thir-

teen-year-old girl and her sister for sympathizing with the ayatollah's political opponents. They weren't the only youngsters to fall victim to
Khomeini's "freedom." When students at a Teheran high school complained because their teachers were being purged for political noncon-
formity, revolutionary officials taught the pupils a lesson. They paraded
four of the instructors into the school courtyard, gathered the student
body, and shot the pedagogues dead. These Teheran pupils were luckier
than many. Of the thousands tortured and executed between June 1981
and September 1983, half were in secondary school or college.

Meanwhile, Gasht-e Thar Allah, "mobile units of the wrath of
God," rode through the streets of Teheran in cars hunting down citizens
with suspicious faces. Gangs of revolutionary thugs broke into homes and
sprayed the inhabitants with machine-gun bullets.

The new Iranian leaders called their harshness "Islamic justice."
Just a short time earlier, these same leaders had condemned far more humane behavior as tyranny. When the ayatollah had cried out steadily for
the shah's ouster in the sixties and seventies, one of the inhumanities he'd
denounced was the shah's execution of Iranians on drug charges. And how
many did the Shah put to death? A handful. Under the ayatollah, on the
other hand, two hundred Iranians were killed for drug trafficking or drug
use.[600] Many of these were executed after trials in which the evidence was
flimsy at best. When the shah had done it, it was an atrocity; but when the
ayatollah did it, it was justice. Justice, then, can be a very relative term.

Four years into the Iranian revolution, its first premier, Mehdi Bazargan (about the only man in Iran who could voice a complaint and live),
said the government of Islamic clergymen had done nothing but create an
"atmosphere of terror, fear, revenge and national disintegration."
"What has the ruling elite done in nearly four years," Bazargan asked,
"besides bringing death and destruction, packing the prisons and the cem-
eteries in every city, creating long queues, shortages, high prices, unem-
ployment, poverty, homeless people, repetitious slogans, and a dark
future?" What the revolutionary government *had* done was simple. It had
rearranged the Iranian pecking order, removing the dominant beast—the
shah—and replacing him with the ayatollah. Along the way, Khomeini
had moved the Islamic clergymen to the top of the hierarchical heap. As
in the case of Vercingetorix, one man's freedom meant another's
oppression.

That principle applies not just to the pecking order within a country but to the pecking order without. The Iranian clerics had wanted not only to move up in the domestic pecking order but to climb in the international hierarchy—the pecking order of nations.

In the days of the shah, Iranian revolutionaries had lectured constantly about the evils of foreign domination. Once the revolution was over, however, those same "freedom fighters" tossed aside the idea that other countries also deserved to be free from foreign control. At Iran's Congress of Muslim Critics of the Constitution, shortly after the revolution, one faction argued that "Islam knows no borders" and passionately insisted on the eventual creation of a unified state of all Moslem nations. The Iranians, of course, would hold the reins of this massive new superstate, directing it with a leaden grip.

Another august debater at this congress ridiculed the idea of self-determination for other countries. This speaker attacked the proposition that Iran "would neither permit itself to be dominated nor seek to dominate others." Islamic governments, he declared, had a duty to spread Islam. "Islamic culture and knowledge," he added, "are by nature domineering."[601] As a result, the preamble of the Iranian constitution established under the ayatollah refers to the "ideological mission of the army and the revolutionary guard to extend the sovereignty of Allah's law throughout the world."

In other words, the day of peace would come only when we heathens have fallen to our knees and embraced the true faith, allowing ourselves to be tucked into the nether regions of a global Islamic order. This is a classic case of a meme driving a superorganism to expand.

The Iranians took their responsibility to conquer the world seriously. They called for the overthrow of Iraq's Saddam Hussein and dreamed of turning Iraq into an Iranian satellite. When 100,000 had been killed in the Iraq-Iran War and two million had been turned to refugees, the Iranians refused to discuss peace. They were determined to fight until Hussein gave way to a "revolutionary" leader under the ayatollah's thumb.[602] By the time the Iran-Iraq War finally entered a cease-fire in August 1988, the conflict had killed a million.[603]

The Iranians stirred up revolutionary movements in the nations of the Persian Gulf. They became a major force in Lebanon, where they exerted considerable influence over the country's violence-prone Shiite population.[604] They broadcast their revolutionary propaganda to the Moslem inhabitants of the Soviet Union.[605] They financed the Palestinian Hamas, which called for the worldwide annihilation of Jews. They incited unrest among Moslem populations a continent away in Malaysia. They helped the Islamic Moro National Liberation Front seize control of thirteen provinces in the largely Catholic Philippines.[606] And by the mid-nineties they were backing African warlords like Somalia's Mohammed Farah Aidid in what pro-Islamic French journalist Thierry Lalevee called a vigorous ''strategy for encirclement of Africa and the Middle East.''[607]

The Iranian ayatollah and the Gallic Vercingetorix both fought for position at the feeding trough like chickens in a barnyard. Both gathered followers to their cause. And both pulled together a hungry superorganism driven by a meme. But these masters of oratory did not bring a lump to the throat and a tear to the eye by comparing themselves to farmyard beasts. Far from it. They rallied their supporters with three misleading words disguised by moral might: *freedom, peace,* and *justice.*

THE RISE AND FALL

OF THE

AMERICAN EMPIRE

THE VICTORIAN DECLINE AND
THE FALL OF AMERICA

≫€

Whether a nation be today mighty and rich or not depends not on the
abundance or security of its power or riches, but principally on whether
its neighbors possess more or less of it.

Philip Von Hornigk, German mercantilist, c. 1690

Violence is not the only way a nation can be beaten in the hierarchical
race. In Lewis Carroll's *Through the Looking Glass,* the Red Queen says that
to stay in place you have to run very, very hard, and to get anywhere, you
have to run even harder. Stalin put it differently: "Those who lag behind
are beaten." That is especially true for superorganisms.

Victorian England forgot the Red Queen's wisdom. In the process,
Britain lost her dominance of the world. The Victorians said that the sun
never set on their empire, and the claim was quite literally true. Under
Victoria's hand, the English dominated 25 percent of the land surface of
the earth.[608] They ruled more than twelve million square miles of territory and a quarter of the globe's population.[609] They produced a massive
22.9 percent of all the world's goods. But that magnificent state of affairs
was not to last forever. The empire that spanned a planet has disappeared.
The British share of world productivity has slipped from almost 23 percent to three.[610] What happened? More important, could the same fate be
overtaking us?

The foundations of Victorian power were laid over a generation
before round-faced Queen Victoria was born. From 1790 to 1815, British
exports skyrocketed. Britain had come up with some startling industrial

innovations. Asians had long since figured out how to turn the fluff produced by a scrubby, low-hung bush into an expensive but extraordinarily comfortable cloth called cotton. But the Anglo-Saxons invented machines that could spit out this fabric at bargain-basement costs, and the result sold like crazy.[611] The English also perfected the art of mass-producing pig iron—another substance in worldwide demand. The British built an international marketing system on a scale that boggled the mind, with ships controlling the sea lanes of every major ocean and colonial footholds that helped them develop markets for their merchandise everywhere from India to South America.

While this explosion in British trade was getting under way, a gentleman on the other side of the English Channel championed the shortsighted proposition that a nation's might depends on military strength. He made occasional disparaging remarks about the "nation of shopkeepers" back in the British Isles. At first, his military power did appear to demonstrate that weapons are more important than trade. The skeptic's name was Napoléon, and from 1796 to 1812 he humbled nearly every country he encountered—Spain, Holland, Prussia, Austria, and even Egypt. But somehow Napoléon couldn't bring the pesky British to their knees. Why? The English kept piling up profits from the worldwide export of their hot new goods and were able to plow those profits into two things: increased industrial innovation and military resistance to Napoléon.

Napoléon, with his military genius, sailed through one battlefield victory after another, but in the countries he conquered, he botched the job of setting industry on the path of innovation. The result: the economies from which he drew the funds for nonstop warfare stagnated. Workers and bosses still mired in obsolete technologies could ill afford to subsidize the little general's exorbitant armies.

Overtaxed populations eventually grew resentful, so when the British "nation of shopkeepers" finally invaded French-held Spain, the Spanish population rose up in arms to support the liberators. Not long after, Napoléon was crushed. He had overlooked the fact that military might depends not just on guns and strategic brilliance but on industrial innovation and marketing smarts.

The next fifty-nine years would be rosy ones for the British. As they were defeating Napoléon at Waterloo, a new set of technologies loomed on the horizon: one was the spread of mechanization beyond mere cotton mills. The second was steam. The British would be the masters of both. English citizens built the most advanced steam-driven ships—and sold them everywhere from South America to Russia.[612] They virtually invented the railroad, then accepted contracts on the construction of millions of miles of railroad track in the unreachable wastes of nearly every continent.[613] They did a spectacular business in the sale of railroad engines, railroad cars, and even motormen's caps.[614]

Back home, the British figured out how to use steam engines to make the goods that artisans had produced painfully by hand. The result: productivity leaped, and costs came tumbling down. One machine-operating British worker could turn out as much cloth as twenty of her old-fashioned competitors. In Queen Victoria's day (1837–1901), productivity per person in Britain rose 2.5 times! The British laborer benefited mightily, and wages rose an astonishing 80 percent in real dollars from 1850 to 1900.

The world clamored to get its hands on inexpensive, high-tech British goods. China and India actually went through productivity declines as their citizens abandoned the expensive handmade items sold by old-fashioned craftsmen at the local bazaar. Instead, Chinese and Indians bought British cloth, pins, and other necessities.[615] By 1860, the English—with a mere 2 percent of the world's population—were turning out a full quarter of the world's wares and a mind-boggling 40 percent of the items that came tumbling from modern industrial plants.

The cornucopia of British products traveled across every sea in an armada of British ships. By the mid-1800s, the English controlled fully a third of the world's merchant marine.[616] Transactions everywhere from Bombay to Bogota were financed by British banks and insured by British insurance companies. Raw materials from the most distant corners of the globe flowed to Britain, were processed in English factories, and were sent back to the bustling bazaars of Borneo and Beirut as finished goods. No wonder one economist, in 1865, called Britain "the trading center of the universe."

In the beginning, the British knew how much their prosperity depended on the fact that they were ahead of any other country in the commercial utilization of technology. As early as 1781, they banned the export of high-tech fabric-making machines and "any . . . tool, press, paper, utensil or implement or any part thereof, which now is or hereafter may be used in the woollen, cotton, linen or silk manufacture of the kingdom.''[617] But as they grew fat with prosperity, British industrialists overlooked three simple facts: (a) every technological breakthrough eventually grows old; (b) new inventions arrive to replace it; and (c) the country that dominates these new technologies often rules the world.[618]

The technologies that made steam look old-fashioned were developed, ironically, in Britain, but British industrialists, blinded by self-satisfaction, seldom tried to turn them into tempting new products. The result would be disastrous.

One of the most important of these cutting-edge developments sprang from a dilemma England encountered in ruling its far-flung empire. Malaria was swatting down British troops stationed in India like flies. So serious was the epidemic that the average British soldier on Indian soil had only half the life expectancy of his compatriots stationed back in the home country. Illness posed an obstacle to Britain's colonial ambitions elsewhere. Africa was a continent every European power longed for a slice of, but white men who traveled a few miles inland from the African coasts almost invariably sickened and died. The key reason was, once again, malaria.[619] Until the disease could be stopped, British entrepreneurs would have to camp out in African port cities, trading with the natives, collecting tales of the dark continent's unexploited resources, yet utterly helpless to go inland and tap those riches themselves.

There was hope. The bark of a Peruvian plant, the chinchona, could be used to make a derivative called quinine. And quinine seemed to be the magic bullet against malarial fever. But British botanists had almost no luck in cultivating enough chinchona plants to make even the smallest amount of quinine, much less enough to serve entire armies.

Meanwhile, the British had learned how to extract a vapor from coal and use this gas for lighting. But the extraction process produced a useless form of trash—coal tar. Though disposing of the stuff was a messy business, scientists poked around in the sludge to get a handle on its chemical properties. At London's Royal College of Chemistry, a German professor suggested to an assistant that the young man see if he could somehow create an artificial quinine from the goo. The assistant, William Perkin, tried hard but missed the mark. Instead of quinine, he ended up with a liquid whose color was a tantalizing shade of mauve. Perkin tried the solution out as a cloth dye, and, sure enough, it worked. Realizing he had a hot property on his hands, the young man dropped his university assistantship, borrowed every penny his father had, and opened a small factory outside London. Not long after, even Queen Victoria was wearing gowns tinted with Perkin's mauve.

The British may have invented the new synthetic dyes, but in the long run, they were not the ones to profit from them. Despite Perkin's rapid rise to millionaire status, most British industrialists turned up their noses at his discovery. The Germans, however, did not. They worked like maniacs to find out what else they could extract from the grunge produced by coal. In 1863, one German researcher came up with a rich shade of green. When the Empress Eugenie wore it to the Paris Opera, it became the fashion rage.

The most impressive theoretical chemical research was still going on in English laboratories. So German industrial firms offered huge amounts of money to German chemists working in Britain. Then they put the British-trained recruits to work making useful new substances in the fatherland. Among those the Germans lured back was the professor whose suggestion had stimulated young Perkin to attempt the synthesis of quinine to begin with.

Perkin himself had made his fortune. At thirty-six, he retired to pursue a life in "pure science." The British dye industry shriveled in his absence, but the German dye business became the first step in a technology that would revolutionize the future. It was the foundation of the chemical industry.[620] That industry would have far-reaching implications for every aspect of human life. One example is the use of chemical fertilizers, with which German farmers were soon able to produce more food per acre than any of the other Great Powers. England had invented, then discarded, one of the keys to the coming age, while the Germans had enthusiastically scooped up what the British had tossed aside.

Chemical products were not the only futuristic goods that would soon make English steam engines look old-fashioned. Laboring in British laboratories were some of the greatest physicists of the age—Michael Faraday and James Clerk Maxwell.[621] These men and a few others were coming up with astonishing discoveries about the properties of a peculiar force that had puzzled men of science for two centuries:[622] electricity. But British industrialists did not buzz around Faraday and Maxwell's labs, anxious to discover what practical uses they could find for the pair's groundbreaking discoveries.

Those who *did* plunge into the electricity business with every scrap of ingenuity they could muster were Americans and Germans.[623] The first electric-generating plant in Britain to sell power to the ordinary householder was built by an unschooled Ohio go-getter named Thomas Alva Edison. And when the Englishman Sir Coutts Lindsay erected a power station to rival the American upstart's, the British aristocrat was forced to import his alternators from the German firm of Siemens. Siemens, in fact, had a good deal of the generating equipment market sewn up. British electric railways, for example, depended on Siemens' machines.

Meanwhile, Britain's Faraday had long since discovered the principle of the alternating-current transformer but hadn't bothered to turn it into anything of practical value. The man who championed A.C.'s development was an American named Westinghouse. Faraday had also shown the principle of an alternating-current electric motor in 1821, but it took a scientist working with Westinghouse in America to make this motor a practical product sixty-seven years later. To top it off, Faraday's experiments inspired a Massachusetts painter to figure out a use for electrical power that would dramatically shrink the world—communication. In 1835, the Yankee portrait artist Samuel Finley Breese Morse built the first telegraph.[624]

Even the French got into the act. They went whole hog for outdoor electric-arc illumination. Paris became the city of light while London—the beacon of Western civilization—was still in the dark.[625]

One consequence: in 1873, Britain went into an economic nosedive. The British called it the "Great Depression."[626] English businessman H. L. Beales wrote, "Everywhere there is a stagnation and a negation of hope. . . . This is not a period like those which followed ordinary panics. It is more likely the beginning of a new era for ourselves and the world." Beales was right. A new era was beginning, and the British wouldn't like it.[627]

What was the Great Depression of the 1870s all about? Steam technology was petering out. Most of the countries interested in buying a railroad from Britain already had one. Most of the factories that could use a British steam engine had already installed a gaggle of them. Meanwhile, all those foreigners who had purchased British steam devices and factory equipment were putting the new contraptions to work. Now many nations were able to turn out the inexpensive fabrics that had once been an indispensable British export. As a consequence, the demand for British cloth was in a nosedive.

There was an enthusiastic clamor for a whole new species of products, but these were not commodities Britain made. They were the tempting new goods churned out by Germany's chemical industry and the delightful electrical devices made by both the Germans and Americans. Britain, the great exporter, found her stores flooded with inexpensive goods from overseas. Native British industries, still clinging to old-fashioned products, were in decline.

British authors raced to the presses with self-help books that told distressed English industrialists and managers how to reorganize their factories along foreign lines. One of these tomes bore a title that testified to just how ubiquitous the Teutonic imports had become: *Made in Germany.*

Meanwhile, Germany's exports tripled from 1890 to 1913, and its share of world manufacturing grew so mighty that it finally passed that of England. By 1913, the German companies Siemens and AEG would dominate the European electrical industry, employing an awesome 142,000 workers. German chemical giants like Bayer and Hoechst would produce 90 percent of the world's industrial dyes.

On the other side of the Atlantic, another power was rising. America was vigorously grasping the new technologies of electricity and steel. In 1902, just one American entrepreneur—Andrew Carnegie—produced more steel than all the factories of England combined.

Until 1870, Britain had been without question the strongest nation on the earth, yet she had spent the least on military hardware. From 1815 to 1865, a minuscule 3 percent of her GNP had gone into military budgets. Her strength had come from the spinning jenny, the steam-driven loom, the Cunard steamship, and the railroad. But Britain forgot that industrial innovation was the key to her power. Floundering British industrial titans dreamed of holding on to their old position by force. From 1880 to 1900, Britain raised her warship tonnage by 64 percent, and she nearly doubled the number of men she kept in arms.

In the last analysis, power comes from the vigor of minds. The English were blind to this fact; the Germans were not. Germany maintained the best school system in the world. By the 1890s, she had 2.5 times as many university students per unit of population as England.[628]

Victorian England, like today's United States, maintained the illusion of prosperity while that prosperity's foundations were being eaten away. In the late 1890s, Britain recovered from the Great Depression, and looked prosperous and expansive. Plants were operating at capacity, and new industries were doing well. The upper classes were making money by the fistful. But appearances were deceiving. Big business was defending itself through counterproductive mergers and takeovers,[629] and the gap between rich and poor was growing ever greater as England was slipping downward in the pecking order of nations.

Meanwhile, Germany was moving up the hierarchical ladder, and the German leaders were gripped by the testosterone high that makes a nation belligerent. Friedrich Naumann was typical of those who gloated over Teutonic good fortune. He said, "The German race brings it. It brings army, navy, money and power. . . . Modern, gigantic instruments of power are possible only when an active people feels the spring-time juices in its organs."[630] Like the Arizona toad in a downpour, the German superorganism was waking up. And sudden prosperity, as we've seen, does not bring peace. The result was the First World War.

Officially, Britain won the Great War. Yet, in the coming years, she would lose her empire. She had already lost her prosperity. The British worker, once the highest paid in the world, would eventually become one of the lowest. The British factories, once the world's most productive, would enter the ranks of the world's most inefficient. Britain would go from the indisputable ruler of world affairs to a second-rate power.

In military terms, the Germans would lose two world wars. Yet Germany would become in 1987 the greatest exporting nation on the globe, overtaking even the Japanese. Economically, the Germans would win.[631]

Today America seems to be following the path that led the British to their downfall. In 1945, the United States produced 40 percent of the world's goods. By the mid-eighties, our share was half of that. Until the early seventies, we were the biggest exporter in the world. Today, we are the biggest importer. Our federal deficits are soaring, and the amount of money we've borrowed from the citizens of foreign countries is so large[632] that we are now the biggest debtors since the prehistoric invention of the loan.[633]

America's educational system has become one of the least effective in the industrial world. The average Taiwanese first grader spends over eight hours a week doing homework; the average American first grader, an hour and nineteen minutes.[634] The American companies that should be making consumer products the whole world wants to buy—Westinghouse, Raytheon, and General Electric—have dramatically reduced their commitment to the manufacture of consumer goods. Instead, they dedicated themselves in the seventies and eighties to living off government welfare, turning out overpriced defense items that guaranteed a fat profit.

When hot new innovations come out of American labs, no American company scoops them up and turns them into the gadgets of tomorrow. Bell Labs created the transistor in the forties, but the people who made a fortune in the sixties and seventies selling us transistorized television sets and radios were Japanese. RCA and Ampex developed the videocassette recorder, but those who raked in over six billion dollars a year selling VCRs to the world were, once again, the Japanese.[635] Yet, these sorry experiences did not teach us any lessons. American scientists were in the forefront of basic research on superconductors—the hot new technology of the nineties—but in 1988, most American companies were asleep at the wheel about applying superconductors in consumer products. The Japanese, on the other hand, were developing ways to use these futuristic materials to revolutionize everything from high-speed computer chips and power generators to simple electric wiring.[636] We also invented, then abandoned, flat panel video screens—an essential element in laptop computers—and amorphous crystal solar panels (the ones that show up on everything from calculators to watches).[637] Meanwhile, throughout the eighties our military budgets climbed dramatically. Like the English under Victoria, we were trying to fool ourselves with the notion that weapons are the real source of strength.

In the 1800s, the British lost their preeminence. They did it by forgetting what counts the most in the pecking order of nations. To stay in place, you have to run. To get anywhere, you have to run even harder![638]

SCAPEGOATS AND SEXUAL HYSTERIA

The social climb—or fall—of a superorganism radically redecorates the psychic interior of the individuals who form its constituent parts. Being bounced from one rung to another reshapes personal emotions, warps the lenses of perception, and twists the course of behavior. In the next few chapters, we'll dig into a few of the more peculiar consequences for the world in which we live today.

When the pecking order status of a national superorganism slides, a frustrated populace looks for someone to blame, preferably a character located conveniently close to home. A declining Victorian England seized on Oscar Wilde, perhaps the most dazzling literary genius of his day. His plays, short stories, fairy tales, and essays scintillated. His wit was exquisite, his cynicism startling. The frenzy that led to Wilde's imprisonment all began with a book.

It was 1893 when Max Nordau published *Degeneration*. England's Great Depression had been dragging on for twenty years.[639] The island kingdom that had led the world into brave new technologies at the turn of the nineteenth century was becoming a technological and industrial backwater. The English knew they were in trouble, but they didn't know why. Then Max Nordau uncovered the real cause. The culprits behind Britain's fall were modern philosophy, modern art, and modern novels. As historian Barbara Tuchman puts it in *The Proud Tower,*

> Through six hundred pages of mounting hysteria, he [Max Nordau] traced the decay lurking impartially in the realism of Zola,

the symbolism of Mallarmé, the mysticism of Maeterlinck, in Wagner's music, Ibsen's dramas, Manet's pictures, Tolstoy's novels, Nietzsche's philosophy, Dr. Jaeger's woolen clothing, in Anarchism, Socialism, women's dress, madness, suicide, nervous diseases, drug addiction, dancing, sexual license, all of which were combining to produce a society without self-control, discipline or shame which was "marching to its certain ruin because it is too worn out and flaccid to perform great tasks."[640]

In the days before television and the compact disc, poetry, plays, and novels were the equivalent of today's electronic mass-consumer fare. Nordau was indicting all of pop culture.

One of the most visible popular artists of the day was Oscar Wilde. In 1895, his play *The Importance of Being Earnest* was a huge success. His books were widely read and his humor was quoted everywhere. But Wilde's sexual habits were exactly the kind that all good citizens knew were destroying England. Oscar was a homosexual.

When the flamboyant author filed a libel case against the marquess of Queensbury, Wilde, rather than the marquess, suddenly became the subject of scrutiny. A series of trials luridly pictured affairs with male prostitutes, a valet, a groom, and even a boat attendant. The newspapers flew into a fit of moral outrage. Cabbies and newsboys derided Oscar's sins. His books were removed from the stores. Two young noblemen implicated in similar activities were quietly let off the hook, but Wilde was sent to prison for two years. The incarceration drained the life from him. A mere thirty-six months after his release from Reading Gaol, Oscar Wilde died. He was forty-six years old.

Oscar Wilde's imprisonment did not save England, nor did the publication of Nordau's irascible book; but both gave the English the comfortable illusion that they had some sort of control over their unpleasant fate, and both distracted Britain from that fate's actual causes.

Since the early 1970s, America has experienced a decline similar to that which afflicted Victorian England. For decades, our exports exceeded our imports. That began to change in 1971. In 1973, we were victimized by an oil embargo that left normally confident American motorists stranded for hours in line waiting for a few gallons of gas. It was our first taste of helplessness.

Presidential adviser Pat Caddell sent a memo to Jimmy Carter in 1979 saying that the United States was in a new, invisible kind of crisis, "a crisis of confidence marked by a dwindling faith in the future, . . . [a crisis that] threatens the political and social fabric of the nation."⁶⁴¹ The year of Caddell's memo, 33 percent of Americans saw their lives going straight downhill.⁶⁴² By 1987, things had gotten worse. According to pollster Louis Harris, a full 60 percent "felt a basic sense of powerlessness" despite the apparent prosperity of the 1980s.⁶⁴³

Then an author came to the rescue. In 1987, America disgorged its own Max Nordau. He was an obscure professor from the University of Chicago named Allan Bloom. Like Nordau, Bloom knew exactly who to blame for America's decline. He did not level a bony finger at the industrialists who ignored the commercial possibilities of the flat-panel video and the VCR, but eerily echoed Nordau. Bloom fulminated against a set of dead German philosophers—Nietzsche, Freud, and Heidegger. And like Nordau, he raged against popular culture, but instead of Oscar Wilde, Bloom attacked rock and roll. "Sex, hate and a smarmy, hypocritical version of brotherly love" are the themes of rock, he declared dogmatically.⁶⁴⁴ "Such polluted sources issue in a muddy stream where only monsters can swim." In MTV videos, Bloom pontificated, "Hitler's image recurs frequently . . . in exciting contexts. . . . Nothing noble, sublime, profound, delicate, tasteful or even decent can find a place in such tableaux." He claimed that rock is a "gutter phenomenon," obsessed with sex, violence, and drugs, ruining "the imagination of young people,"⁶⁴⁵ stealing their zest for learning, impoverishing their emotions, turning them into callow participants in a nation's decline.

One of rock's primary crimes, Bloom claimed, was an overt celebration of sexuality. In Bloom's view, only when sex is driven underground can man create. The pent-up libido, Bloom claimed, is the driving force behind all ennobling accomplishments. (The professor, by the way, was a bachelor.)

Bloom never cited a single fact that would justify his bizarre coupling of sexual gratification with creative sterility. What's more, his view of rock was absurd. Drug lyrics had practically disappeared from rock music over fifteen years before Bloom wrote his book. Hate had never been a major rock theme (though it would later surface in a musical form Bloom was unaware of: rap). And at the time Bloom penned his work,

Hitler's image simply had *never* appeared in an MTV video. (Two years after Bloom's screed hit the stands, Hitler finally showed up in one MTV clip: the führer materialized in Michael Jackson's "Man in the Mirror," a work that urged social responsibility and held up Hitler as an icon of evil.)

But Bloom's singling out of a scapegoat satisfied a deep hunger for someone to blame. His book was wildly successful, and its influence was everywhere. A December 8, 1987, editorial in *The New Republic* picked up the professor's theme. It pointed a prophetic finger at "the prospect of decline that lurks . . . in America," lamented that "our cities have . . . become centers of barbarism," and deplored "the exacerbated cultural degradation of man and environment." The cause of all this? Rock music, with its "numbing norms . . . of random drugs, random sex, and random violence."[646]

The exaggerated—and often false—charges spurred a spate of legal actions. The FCC revised its policies on "obscenity." The new doctrine was worded so murkily that almost anything could be deemed obscene. In response, the listener-supported Pacifica radio stations were forced to suspend their plans to read Allen Ginsberg's classic poem "Howl" on the air.

But that was just the tip of the iceberg. A nineteen-year-old store clerk in Calloway, Florida, was arrested for selling a rap album that contained the word *pussy*. She was taken to jail, and the store she'd worked for was driven out of business. Police broke into the San Francisco home of political rock singer Jello Biafra and arrested him for "selling material harmful to minors." The material in question was a poster by Academy Award–winning designer H. R. Geiger included in a Biafra album. Displayed in numerous galleries, the poster was a surrealistic landscape of penises and vaginas designed to "criticize the standardization of mass consumer society." For nearly two years, Biafra was forced to abandon music and mount a legal defense. By the time he was acquitted, his rock group had disbanded.[647]

In Illinois, a law was introduced before the legislature that would have enabled officials to arbitrarily declare the goods of a bookstore, record store, or video store obscene. Armed with this charge, the government would have been empowered to seize the suspect's property—his store, inventory, bank accounts, and even his home—without a trial.[648] A similar piece of legislation—the so-called Child Protection and Obscenity

Enforcement Act—was introduced in Congress. Though numerous congressmen and senators admitted privately that the bill was thoroughly unconstitutional, it passed both Houses without a dissenting vote.[649] The hysterical search for scapegoats had mounted to such a height that "innocent until proven guilty" was about to be suspended in the case of pop culture.

The pattern was a common one in history. A slide down the ladder of nations brings a search for scapegoats and a rise in sexual hysteria. When Rome was under attack by Hannibal, its citizens looked for a solid, conservative dictator. The man they found pointed out that the traditional religious rituals had been either dropped or carried out with appalling carelessness. The new leader hurriedly restored the old-time worship of the gods.[650] A year later, Hannibal was still ravaging the countryside, so Rome's brave citizens looked for a few *humans* to blame their troubles on. A diligent "inquiry" uncovered the fact that two of the vestal virgins had been less than entirely virginal. To rid the city of its sins, the Romans buried one of the oversexed young women alive. (The other saved her neighbors the trouble. She committed suicide.) Just to be safe, the guardians of respectability interred a few visiting foreigners as well.[651] The return to Rome's old moral shibboleths did not make Hannibal go away.

In the history of our species, the interlocked phenomena of sexual hysteria and the search for scapegoats allow the social beast moving down the pecking order to ignore the forces shoving it toward the bottom. England used Oscar Wilde to seize pop culture by the scruff of the neck and give it a vicious shake. In doing so, Britain forgot the industrialists who had allowed the new chemical and electrical technologies to slip through their fingers. She overlooked the complacency that had eroded the international standing of her schools. She turned away from the siphoning of funds into damaging mergers and takeovers. Max Nordau's denunciation of pop culture did not stop the British economic slump. It did, however, divert England's energies from the tasks that could have saved her.

LABORATORY RATS AND THE
OIL CRISIS

⟫ ⟪

If a laboratory rat is confronted by an artificial beast larger than himself, he cowers. The mechanical bully can abuse the rat all it wants, and the bludgeoned creature will not raise a paw to counterattack. But offer the victimized rodent a smaller rat as a companion and a strange thing happens. The black and blue animal will not turn to his smaller companion for solace. Taking advantage of his new cage-mate's diminutive size, he will turn on the beast and viciously assault it.[652] Rats in groups are even worse than individuals. Put seven or eight of the rodents on an electrified floor, turn on the juice, and what happens? The gang will single out one of its members for punishment and attack him mercilessly.[653]

The rat is not alone in this kind of behavior. Jane Goodall says that one of the most common causes of brutality among chimpanzees is "frustration that leads an individual who has been thwarted by one stronger to turn and vent his aggression on a smaller or weaker bystander."[654] We humans, alas, are built with the same pusillanimous circuitry. When we are battered by forces beyond our control, we look around for someone smaller to punch.

Yale University researchers John Dollard, Neal E. Miller, and a team of colleagues saw this mechanism at work in 1939 as they painstakingly assembled the evidence for their classic "frustration-aggression hypothesis." The scholars reviewed figures on cotton prices in fourteen southern states over a period of forty-eight years. When cotton prices dipped, the white farmers who depended on the crop for their living took a beating. These unfortunate country folk made up for pinched budgets, lowered status, and the humiliation of mounting debts by turning around and tormenting someone even more helpless than themselves: the local blacks. Whenever cotton prices plummeted, lynchings shot up.[655]

The Soviet Union long indulged in the habit of beating up the little guy to soothe its pain. In the late sixties, the Russian economy was in trouble, the country's agricultural system was bogged down, and frustration was running rampant. The best way for a leader to boost his popularity was to mount a foreign adventure, preferably against someone small and helpless. When the Czechoslovakian uprising came in 1968, it was a godsend. Russian leaders sent in the tanks, and the Soviet populace cheered wildly.[656]

More frequently, however, the victim is someone right at home. In the early fifties, the United States suffered seriously from pecking order problems. We had been humiliated in Europe, where the Soviets had grabbed eight countries[657] and walled them off behind what Winston Churchill called an "Iron Curtain."[658] China, the keystone of our plan for Asian security, had fallen to a Marxist revolution. We had lost nearly a quarter of the world's population to our adversaries in only a few short years. Americans were desperate for someone to blame. A slightly alcoholic senator from Wisconsin offered to satisfy their hunger. There had been Communist infiltration of the government, but by the time the pixilated legislator showed up, President Truman had eliminated it.[659] Nonetheless, the senator cooked up fantasies of a Communist plot within the State Department. His claims were so filled with fraud that nearly every major American newspaper revealed their absurdity, but the public couldn't have cared less. They swept the tipsy senator up as a hero. During the next four years, Joseph McCarthy was one of the strongest men in Washington, able to make or break careers with a single word.[660]

Our problems, however, were not in Washington. They were overseas, where Mao Tse-tung had ousted our allies in China and Soviet-sponsored Eastern European "leaders" had seized power in one state after another. A few years earlier, when the Second World War had ended, we disbanded most of our military forces while the Soviets kept theirs at full strength. In what historian William Manchester calls a "near mutiny," servicemen threw massive demonstrations demanding demobilization, and they got it. The result was that by 1950, the Soviets had four times as many soldiers as we did and thirty tank divisions to our one.[661] Confronting the Russians would have been a painful experience. But picking on un-armed fellow Americans couldn't hurt at all.

John F. Kennedy later called this witch-hunt a supreme act of cow-ardice. He said of those who look for scapegoats, "They find treason in our churches, in our highest court, in our treatment of water. . . . Un-willing to face up to the danger from without, [they] are convinced that the real danger is from within."⁶⁶² Kennedy knew what he was talking about. He had just faced the threat of Soviet nuclear missiles in Cuba.

The search for scapegoats impels us to act like the alcoholic who has trouble at work. The drunkard does not direct his anger at his boss or himself, nor does he decide to stop drinking. Instead, he beats his wife. In 1956, the Hungarians rebelled against the Soviets, and the Russians sub-dued the city with tanks and machine guns. We could no more aid the Hungarians than the alcoholic husband could confront the boss at work. Instead, we turned on our friends.

Shortly before the Russians rolled into Budapest, Egyptian President Gamal Abdel Nasser seized the Suez Canal from the French and English. The French had built this engineering wonder in the nineteenth century with over $80 million of European funds—a gargantuan sum at the time. The land for the project had been granted by Egypt's khedive, in exchange for nearly half of the action.⁶⁶³ Now Nasser's move threatened to cut off 80 percent of Europe's oil supply. The British and the French enticed Is-rael into helping them retake the crucial passageway.⁶⁶⁴ They massed troops for an assault, confident we wouldn't stop them. After all, when we had sent our soldiers into Guatemala and Korea, the British and French had backed us!⁶⁶⁵ But we had a collective psychological need to snap at someone. We needed to beat a wife back home. On the morning of Tuesday, November 6, 1956, the French took the Suez Canal's eastern bank. While Hungarians called on their radio stations for our aid, we turned our back on the Russian soldiers who by now were killing the citi-zens of Budapest.⁶⁶⁶ Instead, we scolded our allies for "colonialist aggres-sion" and insisted that our international partners put a European-built canal back in Nasser's hands.⁶⁶⁷

By giving Nasser a victory, we lent tremendous prestige to the Arab revolutionary movement, a political tide that soon planted anti-American autocrats all through the Middle East.⁶⁶⁸ And eventually, with OPEC, those leaders found a way to twist our tail. The resulting rocket rise in the price of oil would give us our first negative international trade balances in the early seventies. It would create a decade of spiraling prices, and it would make a major contribution to the American decline.

At its best, scapegoating is a nasty business. At its worst, it is a sluggish form of suicide. Battering our British allies in the Suez incident, we set the stage for a financial hemorrhage that would later overwhelm us. During the McCarthy era, we attacked talented people in government and the arts—"eggheads," intellectuals, people with insight and brains. Ultimately, *we* were the ones deprived of the fruits of their talents. And in the late eighties, when conservative groups battered away at rock and roll, they were undermining one of the few American fields of endeavor that contributed massively to our economic well-being. In just the first half of 1988 we exported 3.65 million records and compact discs to our arch trading rival, Japan.[669] That's one reason the entertainment industry in 1987 generated a $5.5 billion balance of payments surplus, a surplus second only to that created by U.S. aircraft manufacturers.[670] When we run into pecking order problems, we look around for someone smaller to kick. And in the end, the folks we kick the hardest are ourselves.

WHY NATIONS PRETEND TO BE BLIND

⟿ ⟸

Scapegoating is not the only consequence of pecking order slippage for the operations of the mind. A rise or fall in hierarchical position can radically rearrange the way we see the world.

I own a very large German shepherd who is inordinately fond of pecking order games. When we walk through the park and he spots another canine, he's suddenly all playful eagerness. His ears go up, and his eyes grow bright. He can't wait to run over to the stranger and make friends. One reason for the enthusiasm: in the game of chase and tumble that follows the first sniffed greeting, my dog is certain he will come out on top.

At least that's the way it goes most of the time. The dog that appears on the other side of a meadow is not always some little cocker spaniel or midsized mutt my shepherd can tower over. Now and then, it's a mastiff or Great Dane, a hulking giant that makes my well-built pooch look like a fuzzy midget. When one of these really big dogs appears, a strange thing happens to my pet's enthusiasm: it disappears. Faced with a creature he knows will beat him soundly, my shepherd plays a perceptual game. He trots along staring resolutely ahead, feigning obliviousness to the massive monster in the distance.

Ethologist Frans de Waal has noted the same behavior in chimpanzees. De Waal, who cropped up earlier in this book, spent six years carefully analyzing the nuances of social behavior among the chimps of the Burgers' Zoo in Arnhem, Holland. There, the animals live in a two-acre outdoor compound that approximates as closely as possible the conditions of the wild.[671] Each chimpanzee tribe—whether in the zoo or in the jungle—has its leader. According to de Waals' observations, this dominant male hasn't become king of the chimps by accident. He has waged a long, hard campaign to make it to the top of the pecking order. He's developed his physical strength, learned to make loud and frightening displays, and has become a master of the art of intimidation. What's more, he has become a primate politician, currying favor among the masses—the lower-ranking males and the powerful lady leaders of the female cliques—and carefully building alliances with other strong males.

But chimpanzee leaders, like human power brokers, eventually grow old and weak. In their younger days when a potential rival showed up to challenge them, they reared back on their hind legs and made a dramatic show of brawn and agility. But when strength and swiftness fade, the aging leaders use another tactic. Like my dog, they pretend they do not see. A rival may swagger toward the reigning monarch determined to assert his claims. The muscular youngster jumps up and down. He makes terrifying noises by pounding on any resonant object in sight. He swings huge branches intimidatingly through the air. But the weakened elder deals with this pecking order challenge in a strange way. He turns his head and pretends to be utterly absorbed in examining a banana peel.[672] For a time, the aging leader who refuses to see his rivals retains his top position. His old system of alliances props him up. But if the youngster has played his cards right, he has quietly built up coalitions of his own and gained the favor of the populace. Then the challenger's public humiliations of his elder may one day prove decisive. Eventually, the older statesman will be forced to yield his position, and the young turk will become the new head chimp.[673]

Human superorganisms go through similar pecking order confrontations. They gather allies. They curry support among the common people. They make theatrical shows of strength and power. And when they sense that they may lose, they follow the path of the chimp and the dog: they undergo perceptual shutdown.

In 1931, the Japanese lunged toward a higher position in the hierarchy of nations. They invaded Manchuria, treating the local inhabitants in a manner that was bloody and barbaric, but no one tried to stop them. Six years later, the Japanese made an even more ambitious move. They landed their soldiers in the most populous empire on earth—China. Their goal was to take the country by force.

The United States was determined to avoid a fight. We were suffering from a devastating economic depression. What's worse, Japan was allied with Hitler's Germany. The Germans and Japanese had been building advanced planes, tanks, and warships for years. They had drafted young men in huge numbers and prepared them for battle. We Americans loved to tell each other tales of how we had won the First World War, but we hoped never to be embroiled in a massive conflict again. Our army had been pared down to a force smaller than that of Czechoslovakia,[674] and our military equipment was obsolete. Our men could scarcely have withstood an assault by a determined group of Boy Scouts.

At the end of 1937, the Japanese invaders arrived on the outskirts of the Chinese city of Nanking, which had a small contingent of American citizens. Moving up and down the nearby international waterway of the Yangtze River were American commercial ships. To protect both ships and citizens from the guerrilla bands who roamed the Chinese countryside, the Americans had stationed a small gunboat, the *Panay*, in the Yangtze's waters.

Our government wanted to avoid damage to the gunboat in the brutal fighting between the Chinese and the Japanese. On December 1, 1937, the American ambassador in Tokyo told Japanese officials the gunboat's exact location, explained that it would abide by the terms of American neutrality, and informed the Japanese that the gunboat would be used to evacuate American citizens peacefully if that became necessary. On December 10, evacuation became very necessary indeed. The fighting around Nanking had turned deadly.

The crew of the *Panay* worked for two days and nights loading American journalists, photographers, businessmen, and embassy personnel on board. Then the ship steamed upriver to find a safe haven. The anchorage it chose was not safe enough. At 1:30 in the afternoon, Japanese warplanes came over the horizon and attacked, sinking the neutral USS *Panay*. When the *Panay*'s passengers piled into lifeboats and headed for shore, Japanese planes returned to rake them with machine guns. Three Americans were killed, and another eleven were gravely wounded. A court of inquiry in Shanghai eventually revealed that the attack had been deliberate.

Historian William Manchester feels that the Japanese assault on the *Panay* was an attempt to see how willing America was to resist the Japanese in the struggle for hierarchical dominance. The United States responded like the reluctant dog and the intimidated chimp: we pretended we didn't see. Media scarcely covered the event. Two months earlier, President Franklin Roosevelt had mentioned briefly during an out-of-the-way speech in Chicago that "the epidemic of world lawlessness is spreading." Roosevelt had called on peace-loving nations to combine and stop the violence before it could travel further. The president's words were greeted by a howl of protest. Angry letters and newspaper editorials accused FDR of warmongering. Roosevelt learned his lesson and shut his mouth.

When the *Panay* and her survivors were mauled, isolationist groups went even further. They tried to get footage of the attack banned from newsreels. News films, they said, would have "the unquestioned effect of arousing the American temper." The ban wasn't necessary. Nobody really wanted to pay attention to the killings. Polls revealed that 70 percent of U.S. voters preferred to solve the problem by withdrawing all American citizens from the Far East.[675] Instead of looking the Japanese in the eye and protesting their attack, we pretended we didn't see.

But the Japanese did not melt into oblivion. Lack of opposition, in fact, fed their illusions of invincibility. Four years later, they sank eighteen American ships, wiped out 188 of our planes, and killed 3,400 U.S. citizens at Pearl Harbor.[676]

The nation slipping downward averts its eyes, but the country on the rise is often vigorously alert, looking for the tiniest opportunity to lunge toward the top. Instead of turning their backs and hoping for peace, superorganisms on the move often manufacture confrontations. For an example, let's revisit the days of Victorian England's decline.

You'll recall that in the 1860s, as England was about to enter its slump, two other countries were on the verge of taking over the lead. One was the United States. The other was Germany. As I mentioned earlier, when German industry raked in the deutsch marks hand over fist, it didn't make the citizens of Hamburg and Berlin blissfully content; it made them hunger for more. A testosterone jolt roused Teutonic ambitions and the will to use violence to achieve them.[677]

The man who embodied his country's new mood was a six-foot-tall *Junker* with flaming red hair, Otto von Bismarck. From his earliest years, Bismarck had loved bloodthirsty showdowns. During his university days, he turned seemingly harmless debates into ultimate tests of honor. In just his first nine months at school, young Otto challenged twenty-five students who disagreed with him to duels, and he always emerged from these potentially fatal confrontations on top.[678]

In 1862, eleven years before England's downhill slide revealed itself in the Great Depression, Bismarck became Prussia's premier. Prussia had used its huge coalfields to fuel new iron and steel plants, plants that put it in the forefront of the industrial revolution. The result was "an age of plenty."[679] Otto could sense opportunity in the air, and he wasn't the type to let a winning chance pass him by.

The Prussian parliament was dedicated to peace, but Bismarck was not. He showed his contempt for mild-mannered solutions at a meeting of the parliamentary budget commission in 1862. "The great questions of the day will not be decided by speeches and majority resolutions," he declared in a statement that would be quoted for the next hundred years, "but by iron and blood."[680] So Bismarck circumvented Prussia's parliamentary delegates and furtively used illegal means to siphon money into a military buildup.[681] Bismarck boosted the size of the Prussian standing army dramatically and made enormous investments in new weapons, including high-tech cannons from Krupp that could fire a shell twice as far as any other artillery piece known to man.[682]

Bismarck was not the only Prussian feeling the feistiness that comes to a nation moving up the hierarchical ladder. His policies had overwhelming grassroots support. One observer said that "Before him, the people lay in the dust and adored."[683] To guarantee adoration, Bismarck sidelined his opponents, made vigorous use of a secret police, and stifled complaining voices in the press.[684]

Once Bismarck was confident that his military buildup was complete, he did not flinch from confrontation. Far from it. He sought it out. Bismarck embarked on a plot to force France into war. But he wanted Prussia to look like the victim and France to appear as the aggressor. To achieve this, the Prussian leader worked out a scheme that would corner the French into making the first openly hostile move.

Spain needed a king, and the Spaniards were searching for a European prince who could fill the throne. Bismarck's agents labored secretly in Madrid to make sure the prince the Spanish parliament chose would be a German. Otto knew that the idea of a German grab at the Spanish crown would panic France. If the Prussian move went through, the

French would be surrounded by hostile states: Rubbing up against the northeastern Gallic border would be Germany, and crowding the country's southwestern flank would be a German puppet. The strategic vulnerability would be intolerable.

If Prussia went ahead with the installation of its princeling on the Spanish throne, France said, it would mean war. Princeling on the Spanish throne? said Bismarck, we have nothing to do with that. The choice is a strictly Spanish matter.[685] The Prussian premier's profession of innocence was less than frank. The only man deciding which autocrat would be shoved into Spain's regal chambers was Bismarck himself. The French continued to bluster their discontents. And Bismarck swore he was the innocent victim of French threats. What's more, the world believed him.

Finally, Bismarck was "forced" to go to war in sheer "self-defense." The world saw this as a confrontation between a mighty nation with a strong military tradition, France, and a weak, unprepared principality, Prussia. Europeans and Englishmen were still bitter at France for its astonishing conquests under Napoléon sixty years before. France, they thought, had an army no one could beat.[686] Bismarck knew better. He was quite aware that in the first weeks of a confrontation, France could only mobilize a few hundred thousand men, but in that time, Bismarck could call up well over a million![687] What's more, the Prussian knew that the French still hadn't figured out how to deploy many of the newly developed armaments on which they were counting for their defense. For example, the latest Gallic secret weapon was the *mitrailleuse,* an extremely rapid-fire machine gun. There was only one problem. French generals hadn't worked it into their strategies yet. As a consequence, when war came, they kept this potential battlefield terror in the rear where it was of no earthly value.[688] Bismarck also knew that his extraordinary high-tech artillery could wipe out French forces long before the French would even come within shooting range of Prussian soldiers.[689]

The result was the Franco-Prussian War of 1870. The world was utterly convinced that France would crush Prussia. Instead, Emperor Louis Napoléon was forced to surrender the bulk of his army within the first three weeks of the war. Less than five months later, Prussian squadrons were marching through the streets of Paris.[690]

Bismarck had never flinched or lowered his eyes at threats. In fact, he had been able to control every offensive move. Why? Like the larger of two dogs encountering each other in a field, Otto von Bismarck knew from the beginning just who was on top.

In the 1960s, when we were at the peak of our power, we too were raring for confrontation. We entered the Vietnam War and lost. It was the second conflict we had failed to win since our participation in the defeat of Hitler's Third Reich. The first had been our "police action" in Korea during the early fifties.

By the early seventies, military failures were not our only pecking order problems. We were showing signs of an unaccustomed financial vulnerability. And our willingness to look world events squarely in the eye altered as well.

While we still fought in Vietnam, a guerrilla group called the Khmer Rouge waged war in Cambodia. Aided by the North Vietnamese, these Marxist "freedom fighters" struggled violently to gain control of what had once been a pleasant, peaceful land. Two years after our departure from Vietnam, the Khmer Rouge finally achieved their goals. They ousted the old regime and installed Communist zealot Pol Pot as prime minister. The forty-eight-year-old idealist immediately set to work imposing his vision of freedom, peace, and justice. He executed those who didn't share his dedication to building a new society. He emptied entire cities at gunpoint. He sent his soldiers into hospitals, forcing the patients to pick up their beds and march to the fields. He ordered college professors, doctors, and civil servants into the countryside to help grow crops. He turned schools into torture and extermination centers. He had babies beaten to death, children's throats cut, old women nailed to the walls of their homes and burned alive, and pregnant women shot. Then he heaped the bodies of his countrymen in thousands of mass graves.[691]

By the time North Vietnamese invaders threw Pol Pot out in 1979, roughly three million Cambodians had been killed by bullets, forced labor, or famine. Nearly half the country's population had died.[692]

What was the American reaction? We didn't march in the streets, nor did our newspapers blare the tale in daily headlines. Our moral leaders—our priests and rabbis—failed to protest, and our university students did not demand that we cut off aid that might conceivably trickle into Pol Pot's hands. Instead, *Newsweek* went so far as to deny that there were any atrocities as did the *Sunday Times* of London. When a *Sunday Times* correspondent turned in a story explaining just how bad the plight of the Cambodians really was, the paper refused to print it.[693] And leading "progressive" intellectual Noam Chomsky declared that occasional rumors of unsavory doings were the results of a CIA plot. A holocaust was in motion, and we turned our backs.[694] We are doing the same today toward the ubiquitous genocide in black Africa and the hatred aimed against us by a host of Islamic nations we insist on calling our friends. Why?

For the same reason that one animal passing another it knows to be much bigger stares in another direction. Far from possessing the pugnacity of an Otto von Bismarck, we have taken our knocks in the pecking order and we sense that we are moving down. Like the chimp who hopes he can cling to his preeminence by ignoring a rival, we toy with our equivalent to a banana peel. While events of grisly horror grind down humans by the millions, we simply turn away.

HOW THE PECKING ORDER
RESHAPES THE MIND

≫ ⋐

A rise or fall in the hierarchy of superorganisms has other profound effects on a society's collective psyche. It transforms the emotions and shared values of the human herd. The nation moving up embraces adventure. The country moving down abandons the strange and buries its head in the familiar. It tries to march backward in time. These shifts in attitude are the result of prewired natural strategies. One of the most basic biological mood loops in our animal brain dictates a simple set of alternatives. It makes us conservative in times of difficulty and exploratory when times are good.

When birds are hungry and desperate for food, you might expect them to sample everything they run across that looks edible. After all, with starvation at hand, some peculiar-looking berry the feathered flier has never noticed before could be the key to survival. But hungry birds do not methodically nibble every unusual item to see if it can be turned into a meal; they shy away from food that seems strange. Their fear of the unfamiliar renders them culinary conservatives.

Birds benefiting from a bonanza in the food supply are a very different story. As these creatures strut around with full stomachs, you'd expect them to ignore any unfamiliar morsels they come across. They've already got more than enough to eat, so why try anything new? But well-fed birds are up for every potential adventure in dining.[695]

There is a logic behind this genetically ordained strategy. Exploring the unknown is a risk. A creature in genuine need can't take a chance on being immobilized a day or two by food poisoning;[696] those long hours with nothing to eat could kill him. But the bird who's already stored up his calories can afford to toy with novelty.

Human superorganisms show the same pattern. In times of trouble, they tend to shun the new.[697] When the Turkish empire was crumbling in the sixteenth century, Ottoman authorities were sure that they could recapture former glories by returning to the traditions of the past. Europeans came up with improved methods for preventing plague, but the Turks refused to use them. Why? The foreign techniques departed from the customs that had once made Turkey great. Like hungry birds, the Ottomans sought their comfort in clinging to tradition. When plague did break out in the land, the Turks blamed it on the few foreign innovations they had failed to eradicate. To stop the ravaging illness in 1580, officials destroyed—of all things—an astronomical observatory. The newfangled installation, they were sure, had offended Allah and brought the curse of disease.[698]

Sounds like something that could never happen here, but Allan Bloom's *The Closing of the American Mind* seems to blame our problems on a departure from traditional values. He implies that our abandonment of the sexual and *racial* inequalities of the forties and fifties has a great deal to do with our current plight. His prescription seems to be a return to the standards of our past. In fact, he wants to go very far back indeed, all the way to the values of ancient Greece.

Others, like self-appointed science critic Jeremy Rifkin, want to head backward technologically. Rifkin has garnered great gobs of publicity by trying single-handedly to halt the progress of genetic engineering. This modern Luddite is just one of a new army of American public figures determined to move us resolutely in reverse. Religionists like Moral Majority leader Jerry Falwell and presidential candidate/religious broadcaster Pat Robertson want to turn back the clock and restore the life-style they knew when they were children. Even radical ecologists want to throw the gearbox into reverse—their slogan: "Back to the Pleistocene."

Though religious leaders and ultraconservatives claim that we have strayed from old-fashioned values, the opposite is the case. Since the signs of America's decline first appeared in the late seventies and early eighties, our tendency has been to move toward the old, stable ways. You can see this rearrangement of the American psyche with particular clarity in the pop music of the mid-eighties.

In the late sixties, America benefited from one of the longest periods of peacetime prosperity in the history of the country and simultaneously dominated the pecking order of nations. In those days, pop music was ruled by rock, the music of rebellion, the soundtrack of a generation bent on stepping beyond the boundaries of the traditional and tasting the forbidden. Crosby, Stills, Nash and Young championed the rights of young men to grow long hair as ''a freedom flag'' against ensconced authority. The Beatles expressed views that flew in the face of established institutions and followed the Maharishi instead of a conventional church. John Lennon and Yoko Ono lay naked in a bed and threatened to stay there until the older generation ended the war in Vietnam. The Lennons and many of their fans wanted to wipe away the boundaries of custom. Like the birds with full bellies, they were determined to taste something new.

The phenomenon of the well-fed, adventurous bird showed up in even more subtle ways. To succeed, a rock 'n' roller had to be a young man on his own, totally free of parents and family, a rebel who had bailed out of his childhood home and become a vagabond, roaming the countryside in the company of other young men like himself—his band. The ideal rocker was a hero who had cut himself loose from the old, smothering ways. There was one cardinal rule for rock interviews: never mention the existence of your father and mother. Admitting that you had once been tied to apron strings could instantly kill your appeal.[699]

Then in the early eighties, the pop world was seized by a phenomenon of such proportions that it staggered even the star makers. Up to that point the biggest-selling LP of all time had come from rock guitarist and singer Peter Frampton, who in the late seventies sold an astounding fourteen million copies of *Frampton Comes Alive*. But in 1983, another singer made Frampton's fourteen million discs seem minuscule. With just one album, this crooning, dancing wonder sold over thirty-nine million records. His name was Michael Jackson.

At first glance, Jackson seemed an eccentric. Yet under the surface, he was anything but. For years, the subject of family ties had been taboo, but Michael Jackson was, in fact, the ultimate family figure. Jackson had worked with his brothers onstage since he was five; he had been coached and managed by his father; and he actually lived in a bedroom in his parents' home!

What triggered this about-face between the pop heroes of the sixties and those of the eighties? The Beatles' heyday had come when our stom-achs were full and we were curious to taste the strange, but Michael Jack-son arrived when we were starving. In the mid-eighties, we craved refuge in the familiar. Jackson, in his own eccentric way, brought us back to the values of our past.

Signs that we were trying to wrap ourselves in the safety blanket of tradition showed up outside the world of pop as well. In the sixties, a widespread motto had said that you couldn't trust anyone over thirty, es-pecially parents. But a survey in 1986 showed that a new generation of students felt their ultimate heroes were their fathers and mothers.[700]

When author Lisa Birnbach went out to visit college campuses dur-ing the mid-eighties for an article in *Rolling Stone,* she was shocked at what she saw. Gone were the long hair and rebellious attitudes she had known in her student days. Torn jeans had been replaced by jackets and ties. What's more, students had abandoned majors that would allow them to explore new territory in favor of those that would make them a safe liv-ing. They had tossed aside anthropology and comparative religion, and raced to take accounting and finance. Abandoning the sixties' holy grail of a psychedelic revolution,[701] the collegians Birnbach saw had revived the attitudes that dominated college campuses in the fifties.[702]

Other backward-looking symptoms increased dramatically during the mid-eighties. The number of college students who clung to Christian fundamentalism increased 43 percent from 1978 to 1985.[703] In 1986, the number of American voters who labeled themselves conservatives shot up to an astonishing 39 percent.[704] Political conservatism was particularly high among, of all people, the young.[705] And, in 1993, when some thought that the election of Bill Clinton had dampened the conservative trend, the *New Republic*'s Fred Barnes declared that reaction against the Arkansas Democrat had actually "revived the conservative movement beyond its wildest dreams."[706]

There was a reason. The mid-eighties generation of college kids had experienced the decline of America since they were toddlers. They were in their early days of grammar school when the United States suffered pecking order defeats in Vietnam and the Arab oil embargo of 1973. These students had spent their growing years in families battered by downward mobility. A huge percentage were latch-key children who had seen both of their parents go to work,[707] or had been raised by a single mother or father.

From 1949 to 1973, American wages had moved upward in a seemingly unstoppable surge.[708] But when the mid-eighties crop of college kids were still in day care, all that started to change. In 1971, America experienced its first two months in a row of balance of payments deficits.[709] In 1973, real wages began to fall.[710] When the mid-eighties college kids entered junior high, the older students ahead of them were already showing the first response to the new American realities: an increase in suicide, alcoholism, and drug addiction.[711] To make matters worse, by 1986 when the new collegians were still getting used to dormitory life, real U.S. wages had tumbled back to 1962 levels.[712] And when the former students were beginning careers in 1991, salaries had shrunk even further, moving a full 20 percent below where they'd been two decades earlier.[713]

No wonder the youth of the eighties re-created the attitudes of the fifties when college students were more interested in getting an MBA than exploring life and philosophy. No wonder even young bohemians returned to the traditions of the American ancients, dressing like Woodstock dropouts and reveling in the music of a quintessential sixties band—the Grateful Dead. Like birds on an empty stomach, they were edging away from the unfamiliar and jettisoning the spirit of adventure. They were closing their eyes to the future and burying their heads in the past.

PERCEPTUAL SHUTDOWN AND
THE FUTURE OF AMERICA

Progress is possible only when people believe in the possibilities of growth and change. Races or tribes die out not just when they are conquered and suppressed but when they accept their defeated condition, become despairing, and lose their excitement about the future.

Norman Cousins

Americans threatened by foreign terrorism, battered by ubiquitous crime, victimized by downward mobility, and menaced by the decline of American industry are trying to fight back. The average breadwinner's workweek has gone up from forty-one hours to forty-nine,[714] and the run-of-the-mill family now has two working parents. Yet, despite these adaptations, somewhere deep in the back of their minds, Americans feel trapped. They sense they are being pummeled by forces over which they have no power.

As we've seen earlier in this book, a strange thing happens when humans and other animals are cornered by the uncontrollable. Their perceptions shut down, their thoughts grow more clouded, and they have a harder time generating new solutions to their problems. The outlines of this perceptual shutdown were revealed by an extensive series of experiments we encountered a while back, experiments that have become recognized as classics. In a typical version, two rats are placed in cages next to each other. The cage floors can be electrified, producing a painful shock to the rodents' feet. In one cage is a bar the rat can press to turn off the floor's electrical shock. In the cage next door, there is no such convenient turn-off switch. Both floors, however, are on the same electrical circuit. When the electricity shuts off in the floor of one cage, it shuts off in the other as well.

When the two rats are first put into the cages, they go about their business—scratching, preening, sniffing for food. Soon, the first shock hits. Both rats jump, startled. They race around their cages, searching for relief. Many shocks later, one of the rats discovers the bar that shuts down the current. Now, when a jolt of power hits the floor, he leaps to the switch and bangs it emphatically, giving both himself and his neighbor relief. The rat with the switch in his cage remains alert, doing well on tests of his ability to solve problems and learn.

His neighbor, on the other hand, has no switch allowing him to turn off the source of his pain. This rat at first scrambles around his cage frantically when the charges sizzle his soles, looking for an escape. Eventually, his panic turns to resignation. He huddles in a corner, not moving when the shock arrives. If you open the cage door, he will not even try to escape.[715] The rat who becomes the helpless victim goes through disturbing changes in the way he sees the world. Put him in an unfamiliar maze, and he will have tremendous difficulty learning its intricacies. Other observations indicate that he doesn't seem to focus on the facts around him clearly anymore. To protect him against pain, the rat's body has generated the internal anesthetic endorphin, which makes the stinging sensation of the electricity seem to melt away. But like its artificial equivalent, morphine, endorphin makes reality disappear as well, for this chemical of mercy soothes by blindfolding the senses.[716]

Like the rat, we have been hit with a random burst of punishments beyond our control. Could we be seeking relief by closing our eyes to the world around us? There's a clue that indeed we are. Our collective eyes and ears, our media, still often give us the impression that we are king of the international heap. It is a totally erroneous notion. France's advanced technological programs allowed it to create the smart card and the supertrain. Her Aerospatiale sells far more helicopters in Japan than we do.[717] Germany's BMW holds the lead in developing recyclable auto parts and engines powered by alternative fuels. Europe's Airbus, already ahead of us in the commercialization of electronic controls for passenger planes, is planning a superjet that will make Boeing's 777 look like a minnow. While our space program is bogged down in a comedy of errors, the Japanese are planning to develop hotels and resorts on the moon.[718] Japan has also overtaken us as the nation that files the greatest number of break-through patents[719] and is far outdistancing us in the introduction of products using such American inventions as fuzzy logic.

More ominously, forty-five countries are accumulating the necessary elements to manufacture nuclear weapons. Four of the seven nations that lead in building the bomb—Iran, Libya, North Korea, and Algeria—see America as their major enemy (two of the others—Taiwan and South Korea—regard us with a more tolerant eye). And ballistic missile delivery systems for these weapons are becoming nearly as easy to obtain or manufacture as .22 caliber pistols.

There are other surprising areas in which we have also fallen behind. We think of ourselves as the healthiest nation in the world with the best quality of medical care, but the citizens of eleven other nations can expect to live longer lives than we can.[720] And we often feel we have the highest standard of living, but that ceased to be the case quite a while ago. The citizens of at least nine countries earn higher per capita incomes than we do. And those nations that tower over us in the personal affluence of their inhabitants include such minor powers as Brunei, Canada, Kuwait, Liechtenstein, Luxembourg, Nauru, Qatar, Switzerland, and the United Arab Emirates.[721] We are blind to much of this. And, despite admirable exceptions like the coverage of our lag in education, our media frequently fails to report it.

What's worse, as we saw a few pages ago, some groups want to fossilize us in a featherheaded re-creation of a past we can never recover. Academic champions of multiculturalism idealize the backwardness of third-world cultures and often bluster furiously to prevent a technologically driven movement into the future. Other self-proclaimed champions of the public interest are attempting to stop critical areas of scientific advancement: for roughly twelve years, fundamentalists eliminated the fetal tissue research that may cure patients of Parkinson's disease; those claiming an expertise in "scientific ethics" labor daily to stop the use of biotechnologies designed to feed the starving; animal rights activists bomb labs housing studies that could relieve the permanent impairments caused by head injuries; and African-American interest groups have succeeded in virtually banning conclaves on new empirical approaches to violence. In many intellectual circles, even the concept of progress has been turned into a dirty word.

Biologically-produced endorphins anesthetize a helpless rat to his pain by disabling his senses. Like endorphins, ignorance would help us feel better, for a price. Other countries have attempted the strategy of perceptual shutdown and found it fatal. One of them is China. In 1405, China was geographically enormous, militarily powerful, and internally stable. Her advanced ways put the European world to shame. China experienced unity while the European countries, each a tiny fraction of the size of China, were squabbling like children. Her cities had amenities like specialty restaurants and sidewalk snack vendors.[722] Her citizens were blessed with immense iron, ceramics, and printing industries; cookbooks, encyclopedias, and medical texts; and widespread education.[723] Her government had sponsored a massive—and successful—program of agricultural research and development that nearly doubled the country's production of rice.[724] What's more, Chinese technology was amazing: paddle-wheeled warships 360 feet long carrying 800 men moved up and down her major rivers.[725] When her emperor stepped into the rooms of his palace, automatically operating doors and window blinds opened before him.

The barbarians from the edge of the world, the Europeans, soon arrived with signs that they were catching up. These peculiar-looking people from insignificant lands had solved problems of the calendar that diligent Chinese sages found insoluble, possessed answers to astronomical enigmas that totally baffled the Chinese, and displayed miraculous devices like portable clocks.[726] More ominously, though China's weapons specialists had a head start of fifty-one years over the West in the development of cannon,[727] the Orientals were still making their artillery from iron while Europeans had figured out how to craft a more powerful version using brass.

In the fifteenth century, China regarded itself as literally the center of the world, the only country that counted. And the Chinese had an easy way to remain number one in their own minds: they walled themselves off. For two generations, the Chinese had built boats the size of apartment buildings with which they had plied the seas and exchanged information with the other countries of the eastern world. The emperor's fleet of 317 giant vessels, staffed with a crew of 27,000, had shown up in the harbors of Vietnam, Cambodia, India, the Arab Middle East, and even eastern Africa.[728] But just before Europe could reach out for contact, China's bureaucrats ordered that all such ships be outlawed and that interchange with other countries be stopped. The functionaries actually forbade Chinese merchants from even dealing with outsiders on pain of death.[729] The Chinese spurned Western knowledge, discouraged the spirit of inquiry in their own country, and buried their heads in their ancient texts.[730] Said one scholar, "The Truth has been made manifest to the world. No more writing is needed."[731] In 1839, Lin Tse-hsu wrote contemptuously to Britain's Queen Victoria: "Of all that China exports to foreign countries there is not a single thing which is not beneficial to people. . . . On the other hand, articles coming from outside China can only be used as toys."[732]

Despite the Manchu invasion of 1644, the Chinese maintained the illusion that they were untouchable, invincible, and above challenge.[733] Then in the nineteenth century, the Europeans returned with advanced gunboats, marched troops into China's cities, and showed just what the Chinese had accomplished by wallowing in illusion.[734] In the late 1890s, the Europeans began, in their words, "carving up the Chinese melon." The French took Kwangchow Bay; the Germans snatched the city of Tsingtao; the Russians expropriated the Liaotung Peninsula; the British expanded their hold on Hong Kong and seized the port of Wei-hai-wei; and even the Japanese snatched their share of the fruit.[735]

Foreigners were capable of inflicting any form of atrocity they chose. Kaiser Wilhelm, in fact, claimed that he would deal with the Chinese so savagely that the lesson he taught them about their inferiority would last them fifty generations. Said Wilhelm to his troops when he sent them off in 1900 to attack the Chinese homeland, "Even as a thousand years ago, the Huns under their King Attila made such a name for themselves as still resounds in terror through legend and fable, so may the name of Germany resound through Chinese history a thousand years from now so that never again will a Chinese dare to so much look askance at a German."[736]

By the early twentieth century, the Chinese empire had collapsed in chaos. If the Chinese had kept their eyes open to the dangers of the world around them in the 1500s when the Europeans had first shown up, if they had attempted to absorb the knowledge of the Westerners and outdo it, they might not have experienced this fate. But the Chinese were more interested in the opiate of illusion than the bitter draught of reality. Like the rat who cannot control his fate, they huddled in their corner of the world, indulging in the endorphin strategy, with its dulling of the senses and crippling of the intellect.

Today, we are the ones desperately attempting to remain number one in our own minds. If we are not careful, we may be the next Chinese.

THE MYTH OF STRESS

⇒❦⇐

There is nothing better for men than that they should be happy in their work, for that is what they are here for.

Ecclesiastes

How do we end our downward slide in the pecking order of nations? One helpful step would be to revise the popular misunderstanding that has created the medical myth of stress. Stress, we are told, is one of man's most implacable enemies. Stress causes headaches, back problems, divorce. What's worse, stress kills. And what produces stress? According to magazines and TV news reports, it's the urge to achieve, the desire to compete, the preoccupation with success. To protect ourselves from stress, we must relax.

We try to shield our children from stress's deadly barbs by shunning competitiveness in our schools. We avoid making heavy academic demands, burden our students with only the lightest hint of homework, and banish the concepts of ambition and excellence from the classroom.[737] We fear that if we drive our youngsters too hard, they, like us, will suffer from the demon disease.

But our concept of stress is a fallacy based on a persistent misreading of the medical evidence. As Kenneth R. Pelletier of San Francisco's Langley Neuropsychiatric Institute says, "Both researchers and clinicians have misinterpreted [the] findings."[738] Very few of the studies on so-called stress have dealt with achievement or work. They have centered in most cases on social loss: men who've recently lost their wives, people who've lost their jobs, and folks who've been recently divorced. Sure enough, those who've been sundered from a spouse by death or separation, or those who've been cut off from the vocation that gave their life meaning have suffered a plethora of physical problems. But those problems have not been due to what laypeople call stress. They have not been the product of hard work or the pursuit of excellence. They've been the result of three factors we discussed in the early pages of this book.

Each of us is sewn by invisible threads into the superorganism. We are cells in the beast of family, company, and country. If those social ties are severed we begin to shrivel and die.

There's more. Hard work and the pursuit of challenge have seldom been demonstrated to hurt us, but we can be damaged powerfully by the lack of control. And without striving to achieve, we cannot control our lives.

Position in the pecking order makes an additional contribution to many of the symptoms we blame on stress. With our dream of eliminating competition, we try to wish the pecking order away. But the fact is that we will continue to live in pecking order structures whether we like it or not. As we've seen, social hierarchies are not limited to "capitalist" or "consumer" societies. Not only do they exist among apes, birds, lizards, and lobsters,[739] but pecking orders left their marks in the remains of our Ice Age ancestors, who thrived fifteen thousand years before the birth of agriculture and nearly twenty-five thousand years before the founding of modern industry. In the Ukraine, archaeologists have unearthed Paleolithic palaces of the wealthy—tents with a framework of mammoth skulls and tusks, and a rich covering of fur. The researchers have also dug up the much more modest hovels in which the poor were sheltered.[740]

The brutal fact is that the more we opt out of competition, the lower our position is likely to be. That holds true in our lives as individuals, and it holds even more true in our life as a nation. As the popular expression puts it, "If you snooze, you lose."

Many of the dire consequences supposedly beaten into our lives by stress are the product of pecking order slippage, otherwise known as defeat. Studies show that one of the greatest causes of high blood pressure in humans, for example, is low position in a social order. Raise a human's status, and you reduce his hypertension.[741] Primatologist Robert Sapolsky of Stanford University studied the level of glucocorticoids—stress hormones—among baboons in Kenya's Massai Game Preserve. He discovered that the hormone level was low in males who had a high social position, but stress hormones were alarmingly high in males who were low in the pecking order. The bottom-ranking males stooped when they walked, had bedraggled fur, showed signs of emotional misery, and were in abysmal health. These baboon lackeys were suffering from a pecking order slump like the one that is overtaking America.[742]

Stress is not a product of the desire to achieve the extraordinary. From 1979 to 1982, researchers at the University of Chicago attempted to discover the differences between high-pressure executives who become sick and those who do not. The low-illness executives turned out to be strong in three areas: commitment, control, and *challenge*.[743] In other words, humans need to vigorously pursue goals, to wrestle with problems, and to master them. They need much of what has been popularly interpreted as stress. No wonder Dr. Hans Selye, the pioneering scientist who almost single-handedly put stress on the map, says, "Stress is not something to be avoided. . . . Complete freedom from stress is death."[744]

Excessive relaxation is a slow form of suicide. Take the most primitive, physical level. If you fail to use your organs, your body begins to dispose of them. The phenomenon shows up clearly among women who don't exercise. Internal mechanisms slow the deposit of fresh calcium in the skeletal structure. The result is that women who haven't made demands on their skeletal frame begin to lose it. In their sixties, these women actually begin to shrink.[745] As we saw earlier, unused muscles also atrophy and shrivel away.

The consequences of inactivity are worst for infants. When babies—be they chimps, mice, or humans—do not receive sufficient sensory stimulation, their neural circuitry fails to develop. Their brains should be thickly meshed with nerve cells—microscopically resembling a densely tangled underbrush. Instead, some cerebral areas in the creatures without sensory exercise look more like a desert punctuated only occasionally by the straggle of a plant.[746]

But infants and adults can actually *increase* the density of connections in their cerebral tissue, adding as many as two thousand new synapses per neuron, by mastering new experiences, and seeing and doing new things.[747] To both body and brain, taking it easy is death; vigorous activity, on the other hand, is life itself.

The author of Ecclesiastes showed just how painful the deprivation of anything stimulating can be. The man who penned these biblical passages was apparently one of the most wealthy and powerful in Jerusalem. He had reached the pinnacle of society and had achieved what should have been a delicious leisure,[748] but suddenly, all life bored him. Now that there were no new obstacles to overcome, there was nothing new, he sighed, under the sun.

The ennui that struck the writer of Ecclesiastes can erode and defeat entire civilizations. It is doing that to ours. In 1921, British author G. K. Chesterton traveled the United States by train. He noted that Americans were obsessed with discussing their work while Englishmen talked only about their leisure.[749] That may be one reason America was thriving while England was on the wane. Today, thanks to the popular misunderstanding of stress, it is we who chatter for hours about sports, fishing, or meditation. It is we who are slipping.

The Japanese know what we have forgotten: that work and challenge are the keys to a vigorous life.[750] They have kept alive the essence of two American buzzwords that disappeared from our vocabulary in the early sixties: American ingenuity and American workmanship. The Japanese out-study and out-work us. Midlevel Japanese executives start the business day at nine A.M. and are frequently still at their desks by eight at night, usually putting in six days a week. Many of them even volunteer to work straight through their annual vacations.[751] Contrary to a spate of news stories about "working to death" that appeared in the newspapers of Tokyo and the U.S. in 1993, the "salary man's" dedication seldom maims him with "stress." Far from it. The Japanese spend a staggering 66 percent less on medical care per unit of population than we do, yet they consistently outlive us![752]

Like the Japanese, we have to restore our sense that stimulation can be exhilarating. We have to realize that challenge is not our enemy but our salvation and that the dangers we have interpreted as stress come from something far different than what we've imagined. They do not spring from ambition or the drive of the dedicated. They come from isolation, separation from the social beast, removal from the superorganismic unit that gives our life its meaning. Our pains do not proceed from overactivity but from the loss of control and the feeling that we are allowing ourselves to be shuffled from the pecking order's peak. The solution to our problem is not a good vacation. Our hope *and* our pleasure lie in rolling up our sleeves and going to work.

TENNIS TIME AND THE
MENTAL CLOCK

⇒⋐

*I long for an experiment that would examine, by means of electrodes
attached to a human head, exactly how much of one's life a person
devotes to the present, how much to memories, and how much to the
future. This would let us know who a man really is in relation to his
time. What human time really is. And we could surely define three
basic types of human being, depending on which variety of time was
dominant.*

Milan Kundera

*A turning of our states of consciousness toward the future . . . [makes]
our ideas and sensations succeed one another with greater rapidity; our
movements no longer cost us the same effort.*

Henri Bergson

A man's reach should exceed his grasp, or what's a heaven for?

Robert Browning

Many chapters ago, we visited a peculiar creature: the spadefoot toad of
Arizona. During long dry spells, the toad nursed the stores of moisture
and food packed away in its cellular structure by crawling under the sand,
shutting down its metabolic systems, and slipping into slumber. Lethargy
was the lifesaver that allowed the snoozing beast to go for months—perhaps even years—without a sip of water.

On the other hand, when an infrequent shower soaked the desert floor, the toad shook off its torpor, wriggled to the surface, cried out for company, listened for the croaked sounds of a gathering crowd, then headed for the nearest puddle. There, it leaped into a frenzy of action, wooing females at a rapid rate, then grappled with them in sexual ecstasy. Its manic eroticism was as much a survival mechanism as its former inertia. For only by coupling quickly could the toad sire a new generation that might grow to maturity before the pools of moisture could be sucked away by the desert sun.

The habits of the toad show up in one form or another over and over again. They appear in the hibernation of squirrels and bears, the seasonally fluctuating fat deposits of woodchucks, and in a wide variety of human annual rhythms.[753] They even show up in the mood swings of societies. When there's little to be gained, Nature slows an organism down. When opportunity arrives, she speeds it up. One consequence is the strange gyration of your mental clock.

See if this sounds at all familiar. On a work day when I'm under extreme pressure, I rush through one task, hurry on to the next, then move quickly from that job to the one after it. I see my allotted hours as time in which I can easily accomplish a lot. But on a day when I have less work than usual, my mental clock readjusts. Suddenly, instead of seeing the day as a period in which I can easily perform a plethora of tasks, I get the sense that it'll be a struggle to finish anything at all. My mind has grown sluggish. It's the phenomenon behind the commonsense expression, ''Work expands to fit the time available to it.''

In a person with little to do, the mental clock slows down. In a person with a great deal to accomplish—or a person excited about what he's doing—it speeds up. Take, for example, the athlete who sees every eighteenth of a second[754] of a tennis ball's motion and calculates in a wink exactly where the ball is going to be when he attempts to swat it. For him, every microinstant is filled with meaning. But for the person lying on a beach catching some rays, a whole morning can go by without a single meaningful moment.[755]

For the athlete under high stimulation, there is *more* time. His world is richer, and far more data is processed by his brain.

One difference between a society on the rise and a society in decline may be that the rising society is on the fast clock. It sees each impediment as a challenge, absorbs information quickly, and finds new ways to overcome its obstacles. It operates on tennis time. But the society that has peaked has moved to the slow clock. It has ceased to absorb data rapidly. It is on beach time. Tennis time is the clock of the newly emerged toad, spending energy in a frenzied burst. Beach time is the clock of the dormant toad, hoarding every gram of substance on his bones.

Superorganisms on the trail of growth gravitate toward chemicals that speed the system up to tennis time. The British, when their empire was enthusiastically seizing new possibilities, were fueled by a new import called coffee. The English commercial conquest of the world was planned in the coffeehouses of London in the late 1600s and early 1700s.[756] The Chinese, during the roaringly successful years of the T'ang Dynasty (A.D. 618–907), filled their lives with another beverage that set their mental clock to "fast." They expanded their empire under the influence of tea.[757] The modern Japanese have shown the same predilection for chemicals that turbo-charge the system. The leading drug problem in Tokyo's nightlife neighborhood during the late eighties was not heroin or marijuana—the drugs that slow you down. It was amphetamine.[758]

With perceptual shutdown, we put on our eye protectors and crawl into the stupor of beach time. But exploding societies like those in Japan and Korea today may well be racing on tennis time, a clock that allows them to outrun us as we sit in front of our television sets, cradling a can of beer in our hands, cozy in the low-stimulus, low-challenge life.

How can America put itself back on tennis time? By focusing on the trigger that moves the toad from torpor into overdrive—opportunity. Close to one hundred years ago, the historian Frederick Jackson Turner proposed his highly regarded frontier thesis.[759] The existence of the American frontier, he said, had invigorated the American mind. The possibility of unending resources just over the horizon had filled Americans with zest, imagination, and exuberance.

America was not the only nation thrown into high gear by the presence of a new frontier. England was a puny and somewhat pathetic power up to the time of Henry VIII. Nonetheless, the country had its dreams of glory, and those dreams were associated with the notion of expansion. The single form of expansion the English could imagine, however, was conquering some of the only world they knew—Europe. The British vainly bumped their heads against a brick wall, attempting to saw off pieces of France in the futile bloodbath known as the Hundred Years' War. They had their share of victories, suffered humiliations at the hands of folks like Joan of Arc, and were utterly thwarted in their efforts, eventually losing even the one scrap of territory they'd managed to cut out for themselves—Calais. Meanwhile, they mauled five generations of French peasants innocently trying to plant the next year's crops.[760]

Historian A. L. Rowse, a leading British expert on the Elizabethan age, considers Henry VIII's final failure to conquer the French one of the luckiest embarrassments England ever endured.[761] It forced the English to turn their attentions away from the Continent and made them focus on a sphere in which England would eventually make a fortune—the New World.

The Old World England reluctantly turned its back on was a land of little opportunity. Hungry Italians were reduced to eating songbirds off the trees. (Tourists in southern Europe two hundred years later would be puzzled by the eery silence of the countryside. Melodious warblers like robins, titmice, and wrens had disappeared into the cooking pots of the area's humbler citizens.)[762] Meanwhile, the average French peasant was living so close to starvation that in the fairy tales he recited to his children, the hero was rewarded, not with a pot of gold, but with a decent meal.[763]

At first, the New World looked equally unrewarding. Christopher Columbus was bitterly disappointed by this hulking mass of landscape. He had set off to find the riches of China and ended up in a territory no one ever heard of. The poor sailor insisted for years that this had to be some previously unreported part of the Chinese empire.[764]

But Columbus's disappointment became England's New Frontier. The British missed the easy pickings. The Spanish beat them to the Aztec and Inca lands, where a few hundred Europeans armed with steel swords, muskets, and cannon could overwhelm ten thousand Indians wielding wooden paddles and make off with a king's ransom in gold.[765] But Englishmen settled the North American continent, planted cotton in its southern regions, and set up sugar plantations in the Caribbean islands, bringing back a bounty that boggled the mind.

The new economic horizons utterly transformed England. In the 1400s, the country had been a barbaric backwater, producing almost nothing in the way of literature, art, or science. But charged with the energy unleashed by a whiff of fresh resources, Britain became a cultural dynamo. The era in which she awoke was the age of Elizabeth, and it gave us, among many other things, the plays of Shakespeare.[766]

America has a host of new challenges available to it—biotechnology, nanotechnology (the construction of microscopic machinery from units a few hundred molecules in size), expansion of the cybersphere, and the construction of intelligent self-replicating devices[767] among them. But in the long run, another more daunting frontier awaits us, one that looks as bleak at first glance as our land once seemed to Columbus. Like Britain under Henry VIII, the country that sucks its energy from that next frontier will not necessarily be the one that manages to conquer some small swatch of the world we know. Instead, it may be the one that first finds a way to mine mineral-rich asteroids the size of Manhattan. It may be the country that first "terraforms" another planet, turning its atmosphere into breathable gas and its surface into a place where humans unencumbered by pressurized suits can take a pleasant stroll. It may be the country that relinquishes dreams of conquest in lands where the birds no longer sing and that turns its eyes toward space.

Space could provide a new rain of resources, or it could bankrupt us. But its habitation does offer two other advantages.

The first: international cooperation. No single nation can afford the price of extraterrestrial development. To turn the wastelands of asteroids and planets into lands of plenty would involve consortia including Russia, Europe, and Japan. Those partnerships are already under development, though too often we are not involved in them.

The second, and perhaps more important advantage of following in the footsteps of Captain Kirk: man has as yet invented no way to prevent war. We have found no method for shaking the consequences of our biological curse, our animal brain's addiction to violence. We cannot free ourselves from our nature as cells in a superorganismic beast constantly driven to pecking order tournaments with its neighbors. We have found no technique for evading the fact that those competitions are all too often deadly.

Carl Sagan, Werner Erhard, and the followers of Buckminster Fuller feel that the mere threat of nuclear annihilation will weld us together as one world society. If only the great communicators, they say, can shrill at us loudly enough about the threat of holocaust, all nations will see themselves as brothers, realizing their common stake in the survival of the species. Unfortunately, Sagan, Erhard, and Fuller—much like you and me—have been known to quibble harshly with others who share their goals but differ in beliefs. Even the peacemakers cannot entirely restrain the urge for battle.

Nor can human beings as a species stop their inexorable itch for war. We're like a teenager in the days before the sexual revolution who has been told that masturbation will drive him insane. His guilt makes him feel nearly suicidal, but he still can't stop himself from the unspeakable act. We've found ways to halt illnesses, we've invented means to leapfrog continents in hours, and someday we will find a way to stop war—but only if we survive long enough. Until then our task is to outlast our own impulses. Our task is to outwit the Lucifer Principle.

You could think of us as a species trapped in a car hurtling out of control toward a tree, the steering locked, the brakes frozen. We could sit behind the wheel and pretend that if we felt enough guilt the tree would disappear, or we could throw ourselves out of the auto's door and live. For us, the equivalent of hurling ourselves to safety is moving a few humans off this planet, putting enough of our kind into colonies in space so that if the rest of us down here on earth disappear, those left in the rotating habitations above could keep the species going.[768] Hopefully, the survivors could carry on the knowledge we've acquired so far and with that wisdom and their own fresh discoveries, someday learn to overcome what we could not.

We need a new horizon, a new sense of purpose, a new set of goals, a new frontier to move once again with might and majesty, with a sense of zest that makes life worth living, through the world in which we live. One of the few frontiers left to us hangs above our heads.

THE LUCIFERIAN
PARADOX

THE LUCIFER PRINCIPLE

Over 200 billion red blood cells a day die in the interests of keeping you alive.[769] Do you anguish over their demise? Like those red corpuscles, you and I are cells in a social superorganism whose maintenance and growth sometimes requires our pain or elimination, suppresses our individuality, and restricts our freedom. Why, then, is it of any value to us?

Because the superorganism nourishes every cell within it, allowing a robustness none of its individual components could achieve on its own. Take, for example, the Mediterranean superbeast known as the Roman Empire. Rome was an evil creature with a despicable lust for cruelty. Julius Caesar, according to Plutarch, "took by storm more than 800 cities, subdued 300 nations and fought pitched battles at various times with three million men, of whom he destroyed one million in the actual fighting and took another million prisoners."[770] Caesar did not carry out these deeds with kindliness. When he leveled enemy cities, he occasionally killed off every man, woman, and child[771] just to teach would-be resisters a lesson.

The governors sent to rule the Roman provinces periodically lost their tolerance for nonconformists and punished them brutally. They crucified a backcountry preacher of peace and humility named Jesus, because his views differed from the standard-issue dogmas approved by imperial authority. But the former carpenter was only one of thousands who twisted for hours, hanging by nails from a crude wooden beam.

Even the affluent folks back in the home city of Rome were hungry for the sight of blood. Their favorite recreation was an afternoon at the

Coliseum watching desperate captives disembowel each other in the arena. Roman sports fans took bets on which contestant would manage to live until nightfall.

Rome stamped out or swallowed entire rival civilizations. She even reduced the land she most revered—Greece—to a sleepy, sycophantic occupied territory. Rome, in short, was a vicious society, one whose habits could make anyone with the slightest scrap of moral sensitivity physically ill.

Yet Rome's rise was part of the world's inexorable march to higher levels of form. By force—sometimes sadistic force—she brought an unprecedented mass of squabbling city-states and tribes together. In the process, she allowed an exchange of ideas and goods that radically quickened the pace of progress.

What's more, during the three hundred years between Augustus and the imposition of Christianity under Constantine, she made an additional contribution. She introduced pluralism, an easygoing attitude that allowed wildly diverse cultures to live peacefully side by side.[772]

Just how much the empire had contributed to her sometimes-oppressed citizens could be seen when Rome fell. A set of heroes impelled by ideals of ethnic conquest led their rebel bands against the colonialist power. The mavericks toppled the hegemonic tyrants forever and turned the city of Rome into a ruin.

In the process, they brought despair to Europe. During the next two hundred years, half of the Continent's population would die.[773] Plague ran rampant. Multitudes starved to death, denied the food that had once been transported on Roman ships and roads. Without a stable organizing force, the paved highways on which provisions had traveled sank into disrepair. On land, bandits and warrior chiefs ended the lives of any who might contemplate a trip along the old paths to carry desperately needed supplies. At sea, pirates destroyed the former Mediterranean lanes of trade. The grain that had once sailed from Egypt in fleets of bulging transport hulls no longer came across with the tides. In the Gallic town of Barbagel, the complex of Roman-run mills that had turned the imported wheat into flour for eighty thousand consumers fell into disrepair.[774] And the Gallic citizens who had been freed of the Roman yoke perished by the millions.

Those who survived learned to live as prisoners in self-contained fortress communities, cut off from the ideas and the delicacies that had once made life sweet. The barbarian "freedom fighters" had loosed the chains, not of life, but of death. For Rome had been an oppressor, but Rome had also been the source of nourishment and peace. In her absence came pestilence and war.

The superorganism is often a vile and loathsome beast. But like the body nourishing her constituent cells, the social beast grants us life. Without her, each of us would perish. That knowledge is woven into our biology. It is the reason that the rigidly individualistic Clint Eastwood does not exist. The internal self-destruct devices with which we come equipped at birth ensure that we will live as components of a larger organism, or we simply will not live at all.

Behind these superorganismic imperatives is nature's latest wrinkle in the research and development racket. Despite the claims of individual selectionists, human evolution is propelled not only by competition between single souls but by the forms of their cooperation. It is driven by the games that superorganisms play.

All this lies behind the mystery with which we began—the pattern of violence in Mao's Cultural Revolution. When China lapsed into chaos during the cultural upheaval of the sixties, society did not fragment into 700 million individuals, each fighting for his right to survive. The social fabric ripped, then reknit in a strange new way. Individuals clustered in collaborative clumps. Stitching each gang together was a force with no physical substance—the idea, the meme. In their battles, the Red Guard wolfpacks obeyed a basic commandment of the animal brain—the law of the pecking order. And they drew their energy from emotions that remain repressed in everyday life—the hatreds, frustrations, and hidden cruelty of students who just a month or two before had seemed models of polite obedience.

Behind the writhing of evil is a competition between organizational devices, each trying to harness the universe to its own peculiar pattern, each attempting to hoist the cosmos one step higher on a ladder of increasing complexity. First, there is the molecular replicator, the gene; then, there is its successor, the meme; and working hand in hand with each is the social beast.

Hegel said the ultimate tragedy is not the struggle of an easily recognized good against a clearly loathsome evil. Tragedy, he said, is the battle between two forces, *both* of which are good, a battle in which only one can win. Nature has woven that struggle into the superorganism.

Superorganism, ideas, and the pecking order—these are the primary forces behind much of human creativity and earthly good. They are the holy trinity of the Lucifer Principle.

EPILOGUE

~❦~

Seventeenth-century thinkers described the universe as a decaying creation, a remnant of a paradise corrupted by sin. The world, they said, had been crafted as a perfect and unblemished sphere. Then Adam and Eve stole fruit from the tree of knowledge. A roaring, wrathful God threw the primal pair out of Eden and visited his anger upon the newly conceived planet. He smashed the surface of this perfect place, leaving the wound-like gashes of valleys and the upturned pustules of mountains as eternal signs of his displeasure. Ever since that time, said the philosophers, the earth has been decaying like an ancient ruin, showing only the faintest signs of a beauty that is no more.

Nineteenth-century physicists expressed the old view in a new way. They created the concept of entropy.[775] All matter, said the second law of thermodynamics, tends toward chaos.[776] Leave the most carefully contrived bit of complex form alone, and it will slowly be devoured by decay. The universe, said these scientists, is like a sugar cube. Drop the highly structured block of sweetness into a glass of water, and it will dissolve in a random swarm of glucose molecules, a liquid swirl of chaos. So, too, said the physicists, will the world we know someday disappear, eaten away by entropy.[777]

The Renaissance thinkers and the nineteenth-century scientists were both wrong. The universe has not been drifting from order into chaos. It has, instead, been marching in the opposite direction. Since its first second of being, the cosmos has coughed up fresh forms of creation. From an explosion of energy, it spun one of its first great leaps forward—the atom. Then came another extraordinary innovation—the molecule. Billions of years later, the universe spat forth another brilliant twist—a molecule that could produce copies of itself, the molecule responsible for life. And over three billion years beyond that, the universe conceived an even more revolutionary upward step: intelligence.

But there is a dark side to this movement toward the light. Like a sculptor carving a figure from stone, nature creates by destroying. Her hammer and chisel strike over and over again. Each time, there is a shower of chips. As the splinters pile up on the floor, a new form emerges where the blade has been at work. The sculptor at the end of his day simply sweeps away the heaps of useless dust and shards of stone. So does Nature. But those discarded scraps from the natural workshop are the bodies of creatures who moments before were alive, creatures like you and me.

Nature creates by placing her inventions in competition with each other. In the world of humans, the bloodiest of these contests is between social groups. The voice of our meme's ambition tells us to pound our rivals into submission and force them into servitude—servitude to the cluster of ideas that sits at our culture's core.

The hunger of the superorganism and the ambition of the meme trap us in a moral dilemma. Violence is the most appalling of human expressions. Yet we cannot wish our way to peace. We cannot lash each other with lectures, pound our chests with guilt, and voluntarily throw away our arms. We live in a threatening world—a world of other human beings very much like ourselves. And like us, our fellow humans are dangerous.

There is one small consolation in this grim picture. Snapping and snarling at each other may be automatic, but holding, caring, and collaborating are built into us too. In one Harvard study, a group of experimental subjects was shown a film. Watching the footage actually boosted immune system activity. The nature of the healing cinematic piece? A documentary on Mother Theresa, who has centered her life on helping others. The mere sight of a work focused on kindness triggered a deeply buried response in the human brain.[778] We desperately need each other, and in that need is hope.

We must invent a way in which memes and their superorganismic carriers—nations and subcultures—can compete without carnage. We may find a clue to that path in science. A scientific system is one in which small groups of men and women cohere around an idea, then use the powers of persuasion and politics to establish that idea's dominance in their field, and to drive rival hypotheses—along with those who propound them—to the periphery. In the struggle for control over scientific journals, over the committees that determine what lecturers will be allowed to speak at scientific symposia, over who will get tenure, grants, prizes, and over all the various other power points that determine which ideas and researchers will be admired and which will be shunned, battles can become intense, insults snide and biting. Occasionally, a pitcher of water is even dumped on someone's head.[779] But there is no violence.

In geopolitics, the closest equivalent is pluralistic democracy, a system within which subcultures and the ideas that ride them compete without bloodshed. During the early nineties, it became popular to declare that democracies, unlike more ideologically rigid forms of government, do not make war against each other. The statement was an exaggeration (the War of 1812 pitted democratic England against the equally democratic United States), but it carried a powerful element of validity. According to Yale University political scientist Bruce Russett, "Since 1946, pairs of democratic states have been only one eighth as likely as other kinds of states to threaten to use force against each other and only one tenth as likely to do so."[780] In addition, the Carnegie Commission for Science, Technology, and Government commissioned a 1992 task force that concluded that pluralism is one of the most potentially effective forces for promoting global development.[781] These findings imply strongly that the human race will take a long step forward if it eliminates intolerant autocracies driven by a passion for doctrinal or ethnic cleansing and replaces them with pluralistic democracies.

Even without this step, like the universe itself, the world of *Homo sapiens* has not been marching from utopia toward chaos. The truth, in fact, is quite the opposite. Margaret Mead said that once upon a time men viewed all brothers as people whose lives they should cherish and all outsiders as fair game. In early human groups, said Mead, the number of those whose lives were sacred was fifty to seventy-five. The rest of the earth's population was a target for murder. But today, Mead pointed out, the number a single society prohibits killing extends to 250 million or more.

As we saw earlier, William Divale and Marvin Harris scrutinized data from 561 primitive tribes and discovered that within this sample, 21 percent of the males were killed off violently before they ended adolescence. The percentage of the slaughtered skyrockets if you include the women and children wiped out by indigenous peoples like the South American Taulapang, who burned dozens of families in their huts when trying to eradicate an enemy tribe, then marched home shouting a joyous "hei-hei-hei-hei-hei!"[782] Our techno-capitalist civilization comes nowhere near the resulting grim proportion of butchery. If it did, roughly 720 million modern humans would be blasted to smithereens in wars or homicides every generation. Compare this with the fifty-five million who died in the Second World War, and the bloodlettings of the current century, appalling as they've been, are less than one tenth of what they would amount to under aboriginal conditions. That reduction in violence is a blessing of the superorganism's evolution. It is a result of the growth of social agglomerations from tiny clusters of men and women huddled in jungles and plains to massive nations sprawled across entire continents.

Ironically, however, the partial spread of peace is a product of past battles between superbeasts, the colossal atrocities that accompanied the building of the empires of Alexander, Caesar, and the ancient Chinese, the gore that oozed from the consolidation of the modern European, American, and Russian states.

The movement of humans into social groups, the tendency of one social organism to swallow another, the rise of the meme, the increase in cooperation—all are ways in which the universe has ratcheted upward in degrees of order. But under the natural urge toward more intricate structures, higher planes of wonder, and startlingly new and effective forms of complexity, there is no moral sense. There is no motherly Nature who loves her offspring and protects them from harm. Harm, in fact, is a fundamental tool Nature has used to refine her creations.

No, we are not Clint Eastwoods, nor were we meant to be. We are incidental microbits of a far-larger beast, cells in the superorganism. Like the cells of skin that peel in clustered communities from our arms after a sunburn, we make our contributions to the whole of which we are a part. Sometimes we make that contribution with our life, sometimes we make it with our death.

Superorganisms, ideas, and the pecking order—the triad of human evil—are not recent inventions "programmed" into us by Western society, consumerism, capitalism, television violence, blood-and-guts films, or rock and roll. They are built into our physiology. They have been with us since the dawn of the human race.

But there is hope that we may someday free ourselves of savagery. To our species, evolution has given something new—the imagination. With that gift, we have dreamed of peace. Our task—perhaps the only one that will save us—is to turn what we have dreamed into reality. To fashion a world where violence ceases to be. If we can accomplish this goal, we may yet escape our fate as highly precocious offspring, as fitting inheritors of nature's highest gift and foulest curse, as the ultimate children of the Lucifer Principle.

ACKNOWLEDGMENTS

≫⋘

Many individuals have lent aid and encouragement during *The Lucifer Principle*'s twelve years of research. I owe debts to more than I can mention.

To Dr. Rollin Hotchkiss, professor emeritus at Rockefeller University; Dr. Magda Gabor Hotchkiss; Timothy Ferris; Don Cusic of Middle Tennessee State University; Hester Mundes; Ron Vanwarmer and Randy Vanwarmer; Charles M. Young, Jr.; Bradley and Lib Fisk; and Robin Fox, University Professor of Social Science at Rutgers University, for patiently reviewing the manuscript and making many valuable comments. To Gao Yuan, historian Michael Wood at the BBC, and Cal Tech's John Hopfield for evaluating the chapters pertinent to their work. To the staff of the Ingersoll Library for the thousands of times they helped unearth bits of esoterica in the seven stories of their stacks.

To those who listened patiently as the concepts came spilling out: Patrice Adcroft, former editor of *Omni* (who also provided invaluable Care packages of research materials); Bob Guccione, Jr., publisher and editor-in-chief of *Spin Magazine;* Bob Kubey of Rutgers University; the late Dr. Milton Plesur of the History Department at the University of Buffalo; Timothy White; Stephen Holden of *The New York Times;* Ken Emerson, formerly of *The New York Times Magazine;* Mike Sigman, C.E.O. of *L.A. Weekly;* Jim Henke of *Rolling Stone;* Grace Diekhaus of *60 Minutes;* Liza Wing of *New York Woman;* Alan Weitz of *Details;* Rona Eliot and Carla Morgenstern of *The Today Show;* John Mellencamp (who not only listened, but who brought my attention to the quotation from Ecclesiastes that heads one of the chapters); Paul Simon; Billy Joel; Daryl Hall; and Bette

Midler. To Peter Gabriel, whose suggestion of a piece of reading material eventually triggered this book's title. To Ralph Gardener, who inveigled sociobiology pioneer Lionel Tiger into a three-hour dinner during this book's inception. To my daughter, Noelle Pollet, who transcribed the first two thousand pages of notes, and to the numerous others who transcribed the remaining thousand. To Jim Stein at the William Morris Agency for nursing the book along. And to Leon Uris for his extraordinary moral support in the difficult days between the emergence of the early manuscript and publication.

I am more deeply indebted than I can possibly say for financial and emotional support to Kenny Laguna, Meryl Laguna, and Joan Jett.

I also owe thanks for encouragement and aid of wildly varying kinds to Harrison Salisbury, Walker Percy, James Lovelock, Marilyn Ferguson, Prakash Mishra (founder, the Mountbatten Medical Trust), Judith Gordon, Martin Gardner, Jerome D. Frank (professor emeritus of psychiatry, The Johns Hopkins University School of Medicine), David L. Hull (Department of Philosophy, Northwestern University), Thomas D. Seeley (Department of Neurobiology and Behavior, Cornell University), William L. Rivers (Paul C. Edwards Professor of Communications, Stanford University), Robert B. Cialdini (Regents' Professor of Psychology, Arizona State University), David Barash (professor of psychology and zoology, University of Washington), Dorion Sagan, Herman Golob, Brett Busang, Marion Hyman, Audrey Dawson, Mildred Marmur, Derek Sutton, Mike Gormley, Danny Sugarman, Marie Diane Partie, Kathy Hemingway Jones, Laura Nixon, Jeremy Walker, Geoff Jukes, Deanne Stillman, Martha Hume, Renee Kuker, Deborah Kuker, Len Kuker, Howard Kuker, Professor William T. Greenough, Annette Sbarro, Paul Kaufman, Bob Keating, Ida Langsam, Otto Teitler, Harriette Vidal, Professor Daniel G. Freedman, Professor Mary Douglas, Aaron Tovish, Stephanie Hemmert, Don Gifford, Bruce Bower, John D. Collier, Dr. John E. Sarno, Michael Mendizza, David Krebs, Richard Block, Paul Bresnick, and Lynn Chu.

Finally, a special thanks for her persistence to my agent, Adele Leone, and to my editor, Anton Mueller, a master at exhilarating the brain cells while he digs out the hidden structures of jumbled prose.

NOTES

Who Is Lucifer?

1. Because the early Church was implacable in its wrath against Marcion, not much remains of his work. The best primary sources on his teachings are attacks against him by early Church fathers like Tertullian (e.g., his third-century work *Adversus Marcionem*). Most of the scholarship on Marcion during the last hundred years has appeared in German (e.g., Adolf von Harnack's *Geschichte der altchristlichen Litteratur bis Eusebius*). However, good—though brief—biographies of Marcion appear in W. H. C. Frend, *The Rise of Christianity* (Philadelphia: Fortress Press, 1984), 212–18; James Hastings, ed., *Encyclopaedia of Religion and Ethics* (New York: Charles Scribner's, 1908–27), 8:407–9; Robert R. Wilken, "Marcion," in *The Encyclopedia of Religion,* ed. Mircea Eliade (New York: Macmillan, 1987), 9:194–96; and *The New Encyclopaedia Britannica* (Chicago: Encyclopaedia Britannica, 1986), 7:825–26. For additional details on Marcion, see Roland H. Bainton, *Christianity,* The American Heritage Library (Boston: Houghton Mifflin Co., 1987), 67–68; Robin Lane Fox, *Pagans and Christians,* (San Francisco: Harper & Row, 1986), 332; and Elaine Pagels, *The Gnostic Gospels* (New York: Vintage Books, 1981), 33, 44.

2. The prophet Isaiah used the term "Lucifer" merely to refer poetically to the king of Babylon, not exactly his favorite monarch. Later, Christians like John Milton would take Isaiah's figure of speech and weave an elaborate tale around it, crafting a devil of impressive proportions.

The Clint Eastwood Conundrum

3. Daniel Goleman, *Vital Lies, Simple Truths* (New York: Simon and Schuster, 1985), 161.

4. Bryan Mullen, "Atrocity as a Function of Lynch Mob Composition," *Personality and Social Psychology Bulletin* (June 1986): 187–97.

5. Sigmund Freud, *Group Psychology and the Analysis of the Ego* (New York: Bantam Books, 1965), 13–16.

The Whole Is Bigger Than the Sum of Its Parts

6. The term "entelechy" was introduced into modern scientific discourse over fifty years ago by the German experimental embryologist and philosopher Hans Adolf Eduard Driesch. The version of the concept I've chosen to use is that proposed by Douglas Hofstadter (Douglas R. Hofstadter and Daniel C. Dennet, *The Mind's I: Fantasies and Reflections on Self and Soul,* New York: Bantam Books, 1981, 144–46). For different interpretations of entelechy, see Steven Levy, *Artificial Life,* New York: Vintage Books, 1992, 21; Robert Wright, *Three Scientists and Their Gods: Looking for Meaning in an Age of Information,* New York: Times Books, 1988, 124; and Paul Davies, *The Cosmic Blueprint: New Discoveries in Nature's Creative Ability to Order the Universe,* New York: Simon and Schuster, 1988, 97. Both Hofstadter and I have chosen to use "entelechy" instead of the currently popular alternative phrase "emergent property."

7. Rene A. Spitz, "Hospitalism: An Inquiry into the Genesis of Psychiatric Conditions in Early Childhood," in *The Psychoanalytic Study of the Child* (New York: International Universities Press, 1945), 1:53–74; Rene A. Spitz, M.D., with Katherine M. Wolf, "Anaclitic Depression: An Inquiry into the Genesis of Psychiatric Conditions in Early Childhood," in *The Psychoanalytic Study of the Child* 2:331; Marilyn T. Erickson, *Child Psycho-pathology: Behavior Disorders and Developmental Disabilities* (Englewood Cliffs, N.J.: Prentice-Hall, 1982), 87; Leo Kanner, M.D., *Child Psychiatry,* 4th ed. (Springfield, Ill.: Charles C. Thomas Publisher, 1972), 684–85; Raymond J. Corsini, ed., *Encyclopedia of Psychology,* (New York: John Wiley & Sons, 1984), 1:161; Harry F. Harlow and Margaret Kuenne Harlow, "Social Deprivation in Monkeys," *Scientific American,* November 1962, 136–46; Harry F. Harlow and Gary Griffin, "Induced Mental and Social Deficits in Rhesus Monkeys," in *Biological Basis of Mental Retardation,* ed. Sonia F. Osler and Robert E. Cooke (Baltimore: Johns Hopkins Press, 1965), 87–106; Stephen J. Suomi and Harry F. Harlow, "Production and Alleviation of Depressive Behaviors in Monkeys," in *Psychopathology: Experimental Models,* ed. Jack D. Maser and Martin E. P. Seligman (San Francisco: W. H. Freeman and Co., 1977), 131–73; and Harry F. Harlow, *Learning to Love* (New York: Jason Aronson, 1974), 95.

8. When the crusaders took Jerusalem in 1099, the anonymous author of the *Gesta Francorum* reported, "There was such a massacre that our men were wading up to their ankles in enemy blood" (Rosalind Hill, ed., *Gesta Francorum et Aliorum Hierosolimitanorum—Deeds of the Franks and other Pilgrims to Jerusalem* [London: Thomas Nelson and Sons, 1962], 91; and Stephen Howarth, *The Knights Templar* [New York: Atheneum, 1982], 40). Archbishop William of Tyre described a "spectacle of headless bodies and mutilated limbs strewn in all directions that roused horror in all who looked upon them. Still more dreadful was it to gaze upon the victors themselves, dripping with blood from head to

Notes

foot'' (Aziz S. Atiya, *Crusade, Commerce and Culture* [Bloomington, Ind.: Indiana University Press, 1962], 62).

The Chinese Cultural Revolution

9. Gao Yuan, *Born Red: A Chronicle of the Cultural Revolution* (Stanford, Calif.: Stanford University Press, 1987), 7.

10. O. Edmund Clubb, *20th Century China* (New York: Columbia University Press, 1978), 388–89; K. S. Karol, *The Second Chinese Revolution,* trans. Mervyn Jones (New York: Hill and Wang, 1974), 90–94; and Gargi Dutt and V. P. Dutt, *China's Cultural Revolution* (Bombay: Asia Publishing House, 1970), 9–13.

11. Nancy Makepeace Tanner, *On Becoming Human: A Model of the Transition from Ape to Human & the Reconstruction of Early Human Social Life* (New York: Cambridge University Press, 1981), 104–5.

12. Yukimara Sugiyama, "Social Organization of Hanuman Langurs," in *Social Communication among Primates,* ed. Stuart A. Altmann (Chicago: University of Chicago Press, 1967; Chicago: Midway Press, 1982), 230–31; Kenji Yoshiba, "Local and Inter-troop Variability in Ecology and Social Behavior of Common Indian Langurs," in *Primates: Studies in Adaptation and Variability,* ed. Phyllis C. Jay (New York: Holt, Rinehart and Winston, 1968), 236; Edward O. Wilson, *Sociobiology: The Abridged Edition* (Cambridge: Harvard University, Belknap Press, 1980) 10, 38; David P. Barash, *Sociobiology and Behavior* (New York: Elsevier Scientific Publishing Co., 1977), 99; David P. Barash, *The Whisperings Within: Evolution and the Origin of Human Nature* (New York: Penguin Books, 1979), 102–3; and Laurence Steinberg, "Bound to Bicker; Pubescent Primates Leave Home for Good Reasons. Our Teens Stay with Us and Squabble," *Psychology Today,* September 1987, 38. Barash's two books, by the way, constitute an extremely good introduction to the relatively new field of sociobiology. *The Whisperings Within* is a delightfully written overview for the layman, and *Sociobiology and Behavior* is a comprehensive academic overview. For those willing to tackle the difficult, there is the seminal sociobiology book, an intellectual tour de force, E. O. Wilson's *Sociobiology,* which virtually created the field of sociobiology.

13. Gao Yuan, *Born Red: A Chronicle of the Cultural Revolution,* 53.

14. Mao gloated that "all those who have tried to repress the student movement in China have ended up badly" (Karol, *Second Chinese Revolution,* 112).

15. The entire tale told in this chapter is from Gao Yuan's book *Born Red.* Additional historical background is given in the book's introduction by William A. Joseph of Wellesley College and in Wolfram Eberhard, *A History of China* (London: Routledge &

Kegan Paul, 1977), 347–48. For Mao's success in regaining power, see also Karol, *Second Chinese Revolution,* 345–49; and G. Dutt and V. P. Dutt, *China's Cultural Revolution,* 206–34.

Mother Nature, the Bloody Bitch

16. Rousseau's *Letter to d'Alembert,* his *Discourse on the Origin of Inequality,* his *Social Contract,* and, most important, his *Emile.*

17. Karen Lehrman, literary editor of *Wilson Quarterly,* toured college campuses to survey women's studies programs and concluded that "Most women's studies professors seem to adhere to the following principles in formulating classes: women were and are oppressed; oppression is endemic to our patriarchal social system; men, capitalism and Western values are responsible for women's problems." "Which Way Feminism?" *Wilson Quarterly,* Winter 1994, 135.

18. *Proposal for Endorsement of the Seville Statement on Violence,* Washington, D.C.: American Sociological Society, 1991.

19. Robert G. Wesson, *Beyond Natural Selection* (Cambridge: MIT Press, 1993), 121.

20. John Tyler Bonner, *The Evolution of Culture in Animals* (Princeton, N.J.: Princeton University Press, 1980), 98–99.

21. Mark J. Davis (writer, director, and editor), *The Private Lives of Dolphins,* Nova, (Boston: WGBH, 1992).

22. Wilson, *Sociobiology,* 29.

23. Others had ventured onto the land before the reptiles, but none of them could quite bring themselves to leave the water totally behind. First came the crossopterygians, fish that could gulp air; then the crossopterygians' descendants, the amphibians. Though amphibians spent a good deal of their time ashore, they apparently felt that land was a nice place to visit, but you wouldn't want to raise your kids there. They still laid their eggs in underwater nurseries, where the youngsters stayed until they were old enough to brave the hard, cold facts of life outside the pond. (For a fascinating account of this process, see Lynn Margulis and Dorion Sagan, *Microcosmos: Four Billion Years of Microbial Evolution* [New York: Summit Books, 1986], 197–202.)

24. Paul D. MacLean, *A Triune Concept of the Brain and Behavior* (Toronto: University of Toronto Press, 1973). For more on the triune brain, see Richard Restak, *The Brain* (New York: Bantam Books, 1984), 136; Robert Ornstein and David Sobel, *The Healing Brain* (New York: Simon and Schuster, 1987), 37–38; and Carl Sagan, *The Dragons of Eden: Speculations on the Evolution of Human Intelligence* (New York: Ballantine Books, 1977), 53–83.

25. Richard E. Leakey and Richard Lewin, *People of the Lake: Mankind and Its Beginnings* (New York: Avon Books, 1983). Though this entire book promotes the thesis that

"war is a cultural invention," a summation of the argument can be found on pages 233–36. By the way, Edward O. Wilson points out that "murder is far more common and hence 'normal' in many vertebrate species than in man" (*Sociobiology,* 121).

26. Anthropologist Richard Lee analyzed the data on !Kung homicide and "determined that, within a population of fifteen hundred !Kung, there had in fact been twenty-two killings over five decades—about five more than the same number of New Yorkers would have been expected to commit over the same period." Melvin Konner, "False Idylls," *The Sciences,* September/October 1987, 10; see also Melvin Konner, *The Tangled Wing: Biological Constraints on the Human Spirit* (New York: Holt, Rinehart and Winston, 1982), 9, 109, 204; and Allen W. Johnson and Timothy Earle, *The Evolution of Human Societies: From Foraging Group to Agrarian State* (Stanford, Calif.: Stanford University Press, 1987), 47.

27. Virginia Morell, "Dian Fossey: Field Science and Death in Africa," *Science 86,* April 1986, 17.

28. Dian Fossey, *Gorillas in the Mist* (Boston: Houghton Mifflin Co., 1983), 69. For similar warfare between bands of rhesus macaques, see K. R. L. Hall, "Aggression in Monkey and Ape Societies," in Jay, *Primates,* 155.

29. Fossey, *Gorillas in the Mist,* 75.

30. Jane Goodall, *In the Shadow of Man* (1971; Boston: Houghton Mifflin Co., 1983).

31. Jane Goodall, *Among the Wild Chimpanzees,* ed. and prod. Barbara Jampel, National Geographic Society and WQED/Pittsburgh, National Geographic Special (Stanford, Conn.: Vestron Video, 1987); Jane Goodall, "Life and Death at Gombe," *National Geographic Magazine,* May 1979, 592–620; Michael Patrick Ghiglieri, *The Chimpanzees of Kibale Forest: A Field Study of Ecology and Social Structure* (New York: Columbia University Press, 1984), 3; and Michael Ghiglieri, "War among the Chimps," *Discover,* November 1987, 76.

32. Goodall, *Among the Wild Chimpanzees.*

33. Ghiglieri, "War among the Chimps," 68. When Ghiglieri visited Africa, he was convinced that war among the chimps may have been an indirect human creation. To lure the chimps of Gombe into viewing distance, Jane Goodall had laid out clusters of bananas, a food that soon became the backbone of the animals' diet. Much, much later, Goodall decided to stop the handouts of simian welfare and left the primates to gather food for themselves. A few years after this change in policy, the chimps began to make war. Ghiglieri suspected that the provisioning of food by humans had set the stage for a violence that wouldn't have occurred without it. His years studying unprovisioned chimps in Kibale, however, convinced him he was wrong. Chimpanzees, he concluded, were subject to periodic outbreaks of war, with or without a human lending hand (see Michael P. Ghiglieri, *East of the Mountains of the Moon* [New York: Free Press, 1988], 8–9, 258–59).

Women—Not the Peaceful Creatures You Think

34. For a bit of historical background on the concept of superior feminine morality, see Reay Tannahill, *Sex in History* (New York: Stein and Day, 1980), 390–91. Historian Joan Kelly summed up the prevailing notion when she said, "I know, in the depth of my being and in all my knowledge of history and humanity, I know women will struggle for a social order of peace, equality and joy" (quoted in Antonia Fraser, *The Warrior Queens* [New York: Alfred A. Knopf, 1989], 7).

35. Tina Rosenberg, *Children of Cain: Violence and the Violent in South America* (New York: William Morrow, 1991).

36. Fossey, *Gorillas in the Mist,* 77–78.

37. Jane Goodall's team observed a whole series of similar incidents in the Gombe Reserve. One family of chimpanzee females made a regular habit of killing and eating their rivals' infants (Goodall, "Life and Death at Gombe," 616–20).

38. Graves's historical novels are noted for their solid research. In *Benet's Reader's Encyclopedia* they are lauded for their "scholarly" content (*Benet's Reader's Encyclopedia,* 3d ed. [New York: Harper & Row, 1987], 402).

39. Robert Graves, *I, Claudius,* (New York: Vintage Books, 1934), 13–147.

40. Eberhard, *History of China,* 121. The instances of females who become killers so they can place or maintain their own children on top of the hierarchy is endless. The dominant female Cape hunting dog, for example, establishes herself at the pinnacle of the pack and gives birth to a litter of pups. Then, if a lower-ranking female has the audacity to produce offspring of her own, the top-ranking lady turns killer. She leads her packmates in a puppy-killing frenzy—utterly eliminating the offspring of her rival (Daniel G. Freedman, *Human Sociobiology: A Holistic Approach* [New York: Free Press, 1979], 31).

41. Konrad Lorenz, *On Aggression* (New York: Harcourt Brace Jovanovich, 1974), 58–59, 63–64.

42. For a description of how Yanomamo women egg their men on to battle, see A. W. Johnson and Earle, *Evolution of Human Societies,* 127.

43. Torturing fellow Christians and plundering their villages was actually a common practice among the crusaders. See Frederic Duncalf, "The First Crusade: Clermont to Constantinople," in *A History of the Crusades: The First Hundred Years,* ed. Marshall W. Baldwin (Philadelphia: University of Pennsylvania Press, 1955), 263, 265, 269, 271, 282. The worst example was in 1204 when the knightly "saviors of the faith" actually sacked one of the two most important capitals of the Christian world, Constantinople (J. M. Roberts, *Pelican History of the World* [Harmondsworth, Middlesex: Penguin Books, 1983], 349).

44. For a more detailed explanation of the why's behind peacock vanity, see Matt Ridley, "Swallows and Scorpionflies Find Symmetry Beautiful," *Science,* 17 July 1992, 327–28.

45. William R. Polk and William J. Mares, *Passing Brave* (New York: Ballantine Books, 1973), 33–36.

46. Susan Walton, "How to Watch Monkeys," *Science 86,* June 1986, 23–27.

Fighting for the Privilege to Procreate

47. Sugiyama, "Social Organization of Hanuman Langurs," 230–31; and Yoshiba, "Common Indian Langurs," 236. By the way, a good many other creatures engage in this sort of infanticide, including male lions and chimpanzees. See Bonner, *Culture in Animals,* 31; Ghiglieri, *Chimpanzees of Kibale Forest,* 182; Ghiglieri, *Mountains of the Moon,* 255; and Wilson, *Sociobiology,* 42, 72.

48. David P. Barash, *The Hare and the Tortoise: Culture, Biology, and Human Nature* (New York: Penguin Books, 1987), 108.

49. Napoleon Chagnon, *Yanomamo: The Fierce People* (New York: Holt, Rinehart and Winston, 1968), 82–83; and Marvin Harris, *Cows, Pigs, Wars and Witches: The Riddles of Culture* (New York: Vintage Books, 1977), 75–78.

50. This description, by the way, disagrees with the standard account of Yanomamo warfare in Napoleon Chagnon's *Yanomamo: The Fierce People.* Chagnon shows his raiders surprising an enemy village, killing a man or two, and stealing any woman they happen to be lucky enough to find unprotected. My information comes from a different source. Several years ago, anthropological film maker Jean Claude Luyat showed me a motion picture he had made of the Massai warriors of Africa at war. The Massai stood on a dusty plain about the size of a high school football field, facing off in a loosely organized mass. Occasionally, a warrior would hurl a spear or stone. It seldom hit anyone, yet killing was definitely the business of the day. Luyat said that all he could think of as he filmed was the army of the Greeks on the dusty plain outside the walls of Troy. The Homeric heroes, Luyat was certain, must have made war very much like this. Then, his mind still on primitive war, the film maker said, "There is a book you must read. I will send it to you tomorrow." The next day, there arrived a volume, in French, called *Yanoama.* It was the first-person account of a European girl, Helena Valero, living with her parents on the Rio Negro, who had been kidnapped by the Yanomamo when she was still small. The Yanomamo had attacked Valero's parents, riddled her father with arrows, then taken the little girl back into the forests and adopted her. After all, soon she would be valuable as a wife. The writer spent her teens and some of her adult years among the Yanomamo, experiencing far more of their brutal ways than Chagnon in his landmark fieldwork was ever able to see. It is from her account that I take my description of Yanomamo assaults. (Helena Valero's story is available in English as *Yanoama: The Narrative of a White Girl Kidnapped by Amazonian Indians,* as told to Italian anthropologist Et-

tore Biocca, trans. Dennis Rhodes [New York: E. P. Dutton, 1970]. The tale of a Yanomamo raid and the brutal massacre of children appears on pages 34–37. See also A. W. Johnson and Earle, *Evolution of Human Societies,* 124–26.)

51. According to anthropologist Judith Shapiro of the University of Chicago, quoted in M. Harris, *Cows, Pigs, Wars and Witches,* 77.

52. Here's how Eusebius put it: "Writers of history record the victories of war and trophies won from enemies, the skill of generals, and the manly bravery of soldiers, defiled with blood and with innumerable slaughters for the sake of children and country" (quoted in Daniel J. Boorstin, *The Discoverers: A History of Man's Search to Know His World and Himself* [New York: Vintage Books, 1985], 573).

53. Anthropologists describing this ploy among the Yanomamo call it "the treacherous feast." See A. W. Johnson and Earle, *Evolution of Human Societies,* 121.

54. Michael Grant and John Hazel, *Gods and Mortals: Classical Mythology, a Dictionary* (New York: Dorset Press, 1985), 303; Robert J. Gula and Thomas H. Carpenter, *Mythology, Greek and Roman* (Wellesley Hills, Mass.: Independent School Press, 1977), 232; and *New Encyclopaedia Britannica,* 10:281.

55. Homer, *The Iliad,* trans. Richard Lattimore (Chicago: University of Chicago Press, 1961), 494.

56. The habit of raiding a town, killing the men, then making off with the women was so common in classical times that Odysseus and his merry band, after the burning of Troy, pulled off more of these despicable attacks as they wended their way home. The wiley hero of *The Odyssey* brags, "From Ilion the wind drove me along and brought me to Ismaros, in the land of the Ciconians. There I sacked the city and put the men to death. We captured from the city their wives and much treasure and divvied it all among us" (Homer, *Odyssey,* 9:39–42, quoted in M. M. Austin and P. Vidal-Naquet, *Economic and Social History of Ancient Greece: An Introduction* [Berkeley: University of California Press, 1977], 42). The practice of raiding to steal women has been almost universal. Twenty-four hundred years after Odysseus, Mongol heroes found warfare a convenient way of acquiring new wives (James Chambers, *The Devil's Horsemen: The Mongol Invasion of Europe* [New York: Atheneum, 1979], 53). And until the white man arrived, even the Kwakiutl Indians of the Pacific Northwest made war in the hope of enslaving a rival tribe's females (A. W. Johnson and Earle, *Evolution of Human Societies,* 171).

57. For recent arguments between scientists who view genes as the driving force behind war and their theoretical adversaries who regard the cause of conflict as a struggle for territory and resources, see Ann Gibbons, "Evolutionists Take the Long View on Sex and Violence: Warring over Women," *Science,* 20 August 1993, 987–88.

The Greed of Genes

58. Steven Frautschi, "Entropy in an Expanding Universe," in *Entropy, Information and Evolution: New Perspectives on Physical and Biological Evolution,* ed. Bruce H. Weber,

David J. Depew, and James D. Smith (Cambridge, Mass.: MIT Press, Bradford Book, 1988), 11; and George Gamow, *One, Two, Three—Infinity* (New York: Dover Publications, 1988), 298–313.

59. Richard Dawkins, *The Selfish Gene* (New York: Oxford University Press, 1976), 13–22.

60. For a vision of the rise of replicators that differs in interesting ways from Dawkins's, see Jeffrey S. Wicken, "Thermodynamics, Evolution and Emergence: Ingredients for a New Synthesis," in Weber, Depew, and J. D. Smith, *Entropy, Information and Evolution*, 160–63.

61. Don't console yourself with the notion that Neanderthals were too primitive to mark the loss. Neanderthals had an aesthetic sensibility: they buried their dead with flowers and used ocher dyes (Leakey and Lewin, *People of the Lake*, 154). They carried out elaborate rituals, crafted tools and weapons, cooked their food, and made fur clothing with bone needles (J. B. Birdsell, *Human Evolution: An Introduction to the New Physical Anthropology* [Chicago: Rand McNally & Co., 1972], 282–83). Archaeologists in China have even found the remains of Neanderthal houses (E. N. Anderson, *The Food of China*, [New Haven, Conn.: Yale University Press, 1988], 9).

62. Lest you think that man in the paradisal state that preceded civilization would never have stooped to the barbarity of massacring a near cousin, consider the case of the chimpanzee. When a hungry chimp is looking for meat, he is quite likely to satisfy his craving by murdering another primate. Jane Goodall describes in great detail how one carnivorous chimp managed to supply himself with cold cuts by creeping up on a group of baboons, grabbing a juvenile, swinging the victim over his head, then thwacking its skull on the rocks until it was dead. This particular killer chimpanzee was not the only one who enjoyed a bit of baboon flesh. His troop-mates crowded around him all day begging for a taste. They even avidly licked the leaves on which tiny drops of blood had fallen. Goodall's chimps, in fact, feasted fairly frequently on slaughtered baboons and colobus monkeys (Goodall, *In the Shadow of Man*, 200). *Groups* of chimpanzees also hunt colobus monkeys (see Christophe Boesch and Hedwige Boesch-Acherman, "Dim Forest, Bright Chimps," *Natural History*, September 1991, 50). For similar behavior among baboons, see S. L. Washburn and D. A. Hamburg, "Aggressive Behavior in Old World Monkeys and Apes," in Jay, *Primates*, 469. For the relationship between early humans and Neanderthal, see Jared Diamond, "The Great Leap Forward," *Discover*, May 1989, 58.

The Theory of Individual Selection and Its Flaws

63. Suicide was so popular among Valentino's bereaved fans that even two years after his death, women were still sending letters that read like this one: "How can we go on in this life when you are in the hereafter? My life is empty, a void, send me a sign that you want me in heaven and I will join you there" (Irving Shulman, *Valentino* [New York: Simon and Schuster, Trident Press, 1967], 25, 370). See also *New Encyclopaedia Britannica* 12:243.

64. For the manner in which the Japanese viewed their superiority as a blessing from their gods, see Edwin O. Reischauer, *The Japanese* (Cambridge: Harvard University Press, Belknap Press, 1981), 217–19; and W. G. Beasley, *The Meiji Restoration* (Stanford, Calif.: Stanford University Press, 1972), 75.

65. John Toland, *The Rising Sun: The Decline and Fall of the Japanese Empire* (New York: Random House, 1970).

66. John Kenneth Galbraith, *The Great Crash: 1929* (1954; Boston: Houghton Mifflin Co., 1988), 128–30.

67. William Manchester, *The Glory and the Dream: A Narrative History of America— 1932–1972* (New York: Bantam Books, 1974), 55.

68. Emile Durkheim, *Suicide: A Study in Sociology,* trans. John A. Spaulding and George Simpson (New York: Free Press, 1951), 217, 241; Walter T. Martin, "Theories of Variation in the Suicide Rate," in *Suicide,* ed. Jack Gibbs (New York: Harper & Row, 1968), 76–77; and T. O. Beidelman, "Emile Durkheim," in *Academic American Encyclopedia* (Danbury, Conn.: Groher, 1985), 6:306.

69. Marcel Mauss, *Sociology and Psychology: Essays by Marcel Mauss,* trans. Ben Brewster (London: Routledge & Kegan Paul, 1979), 19–20. These essays were delivered in the 1920s.

70. P. Diamandopoulos, "Thales of Miletus," in *Encyclopedia of Philosophy,* ed. Paul Edwards (New York: Macmillan, 1967), 8:97.

71. Alan Moorehead, *Darwin and the Beagle* (Newport Beach, Calif.: Books on Tape, 1969).

72. Charles Darwin, *The Origin of Species by Means of Natural Selection or the Preservation of Favoured Races in the Struggle for Life,* ed. J. W. Burrow (London: Penguin Books, 1968), 257. (Originally published in 1859.)

73. Charles Darwin, *The Descent of Man and Selection in Relation to Sex* (London: John Murray, 1871), 93.

74. V. C. Wynne-Edwards, *Animal Dispersion in Relation to Social Behavior* (New York: Hafner, 1962).

75. David L. Hull, *Science as a Process: An evolutionary Account of the Social and Conceptual Development of Science* (Chicago: University of Chicago Press, 1988), 210. For a further summation of the attacks on Wynne-Edwards and individual selection, see Eric Alden Smith and Bruce Winterhalder, "Natural Selection and Decision-Making: Some Fundamental Principles," in *Evolutionary Ecology and Human Behavior,* ed. Eric Alden Smith and Bruce Winterhalder (New York: Aldine de Gruyter, 1992), 29–32. One of the primary arguments used to dismiss Wynne-Edwards and group selection has been that competition between groups is not sufficiently frequent to be statistically significant.

Yet Charles Janson, associate professor of ecology and evolution at SUNY Stony Brook, cites "the frequency of between-group contests" as one of "the two major ecological benefits of large social groups" among primates. And he makes this assertion in a book that repeatedly expresses the obligatory skepticism about group selection (Charles Janson, "Evolutionary Ecology of Primate Social Structure," in E. A. Smith and Winterhalder, *Evolutionary Ecology and Human Behavior,* 106, 109).

76. For J. B. S. Haldane's early suggestions of the concept of kin selection, see Hull, *Science as a Process,* 60.

77. V. C. Wynne-Edwards, personal correspondence with the author.

78. Stephen Jay Gould, *Hen's Teeth and Horses' Toes* (New York: W. W. Norton, 1984).

79. For Depew and Weber's assertions that "there is a plurality of biological units and levels at which and between which . . . evolutionary processes can act," see David J. Depew and Bruce H. Weber, "Consequences of Nonequilibrium Thermodynamics for Darwinism," in Weber, Depew, and J. D. Smith, *Entropy, Information, and Evolution,* 318, 326, 334–35, 338–39. For other tentative approaches to these ideas, see also: Leo W. Buss, *The Evolution of Individuality* (Princeton, N.J.: Princeton University Press, 1987), viii, 171; Dorion Sagan, "What Narcissus Saw: The Oceanic 'I'/'Eye,' " in *The Reality Club,* ed. John Brockman (New York: Lynx Books, 1988), 204–6; and Hull, *Science as a Process,* 59, 402. Meanwhile, David P. Barash sums up the state of mainstream scientific thought on group versus individual selection in *Sociobiology and Behavior,* 70–79.

80. David C. Queller, Joan E. Strassman, and Colin R. Hughes, "Genetic Relatedness in Colonies of Tropical Wasps with Multiple Queens," *Science,* November 1988, 1155–57.

81. Donald T. Lunde, *Murder and Madness* (San Francisco: San Francisco Book Co., 1976), 5.

82. Lunde, *Murder and Madness,* 5; see also Lunde, *Murder and Madness,* 98–99.

83. Lunde, *Murder and Madness,* 45. Martin Daly and Margo Wilson attempt to address this problem with a model based on kin and individual selection in "Evolutionary Social Psychology and Family Homicide" (*Science,* 28 October 1988, 519–23). Unfortunately, their hypothesis is tortuous and assiduously avoids the frequency with which women kill their own children. Daniel G. Freedman presents a much more convincing approach in *Human Sociobiology,* 22.

84. Douglass H. Morse, *Behavioral Mechanisms in Ecology* (Cambridge: Harvard University Press, 1980), 123–24.

85. Donald R. Griffin, *Animal Thinking* (Cambridge: Harvard University Press, 1984), 78–82; and Bernhard Grzimek, *Grzimek's Animal Life Encyclopedia* (New York: Van Nostrand Reinhold Co., 1972), 13:295.

86. "Little precise information exists about the relatedness of the members of most natural populations" (Morse, *Behavioral Mechanisms in Ecology,* 119–20); "Temporary seasonal groups, such as wintering flocks of migratory sparrows . . . are unlikely to be composed of closely related individuals" (Morse, *Behavioral Mechanisms in Ecology,* 122).

87. For a different approach to the problem of self-sacrifice, see Herbert A. Simon, "A Mechanism for Social Selection and Successful Altruism," *Science,* 21 December 1990, 1665–68.

88. H. F. Harlow and M. K. Harlow, "Social Deprivation in Monkeys," 138; H. F. Harlow and G. Griffin, "Rhesus Monkeys," 99–105; H. F. Harlow, *Learning to Love,* 113; Ernest R. Hilgard, *Psychology in America: A Historical Survey* (San Diego: Harcourt Brace Jovanovich, 1987), 400; and Allan M. Schrier, "Harry F. Harlow," *Academic American Encyclopedia* 10: 50–51.

89. For a detailed description of the festival of Muharram, see Elias Canetti, *Crowds and Power,* trans. Carol Stewart (New York: Farrar, Straus and Giroux, 1984), 146–54.

90. For a thorough history of early Christianity's love affair with celibacy, see Peter Brown, *The Body and Society: Men, Women, and Sexual Renunciation in Early Christianity* (New York: Columbia University Press, 1988).

91. V. C. Wynne-Edwards, *Evolution through Group Selection* (Oxford: Blackwell Scientific Publications, 1986), 87, 91–93.

92. Marcia Barinaga, "Cell Suicide: By ICE Not Fire," *Science,* 11 February 1994, 754–56; Martin C. Raff et al., "Programmed Cell Death and the Control of Cell Survival: Lessons from the Nervous System," *Science,* 29 October 1993, 695–99; M. Stroh, "Genes Determine When Cells Live or Die," *Science News,* 11 April 1992, 230; and Gabrielle Strobel, "Guardian Genes," *Science News,* 15 January 1994, 44–45.

93. Danny A. Riley of the Medical College of Wisconsin, who ran an experiment with five rats in the Soviet Cosmos satellite, discovered that when weightlessness renders excess musculature unnecessary, "muscles not only shrink but also lose blood vessels, nerve connections and even their own cells." These disturbingly deleterious effects showed up in only two weeks ("Muscles in Space Forfeit More than Fibers," *Science News,* 29 October 1988, 277).

94. According to Soviet research, confirmed in the author's personal communication with NASA.

95. The Native American Crow tribe ritually lopped off the joint of one finger. The Sioux ran thongs under their pectoral muscles, then were hoisted toward the sky until a muscle tore. African tribes have mutilated themselves with ritual scarifications, while the primitive tribes of Malaysia have indulged in painful piercing and distension of the earlobes. How could there possibly be an adaptive value to these practices? Quite simple.

By opening wounds in the body, the rituals invited infection. Those who survived the deliberate breach in the body's protective barriers and overcame the resultant microbial invasion had immune systems that would stand their progeny in good stead. Most of the rites that sliced through the fortress of the skin took place as part of the ceremony that allowed young males or females to pass into adulthood, when sexual activity is permitted and reproduction becomes a possibility. Those who didn't survive the ordeal did not get to reproduce. For the individual, self-mutilation was not a great way to ensure survival, but it was an effective way to raise the overall health of the group (for information on the Crow Indians, see Dudley Young, *Origins of the Sacred: The Ecstasies of Love and War* [New York: St. Martin's Press, 1991], 223; for Malaysia, see Redmond O'Hanlon, *Into the Heart of Borneo* [Edinburgh: Salamander Press Edinburgh, 1984]).

Superorganism

96. Richard Bergland, *The Fabric of Mind* (Harmondsworth, Middlesex: Viking Penguin Books, 1986), 64.

97. Quoted in Bergland, *Fabric of Mind,* 64.

98. George Ordish, *The Year of the Ant* (New York: Charles Scribner's Sons, 1978), 61–62. Ordish is an economic entomologist.

99. William Morton Wheeler, "The Ant Colony as an Organism," *Journal of Morphology* 22 (1911): 307–25; and Edward O. Wilson, *The Insect Societies* (Cambridge: Harvard University Press, Belknap Press, 1971), 317. James Lovelock, co-creator of the controversial Gaea hypothesis, feels that "the first person to use the concept 'superorganism' " was probably "James Hutton, the father of geology," who applied the notion to the earth itself in 1789 (James Lovelock, personal communication to the author, 10 September 1990).

100. Lewis Thomas, *Lives of a Cell: Notes of a Biology Watcher* (New York: Bantam Books, 1975), 149–55.

101. Ilya Prigogine and Isabelle Stengers, *Order out of Chaos: Man's New Dialogue with Nature* (New York: Bantam Books, 1984), 156–59 (Prigogine is a holder of a Nobel Prize for his work on mathematical patterns that may someday help explain the agglomeration of superorganisms); Bonner, *The Evolution of Culture in Animals,* 78, 80, 103 (these pages include an extra bonus—illustrations of the slime mold's life cycle); L. Thomas, *Lives of a Cell,* 14–15; Norman S. Kerr, "Slime Mold," in *Academic American Encyclopedia* 17:362; *New Encyclopaedia Britannica* 10:879; and Buss, *Evolution of Individuality,* 70–73.

Isolation—the Ultimate Poison

102. Spitz, "Hospitalism," 53–74; Spitz with Wolf, "Anaclitic Depression," 331; M. T. Erickson, *Child Psycho-pathology,* 87; Kanner, *Child Psychiatry,* 684–85; and Corsini, *Encyclopedia of Psychology,* 1:161.

103. Jon Franklin, *Molecules of the Mind: The Brave New Science of Molecular Psychology* (New York: Atheneum, 1987), 161.

104. For a thorough review of these consequences, see James S. House, Karl R. Landis, and Debra Umberson, "Social Relationships and Health," *Science,* July 1988, 540–45.

105. Jay R. Kaplan et al., "Social Stress and Atherosclerosis in Normocholesterolemic Monkeys," *Science,* 13 May 1983, 733–35.

106. Restak, *Mind,* 152.

107. Bertram H. Raven and Jeffrey Z. Rubin, *Social Psychology* (New York: John Wiley & Sons, 1983), 56–57. The number of studies that demonstrate the damage of disrupted social ties is now enormous. See, for example, Kenneth R. Pelletier's references to the impact of bereavement and job loss on mortality in his article "Stress: Etiology, Assessment, and Management in Holistic Medicine," in *Selye's Guide to Stress Research,* ed. Hans Selye, Scientific and Academic Editions (New York: Van Nostrand Reinhold Co., 1983), 3:51–53. See also I. G. Sarason, B. R. Sarason, and G. R. Pierce, "Social Support, Personality, and Health," in *Topics in Health Psychology,* ed. S. Maes et al. (New York: John Wiley & Sons, 1988), 245–56; and Sheldon Cohen et al., "Chronic Social Stress Affiliation and Cellular Immune Response in Non-Human Primates," *Psychological Science* (September 1992): 301. Early researchers like Durkheim and Halbwachs also saw a clear relationship between isolation and suicide (Martin, "Theories of Variation in the Suicide Rate," 76–80). More recent research has shown that suicides actually go down on holidays that stress "social integration" by bringing families together (David P. Phillips, "A Dip in Deaths before Ceremonial Occasions: Some New Relationships between Social Integration and Mortality," *American Journal of Sociology* 84 [1979]: 1150–74; David P. Phillips and Judith Lu, "The Frequency of Suicides around Major Public Holidays: Some Surprising Findings," *Suicide and Life Threatening Behavior* [Spring 1980]: 41–50).

108. Goodall, *In the Shadow of Man,* 99, 232–36.

109. Flo and Flint's deaths occurred after Goodall's book *In the Shadow of Man* was written. They are chronicled in *Among the Wild Chimpanzees,* a National Geographic Special made by the researcher. For additional details, see Goodall, "Life and Death at Gombe," 605, 614.

110. Pelletier, "Stress," in Selye, *Selye's Guide to Stress Research,* 3:53.

111. Even infants treat inanimate objects as if they were people. Psychologist John Watson built a number of contraptions that rotated mobiles above the heads of babies when the youngsters put pressure on a pillow. Once the infants got the hang of the apparatus, they tended to smile and coo conversationally whenever the whirligigs began to turn (Herbert M. Lefcourt, *Locus of Control: Current Trends in Theory and Research,* 2d. ed. [Hillsdale, N.J.: Lawrence Erlbaum, 1982], 144).

112. Rhesus monkeys share this need with us. A simian subject isolated in a box will pull a lever over and over again just to get a glimpse of another monkey (Wilson, *Sociobiology,* 7).

113. Bruce Bower, "Personality Linked to Immunity," *Science News,* 15 November 1986, 310. Isolation and the personality test scores correlated with it also showed up as factors increasing the risk of cancer in a series of studies surveyed in *Science News* (Bruce Bower, "The Character of Cancer," *Science News,* 21 February 1987, 120–21. See also Bruce Bower, "Heart Attack Victims Show Fatal Depression," *Science News,* 23 October 1993, 263).

114. Friedman made the statement on "The Phil Donahue Show," 16 May 1983.

115. Manchester, *Glory and the Dream,* 755–56.

116. T. E. Lawrence, *Seven Pillars of Wisdom* (New York: Doubleday & Co., 1926; New York: Dell Publishing, 1962).

117. Desmond Stewart, *T. E. Lawrence* (New York: Harper & Row, 1977), 293.

118. Philip Knightley and Colin Simpson, *The Secret Life of Lawrence of Arabia* (New York: McGraw-Hill Book Co., 1969), 175.

119. *The Biography of Thomas Edward Lawrence, Lawrence of Arabia, 1883–1935* (Pasadena, Calif.: Cassette Book Co., 1983). In fact, Lawrence had tried suicide once before. Eight years after the war ended, he had taken out a pistol, put it to his head, and squeezed the trigger. Fortunately, a friend had guessed Lawrence's intentions and emptied the chamber of bullets (Knightley and Simpson, *The Secret Life of Lawrence of Arabia,* 224–25).

120. Lawrence's profound sense of purposelessness led biographer John E. Mack to consider the question of suicide and to conclude that as Lawrence took his motorcycle on its last ride, he was "less vigorous in preserving his own life than he might once have been" (for the quotations from Lawrence's letter to Eric Kennington and from Mack's biography, see Stewart, *T. E. Lawrence,* 292).

121. Robert B. Cialdini, *Influence: How and Why People Agree to Things* (New York: William Morrow, 1984), 145. Despite its breezy title, this book is an outstanding summary of findings in the field of social psychology. Its author, Regents Professor of Psychology at Arizona State University, is a former associate editor of the *Journal of Personality and Social Psychology.*

Even Heroes Are Insecure

122. Marilyn Machlowitz, *Whiz Kids: Success at an Early Age* (Newport Beach, Calif.: Books on Tape).

123. Ernle Dusgate Selby Bradford, *Hannibal* (New York: McGraw-Hill Book Co., 1981), 39, 44.

Notes

124. Bradford, *Hannibal,* 208–9.

125. Lionel Tiger and Robin Fox, *The Imperial Animal* (New York: Holt, Rinehart and Winston, 1971), 30. Tiger and Fox, whose background was originally in sociology and anthropology, have become two of the best-known proponents of the sociobiological mode of thought. See also Jane Van Lawick-Goodall, "A Preliminary Report on Expressive Movements and Communication in the Gombe Stream Chimpanzees," in Jay, *Primates,* 323.

126. Frans de Waal, *Chimpanzee Politics: Power & Sex among Apes* (New York: Harper Colophon Books, 1984), 133.

127. Albert Speer, *Inside the Third Reich—Memoirs,* trans. Richard Winston and Clara Winston (New York: Collier Books, 1970). Several visitors to the führer, among them physicist Max Planck, described how Hitler would lapse into an incoherent frenzy if someone hit on one of his weak points (Robert G. L. Waite, *The Psychopathic God: Adolf Hitler* [New York: New American Library, 1978], 10, 49, 454).

128. Lawrence, *Seven Pillars of Wisdom,* 526.

Loving the Child Within Is Not Enough

129. "Phil Donahue Show," 16 May 1983.

130. Jesse Roth and Derek LeRoith, "Chemical Cross Talk: Why Human Cells Understand the Molecular Messages of Plants," *The Sciences,* May/June 1987, 54; and Robert Wright, "The Information Age: The Life of Meaning," *The Sciences,* May/June 1988, 12.

131. Andrew Liebman (writer, director, and producer), *The Secret of Life: Conquering Cancer* (London: BBC-TV; Boston: WGBH, 1993).

Us versus Them

132. Harold M. Schmeck, Jr., *Immunology* (New York: George Braziller, 1974), 44–45; Carla Reiter, "Toy Universes," *Science 86,* June 1986, 56; and L. Thomas, *Lives of a Cell,* 43–48.

133. Herman Harvey, interview with Margaret Mead, audiotape series *Sum and Substance;* see also Lorenz, *On Aggression,* 83; and Ruth Benedict, *Patterns of Culture* (1934; New York: New American Library, Mentor Book, 1950), 6.

134. For a summary of social psychological research on the ease with which humans fall into us versus them patterns, see Raven and Rubin, *Social Psychology,* 639–50.

135. Lewis Thomas and Robin Bates, *Notes of a Biology Watcher,* prod. and dir. Robin Bates, Nova, no. 818 (Boston: WGBH, 1981).

136. Barash, *Hare and the Tortoise,* 279.

137. "Avian dialects permit a group cohesiveness and . . . tend to isolate . . . groups into separate geographic regions" (Bonner, *The Evolution of Culture in Animals,* 179). See also Wilson, *Sociobiology,* 80; Harold E. Burtt, *The Psychology of Birds: An Interpretation of Bird Behavior* (New York: Macmillan, 1967), 174; and James W. Grier, *Biology of Animal Behavior* (St. Louis, Mo.: Times-Mirror, 1984), 575. For dialects that serve the same purpose among frogs, see M. J. Ryan and W. Wilczynski, "Coevolution of Sender and Receiver: Effect on Local Mate Preference in Cricket Frogs," *Science,* 24 June 1988, 1786–87. For similar dialects among whales, see Constance Holden, "Do Whales Speak in Many Tongues?" *Science,* 11 February 1994, 753.

138. Wilson, *Insect Societies,* 272; and Ordish, *Year of the Ant,* 43.

139. The leading expert on social distance as a cultural marker is anthropologist Edward T. Hall. See his *Beyond Culture* (New York: Anchor Books, 1977).

140. Deuteronomy 6:7–9.

141. Harrison E. Salisbury, *Black Night, White Snow: Russia's Revolutions, 1905–1917* (New York: Plenum Publishing, Da Capo, 1981), and Alan Brien, letter, *New York Times Book Review,* 1 January 1989, 2.

142. D. S. Roberts, *Islam: A Concise Introduction* (New York: Harper & Row, 1981), 102–3. Ayatollah Sayyed Ruhollah Mousavi Khomeini, *A Clarification of Questions: An Unabridged Translation of Resaleh Towzih al-Masael,* trans. J. Borujerdi (Boulder, Colo., Westview Press, 1984), 38–39.

143. William Manchester, *The Arms of Krupp: 1587–1968* (New York: Bantam Books, 1978), 276, 595, 836.

144. Manchester, *Arms of Krupp,* 540.

The Value of Having an Enemy

145. Edward Sagarin and Robert J. Kelly, "Collective and Formal Promotion of Deviance," in *The Sociology of Deviance,* ed. M. Michael Rosenberg, Robert A. Stebbins, and Allan Turowetz (New York: St. Martin's Press, 1982), 214.

146. Manchester, *Glory and the Dream,* 799–809.

147. Tad Szulc, *Fidel: A Critical Portrait* (New York: William Morrow, 1986), 432. Szulc, the *New York Times* reporter who broke the Bay of Pigs story, wrote this book with Fidel Castro's cooperation. Castro gave Szulc interview time and access to Cuban officials and documents normally off-limits to foreigners.

148. Szulc, *Fidel,* 488–507.

149. Szulc, *Fidel,* 534.

The Perceptual Trick That Manufactures Devils

150. Elizabeth Loftus, *Memory* (Reading, Mass.: Addison-Wesley Publishing Co., 1980), 122–23.

151. Loftus, *Memory,* 135–44.

152. Donn Byrne, *An Introduction to Personality* (Englewood Cliffs, N.J.: Prentice-Hall, 1966), 239–83; and Bob Altemeyer, "Marching in Step: A Psychological Explanation of State Terror," *The Sciences,* March/April 1988, 30.

153. Figures from a presentation by Skipp Porteous, former fundamentalist minister and publisher of the newsletter *Freedom Writer*. Porteous, an authority on the religious right, delivered his presentation at a panel on censorship organized by the author. According to Antony Thomas's documentary *Thy Kingdom Come,* fundamentalist ministers claim to reach a daily TV and radio audience of forty million (*Thy Kingdom Come, Thy Will Be Done* [London: Central Television Enterprises, 1987]).

154. For details on the computerized mailing organization that helps the fundamentalists and their political allies send out a staggering seventy-five million mailing pieces per year, see Antony Thomas's documentary film *Thy Kingdom Come, Thy Will Be Done.*

155. The leading proponents of a total fundamentalist government takeover are the preachers of Dominion Theology and Christian Reconstructionism. For a profile of these men and their movements, see the television documentary *Moyers: God & Politics—The Battle for the Bible,* prod. Gail Pellett, 16 December 1987 (New York: Public Affairs Television).

156. Jimmy Swaggart, "Rock 'n' Roll Music in the Church," *The Evangelist,* January 1987, 8; *Year of Action* (Lakemont, N.Y.: Freedom Village [a fundamentalist organization] 1985); Jimmy Swaggart, *A Letter to My Catholic Friends,* cited in "TV Evangelist Denies Charges by Mondale," *New York Times,* 26 September 1984; Kenneth L. Woodward with Vincent Coppola, "King of Honky-Tonk Heaven," *Newsweek,* 30 May 1983, 89–90; and "Jerry Falwell; Circuit Rider to Controversy," *U.S. News and World Report,* 2 September 1985, 11.

157. Michael Kramer, "Are You Running with Me Jesus? Televangelist Pat Robertson Goes for the White House," *New York* magazine, 18 August 1986, 24; and Tim LaHaye, *Has the Church Been Deceived?* (Washington, D.C.: American Coalition for Traditional Values). For additional information, see the congregational bulletins of grassroots figures like the Reverend Paul McGechie of Goshen, Indiana, and the American Family Association's *AFA Journal.* These were collected in the files of Music in Action, an anti-censorship group of which the author was a co-founder.

How Hatred Builds the Walls of Society's Bungalow

158. Humans are so addicted to officially authorized hatred that in the mid-eighteenth century, when the British government proposed lifting legal sanctions against Catholics, the outraged population of London rioted in protest and burned down parts of the city (Dero A. Saunders, ed., introduction to *The Decline and Fall of the Roman Empire* [abridged] by Edward Gibbon [New York: Penguin Classics, 1985], 4).

159. Leonard Berkowitz, "The Frustration-Aggression Hypothesis Revisited," in *Aggression: A Re-Examination of the Frustration-Aggression Hypothesis*, ed. Leonard Berkowitz (New York: Atherton Press, 1969), 4, 7, 8, 19, 22. For similar experiments with pigeons, squirrel monkeys, rhesus monkeys, and humans, see N. H. Azrin, R. R. Hutchinson, and D. F. Drake, "Extinction Induced Aggression," in Berkowitz, *Aggression*, 34, 41, 42. Frustration is not the only experience that can make a rat or human turn on his fellows; pain also does the trick. See R. F. Ulrich and N. H. Azrin, "Reflexive Fighting in Response to Aversive Stimulation," *Journal of the Experimental Analysis of Behavior* (October 1962): 511–20. The classic work on the subject of frustration and aggression, which is examined later in this work, is John Dollard et al., *Frustration and Aggression* (New Haven, Conn.: Yale University Press, 1957). It would be pointless to give specific page numbers, since virtually the entire book is dedicated to this thesis. See also Hilgard, *Psychology in America*, 371–72; Raven and Rubin, *Social Psychology*, 271–73; and Goodall, "Life and Death at Gombe," 598–99.

160. Margulis and D. Sagan, *Microcosmos*, 75.

161. Barash, *Hare and the Tortoise*, 71.

162. de Waal, *Chimpanzee Politics*, 49, 167–68, 175, 179.

163. Wilson, *Insect Societies*, 147–52.

164. Food is not the only factor determining which form a growing ant will take. The others include the amount of winter chilling the ant goes through while it's still an egg, the size of the egg it hatches from, the temperature of its nursery when the ant is still an infant, and the age and condition of its mother (Wilson, *Insect Societies*, 152).

165. Ordish, *Year of the Ant*, 114.

166. For a brilliant evocation of Grant's years of shame, see MacKinlay Kantor's short story "Then Came the Legions," reprinted in Roger B. Goodman, ed., *75 Short Story Masterpieces: Stories From the World's Literature* (New York: Bantam Books, 1961), 160–64. For a slightly more charitable version of the facts, see Ishbel Ross, *The General's Wife: The Life of Mrs. Ulysses S. Grant* (New York: Dodd, Mead & Co., 1959), 88–105; and *New Encyclopaedia Britannica* 5:425. But for the bottom line on Grant's drinking (he didn't do it often, but when he did, his benders were spectacular), see John Keegan, *The Mask of Command* (New York: Viking, Elisabeth Sifton Books, 1987), 204.

167. For a view of how our lives are arbitrarily limited by role playing, see Ervin Goffman, *The Presentation of Self in Everyday Life* (New York: Doubleday, Anchor Books, 1959).

168. Alvin Toffler, introduction to *Order out of Chaos,* by Prigogine and Stengers, xxiv.

169. William K. Purves and Gordon H. Orians, *Life: The Science of Biology* (Sunderland, Mass.: Sinauer, 1987), 403; and C. C. Ford, "Development," in *Academic American Encyclopedia* 6:137–39.

170. D. G. Freedman, *Human Sociobiology,* 46, 169–70. By the way, female campers also sorted themselves out in a hierarchy; but the process by which they arrived at their social arrangement was a bit different than that of the boys. It involved more vicious backbiting and less physical forms of cruelty. Yet the cruelty was so potent that at one time or another, it reduced the camp counselors to tears. Said one of these counselors, "Now I know why no one studies junior high school girls! They are so cruel and horrible that no one can stand them" (D. G. Freedman, *Human Sociobiology,* 47–49).

171. Boorstin, *Discoverers,* 126.

172. Tannahill, *Sex in History,* 239–40; and Robert K. Massie, *Peter the Great* (New York: Ballantine Books, 1986), 553.

173. Margulis and D. Sagan, *Microcosmos,* 179–86.

174. For an interesting description of how the early Christians managed to turn the pagan gods into demons, see Gibbon, *The Decline and Fall of the Roman Empire* (Penguin Classics), 270. See also R. L. Fox, *Pagans and Christians,* 137.

From Genes to Memes

175. Anne Givens, "Chimps, More Diverse Than a Barrel of Monkeys," *Science,* 17 January 1992, 287. The twenty-two-million-year estimate can be pushed even further back if we take into account the remarkable ability of birds to develop dialects and food-tapping techniques, which research shows they literally invent, then pass on through learning, not instinct (Bonner, *The Evolution of Culture in Animals,* 183–85).

176. *New Encyclopaedia Britannica* 23: 575–76.

177. This estimate of Marx's weight is based on the report of a Prussian police spy who visited Marx's London home in 1853. The agent said, "Marx is of medium height . . . his figure is powerful" (Saul K. Padover, *Karl Marx: An Intimate Biography* [New York: McGraw-Hill Book Co., 1978], 291). In the nineteenth century, average height was several inches shorter than it is today.

178. David McLellan, *Karl Marx: His Life and Thought* (New York: Harper Colophon Books, 1973), 32, 33, 53, 102–3. See also Boorstin, *Discoverers,* 617–21. The anar-

chist leader Mikhail Bakunin, with whom Marx had an extremely acrimonious relationship, painted an even more damning picture of his adversary. Bakunin said that Marx is "vengeful to the point of madness. There is no lie or calumny that he is not capable of inventing against anyone who has had the misfortune of arousing his jealousy, or, which is the same thing, his hatred" (Padover, *Karl Marx,* 180).

179. The censor said, "Few people in Russia will read it, and still fewer will understand it" (Boorstin, *Discoverers,* 618).

180. Before these fathers of the Russian Revolution came along, the Russian Marxist movement was pitifully small. In 1872, only three thousand Russian readers had purchased the first Russian edition of *Das Kapital* and waded painfully through its turgid prose (Boorstin, *Discoverers,* 618). Among these was Georgy Valentinovich Plekhanov, the son of a wealthy country gentleman, who started a Russian Marxist movement in 1883. Through much of his life, Plekhanov led a struggling group of underground cells from his exile in Geneva. Plekhanov's efforts planted the seeds that would come to fruition under Lenin and Stalin (Alan Moorehead, *The Russian Revolution* [New York: Bantam Books, 1959], 34–38).

181. Salisbury, *Black Night, White Snow* 325. Exile was not the only reason that Lenin found fewer followers than he would have liked. Like Marx, Lenin was quarrelsome in the extreme. Those who started out liking him generally changed their minds pretty fast. One of his acquaintances, Vera Zasulich, said Lenin was like a bulldog with a "deadly bite" (Salisbury, *Black Night, White Snow,* 94, 143–46).

182. Salisbury, *Black Night, White Snow,* 324–25.

183. In Edwin O. Reischauer's opinion, the fleet was "annihilated." Three badly battered ships did manage to flee their battle with the Japanese and hobble into port. But even Alan Moorehead, who reports on the survival of this tattered trio, calls the event a "massacre" (Edwin O. Reischauer, *Japan: Past and Present,* 3d ed. [Tokyo: Charles E. Tuttle Co., 1981], 139; Moorehead, *Russian Revolution,* 27–28; and Salisbury, *Black Night, White Snow,* 96–97).

184. By the early eighties, Marxism controlled 39.7 percent of the earth's population and 27.5 percent of its land mass (Mikhail Heller and Aleksandr M. Nekrich, *Utopia in Power: The History of the Soviet Union from 1917 to the Present,* trans. Phyllis B. Carlos [New York: Summit Books, 1986], 717).

185. Bonner, *Culture in Animals,* 57.

The Nose of a Rat and the Human Mind—a Brief History of the Rise of Memes

186. All of this information on rats and most in the following paragraphs come from Konrad Lorenz's ground-breaking *On Aggression,* 157–63.

187. Smell is such a ubiquitous clue to genetic relatedness that it is used by an extraordinary variety of animals, from ants to goats. Mother goats will let their own children starve if the youngsters don't exude the correct aroma (Wilson, *Sociobiology,* 102).

188. Lorenz, *On Aggression,* 162.

189. Buss, *Evolution of Individuality,* viii. Though Buss never attempts to extend this idea to humans and their societies, his principle applies extremely well to memes and their role in the construction of superorganisms.

190. "Now those who were scattered after the persecution that arose over Stephen traveled as far as Phoenicia, Cyprus, and Antioch, preaching the word to no one but the Jews only" (Acts 11:19). "The original apostles . . . at first . . . were of no mind even to consort with the Gentiles" (Bainton, *Christianity,* 50).

191. According to the Acts of the Apostles, shortly after Jesus' death, his disciple Peter made a stirring speech. In it, the apostle spoke of the Lord's promise to "make Your enemies Your footstool" and declared that "the promise is to you [the house of Israel] and to your children" (Acts 2:34–39).

192. In Acts of the Apostles, it is stated: "The Jews . . . opposed the things spoken by Paul. Then Paul . . . grew bold and said, 'It was necessary that the word of God should be spoken to you first; but since you reject it, and judge yourselves unworthy of everlasting life, behold, we turn to the Gentiles' " (Acts 13:45–46).

193. Sources for the story of Paul are: Acts 8–13; Joseph Klausner, *From Jesus to Paul,* trans. W. F. Stinespring (New York: Macmillan, 1943); A. Powell Davies, *The First Christian* (New York: New American Library Mentor Books, 1959); Bainton, *Christianity,* 48–54; and H. G. Wells, *The Outline of History* (New York: Macmillan, 1926), 332. For a map of Saint Paul's extensive travels in search of converts, see Robert Jewett, "Saint Paul," *Academic American Encyclopedia,* 15:117.

194. Alexander the Great also deserves a certain amount of credit for freeing memes from genes. Though Alexander did not force Greek gods down the throats of all who fell under his power, in the fourth century B.C., he carried Hellenic ideas to the old empires of Persia, Egypt, and India, hopscotching over genetic boundaries as he went. A few hundred years later, the Romans would do the same with their concepts. In the east, K'ung Fu-tse—Confucius—fashioned a gene-free philosophy as early as 500 B.C.

How Wrong Ideas Can Be Right

195. Leon Festinger, Henry W. Riecken, and Stanley Schachter, *When Prophecy Fails: A Social and Psychological Study of a Modern Group That Predicted the Destruction of the World* (New York: Harper Torchbooks, 1966). See also the summaries and descriptions of the background behind Festinger, Riecken, and Schachter's study in Cialdini, *Influence,* 121–27; and Raven and Rubin, *Social Psychology,* 6–12.

196. There is considerable disagreement among historians about the dates Miller predicted for the world's demise. I've used those given in the *Encyclopaedia Britannica,* 1986 edition, 8:136. My remaining sources for the story of William Miller and the subsequent rise of Seventh Day Adventism include: R. Laurence Moore, *Religious Outsiders and the Making of America* (New York: Oxford University Press, 1986), 131–32; Jack Gratus, *The False Messiahs* (New York: Taplinger Publishing Co., 1975), 50–52; William Joseph Whalen, *Minority Religions in America* (New York: Alba House, 1981), 8; Conrad Wright, "Adventists," in *Academic American Encyclopedia* 1:111; and Clarke F. Ansley, ed., *The Columbia Encyclopedia in One Volume* (New York: Columbia University Press, 1940), 21, 1173.

197. Barbara W. Tuchman, *The Proud Tower: A Portrait of the World before the War, 1890–1914* (New York: Bantam Books, 1967), 481.

198. For a paradigmatic experimental demonstration of how ideas can pull together groups that instantly compete, see Giyoo Hatano and Kayoko Inagaki, "Sharing Cognition through Collective Comprehension Activity," in *Perspectives on Socially Shared Cognition,* ed. Lauren B. Resnick, John M. Levine, and Stephanie D. Teasley (Washington, D.C.: American Psychological Association, 1991), 339–40.

The Village of the Sorcerers and the Riddle of Control

199. Melville J. Herskovits, *Economic Anthropology: The Economic Life of Primitive Peoples* (New York: W. W. Norton & Co., 1940; reprint 1965), 157–59.

200. Herskovits, *Economic Anthropology,* 157–59; and John Reader, *Man on Earth* (Austin, Tex.: University of Texas Press, 1988), 176–78.

201. Sigmund Freud, *The Future of an Illusion* (New York: W. W. Norton & Co., 1989), 38–43.

202. For a good review of animal experiments on control, see Lefcourt, *Locus of Control* 8–18. See also: William R. Miller, Robert A. Rosellini, and Martin E. P. Seligman, "Learned Helplessness and Depression," in *Psychopathology: Experimental Models,* ed. Jack D. Maser and Martin E. P. Seligman (San Francisco: W. H. Freeman and Co., 1977), 104–130; and T. J. Shors et al., "Inescapable versus Escapable Shock Modulates Long-Term Potentiation in the Rat Hippocampus," *Science,* 14 April 1989, 224–26.

203. Restak, *Brain,* 167–69.

204. Goleman, *Vital Lies, Simple Truths,* 38. At the University of Pennsylvania, Martin E. Seligman and Steven Maier achieved the same results reported by Goleman. (Richard M. Restak, *The Mind* [New York: Bantam Books, 1988], 152.)

205. In fact, the word *endorphin* is a contraction of the phrase "endogenous morphine" (Floyd E. Bloom, "Endorphins," in *Encyclopedia of Neuroscience,* ed. George Adelman [Boston: Birkhauser, 1987], 1:393). See also Roth and LeRoith, "Chemical

Cross Talk,'' 53; *McGraw-Hill Encyclopedia of Science and Technology* (New York: McGraw-Hill Book Co., 1982), 5:72; Goleman, *Vital Lies, Simple Truths,* 30–31; Franklin, *Molecules of the Mind,* 78; and O. T. Phillipson, ''Endorphins,'' in *The Oxford Companion to the Mind,* ed. Richard L. Gregory (New York: Oxford University Press, 1987), 221–23.

206. Goleman, *Vital Lies, Simple Truths,* 34–36, 38.

207. I've taken the liberty of combining the results of an extensive series of experiments. UCLA researchers found that subjecting rats to uncontrollable foot shocks raised their endorphin levels, but giving them shocks they could control did not tweak endorphin levels in the least. To the contrary, the controllable shocks upped the release of ''non-opioids''—presumably chemicals like ACTH, a substance that heightens attention (Goleman, *Vital Lies, Simple Truths,* 38). Meanwhile, scientists at the University of Pennsylvania and numerous other institutions also subjected rats, dogs, and other animals to uncontrollable punishment. The investigators discovered that this treatment—the very same form of torture that had been proven to boost endorphin levels—had a devastating impact on vertebrate learning and behavior. The lab animals lost interest in food and sex, and when they were taught to run mazes, their learning rate was far, far lower than that of their normal cousins. They showed an appalling mental sluggishness (Franklin, *Molecules of the Mind,* 131; and Leonard A. Sagan, ''Family Ties: The Real Reason People Are Living Longer,'' *The Sciences,* March/April 1988, 28). Experiments with the endorphin-blocking chemical naltrexone indicated that the substance which had muffled the creatures' brains was almost certainly endorphin.

208. Beth Livermore, ''At Least Take a Deep Breath,'' *Psychology Today,* September 1992, 44.

209. Lefcourt, *Locus of Control,* 3–6.

The Modern Medical Shaman

210. Norman Cousins, *Human Options* (New York: W. W. Norton & Co., 1981).

211. John Pfeiffer, ''Listening for Emotions: Videotapes Show that Many Doctors Aren't—and Patients Suffer,'' *Science 86,* June 1986, 16.

212. Dorothy W. Smith and Carol P. Hanley Germain, *Care of the Adult Patient: Medical, Surgical Nursing,* 4th ed. (Philadelphia: J. B. Lipincott Co., 1975), 398; and William A. R. Thomson, M.D., *Black's Medical Dictionary* (Totowa, N.J.: Barnes & Noble Books, 1984), 519.

213. Ornstein and Sobel, *Healing Brain,* 21–24. David Sobel, M.D., is director of patient education and health promotion for Kaiser Permanente Medical Care Program in northern California and chief of preventive medicine at Kaiser Permanente Medical Center in San Jose. Robert Ornstein, Ph.D., teaches at the University of California Medical Center in San Francisco and at Stanford University. Leonard A. Sagan covers similar

ground in "Family Ties," 22. Sagan is an epidemiologist at the Electric Power Research Institute in Palo Alto, California, and author of *The Health of Nations: True Causes of Sickness and Well-Being.*

214. Marjory Roberts, "Patient Knows Best," *Psychology Today,* June 1987, 10.

215. Jerome S. Bruner, *Beyond the Information Given: Studies in the Psychology of Knowing,* ed. Jeremy M. Anglin (New York: W. W. Norton, 1973), 33–37; Spencer A. Rathus, *Psychology* (New York: Holt, Rinehart and Winston, 1987), 418; Albert R. Gilgen, *American Psychology since World War I: A Profile of the Discipline* (Westport, Conn: Greenwood Press, 1982), 121; and Morris Eagle and David Wolitzky, "Perceptual Defense," in *International Encyclopedia of Psychiatry, Psychology, Psychoanalysis & Neurology,* ed. Benjamin B. Wolman (New York: Van Nostrand Reinhold Co., 1977), 8:260–65.

216. San Francisco internist R. Dennis Collins confessed to a national medical conference in 1989 that "there is probably some degree of arrogance among physicians who feel that if you haven't learned about it in medical school or training, then it doesn't exist. Physicians don't feel comfortable saying 'I don't know,' so they may prefer not to deal with it" (Sari Staver, "Conference Shows One Skeptic: 'It's Clear We Have a Real Syndrome,' " *American Medical News,* 26 May 1989, 9).

217. Joseph Alper, "Depression at an Early Age," *Science 86,* May 1986, 45–46. See also John F. McDermott, Jr., "Child Psychiatry," in Wolman, *International Encyclopedia of Psychiatry, Psychology, Psychoanalysis and Neurology* 3:112.

218. For a history of the development of antidepressants, see Franz G. Alexander and Sheldon T. Selesnick, *The History of Psychiatry: An Evaluation of Psychiatric Thought and Practice from Prehistoric Times to the Present* (New York: Harper & Row, 1966), 289; Daniel X. Freedman, "Psychic Energizer," *McGraw-Hill Encyclopedia of Science and Technology* 11:65–66; Eliott Richelson, "Antidepressants," *Encyclopedia of Neuroscience* 1:52; Leonard Cammer, *Up from Depression* (New York: Simon and Schuster, 1969), 140; Herman C. B. Denber, "Depression: Pharmacological Treatment," in Wolman, *International Encyclopedia of Psychiatry, Psychology, Psychoanalysis and Neurology* 4:55–58.

Control and the Urge to Pray

219. For a detailed analysis of the background behind this dispute, see Geoffrey Barraclough, *The Origins of Modern Germany* (New York: W. W. Norton, 1984), 30–125. See also Bainton, *Christianity,* 159–63; and J. M. Roberts, *Pelican History of the World,* 470–72.

220. Barbara W. Tuchman, *A Distant Mirror: The Calamitous 14th Century* (New York: Ballantine Books, 1979), 27–34.

221. Bruce Bower, "Taking Hopelessness to Heart," *Science News,* 31 July 1993, 79.

222. Barbara Tuchman, *Distant Mirror.*

Power and the Invisible World

223. David Holzman, "Medicine Minus a Cost Tourniquet," *Insight,* 8 August 1988, 9–16.

224. Senator Lawton Chiles, press conference, "MacNeil/Lehrer Newshour," 4 August 1988.

225. J. D. Mackie, *Oxford History of England: The Earlier Tudors, 1485–1558* (London: Oxford University Press, 1962), 372.

226. For a vivid sense of how powerful the Tibetan religious authorities were, see Heinrich Harrer, *Seven Years in Tibet,* trans. Richard Graves (Los Angeles: Jeremy P. Tarcher, 1982). Harrer was one of the few foreigners in modern times to become a member of Tibet's preinvasion royal inner circle. See also: H. E. Richardson, *A Short History of Tibet* (New York: E. P. Dutton & Co., 1962), 11; Anna Louise Strong, *When Serfs Stood Up* (San Francisco: Red Sun Publishers, 1976), 12; and "Tibetan Buddhism," in *New Encyclopaedia Britannica* 11:756.

227. John H. Campbell, "Evolution as Nonequilibrium Thermodynamics: Halfway There?" in Weber, Depew, and J. D. Smith, *Entropy, Information, and Evolution,* 278.

228. Boorstin, *Discoverers,* 412–16. Boorstin's book, by the way, is a delightful excursion through nearly five thousand years of human progress. For lovers of history, it is a must.

229. J. M. Roberts, *Pelican History of the World,* 79; and Boorstin, *Discoverers,* 5, 17.

230. Boorstin, *Discoverers,* 73.

231. Linda Schele and David Freidel, *A Forest of Kings: The Untold Story of the Ancient Maya* (New York: William Morrow and Company, 1990), 73–81. T. Patrick Culbert, "The Collapse of Classic Maya Civilization," in *The Collapse of Ancient States and Civilizations,* ed. Norman Yoffee and George L. Cowgill (Tucson: University of Arizona Press, 1988), 44–68.

232. Dr. Woodrow Borah, quoted in *The New York Times,* 19 February 1977. Cited in Tannahil, *Sex in History,* 304.

233. John G. Neihardt, *Black Elk Speaks: Being the Life Story of a Holy Man of the Oglala Sioux* (New York: Pocket Books, 1972). See pages 17–39 for particular instances of these concepts. The same imagery, however, reappears throughout the book.

234. Erik H. Erikson, *Childhood and Society,* 2d ed., (New York: W. W. Norton & Co., 1953), 127.

235. Quoted in Konner, *Tangled Wing,* 311. When he wrote this encyclopedic volume on the biological underpinnings of human behavior, Melvin Konner was an associate professor at Harvard University. He holds degrees in biological anthropology and medicine. I owe the entire comparison of Watson's attitudes with those of the !Kung to Konner's work.

236. Konner, *Tangled Wing,* 313.

Einstein and the Eskimos

237. Barash, *Whisperings Within,* 43.

238. Marvin Harris, "India's Sacred Cow," in *Conformity and Conflict: Readings in Cultural Anthropology,* ed. James P. Spradley and David W. McCurdy (Boston: Little, Brown and Co., 1986), 208–19; Marvin Harris, *Cannibals and Kings: The Origins of Cultures* (New York: Vintage Books, 1977), 211–32; and M. Harris, *Cows, Pigs, Wars and Witches,* 6–27. Actually, the Arab scholar Abu Raihan Muhammad ibn Ahmad al-Biruni, who traveled extensively in India during the eleventh century, anticipated Harris's economic explanation of cow worship by over nine hundred years. See Abu Raihan Muhammad ibn Ahmad al-Biruni, *Albiruni's India,* trans. Edward Sachau and ed. Ainslie T. Embree (New York: W. W. Norton & Co., 1971), 152. For al-Biruni's background, see Ainslie Thomas Embree, ed., *Encyclopedia of Asian History* (New York: Charles Scribner's Sons, 1988), 1:164.

239. *The Three Worlds of Bali,* written by and based on the research of J. Stephen Lansing, prod. and dir. Ira R. Abrams, Odyssey television series, co-prod. Public Broadcasting Associates and the University of Southern California (1981); Reader, *Man on Earth,* 69–72; and Jane E. Stevens, "Growing Rice the Old-Fashioned Way, with Computer Assist," *Technology Review,* January 1994, 16–18.

240. Leibnitz felt that the process of working out the structure of possible worlds was the very essence of mathematics (Heinz Pagels, *The Dreams of Reason: The Computer and the Rise of the Sciences of Complexity* [New York: Simon and Schuster, 1988], 302). For an interesting sense of what a curved four-dimensional world is like, see Edwin A. Abbott's nineteenth-century classic *Flatland: A Romance of Many Dimensions* (New York: Barnes & Noble Books, 1983).

241. Albert Einstein, *The Meaning of Relativity,* 5th ed. (Princeton, N.J.: Princeton University Press, 1955), 64, 103–104; Max Jammer, *The History of Theories of Space in Physics* (Cambridge, Mass.: Harvard University Press, 1954), 143, 149; G. J. Whitrow, *Einstein: The Man and His Achievement* (New York: Dover Publications, 1973); "Relativity," in *McGraw-Hill Encyclopedia of Science and Technology,* 11:492–93; "Riemannian Geometry," *McGraw Hill Encyclopedia of Science and Technology,* 11:671; "Bernhard Riemann," *New Encyclopaedia Britannica,* 10:62; and Michael Guillen, *Bridges to Infinity: The Human Side of Mathematics* (Los Angeles: Jeremy P. Tarcher, 1983), 84–87, 110–11.

242. Robert Jastrow, *The Enchanted Loom: Mind in the Universe* (New York: Simon and Schuster, Touchstone Book, 1983), 67–70. Jastrow is founder of NASA's Goddard Institute.

243. For more on animals' internal models of the world, see British Royal Society member Janos Szentagothai's "The Brain-Mind Relation: A Pseudoproblem?" in *Mindwaves,* ed. Colin Blakemore and Susan Greenfield (Oxford: Basil Blackwell, 1959), 324.

244. For a different, but extraordinary, outline of the relationship between metaphor, mind, science, and mathematics, see Julian Jaynes, *The Origin of Consciousness in the Breakdown of the Bicameral Mind* (Boston: Houghton Mifflin Co., 1976), 50–54. Also see Peter Hacker, "Languages, Minds and Brains," in Blakemore and Greenfield, *Mindwaves,* 485–88.

245. Boorstin, *Discoverers,* 226.

The Connectionist Explanation of the Mass Mind's Dreams

246. William F. Allman, "Mindworks," *Science 86,* May 1986, 23–31. Additional information for this chapter comes from James L. McClelland, David E. Rumelhart, and the PDP Research Group, *Parallel Distributed Processing: Explorations in the Microstructure of Cognition* (Cambridge: MIT Press, Bradford Book, 1986), vol. 2, *Psychological and Biological Models;* and Doyne Farmer et al., eds. *Evolution, Games and Learning: Models for Adaptation in Machines and Nature, Proceedings of the Fifth Annual International Conference of the Center for Nonlinear Studies, Los Alamos, NM 87545, USA, May 20–24, 1985* (Amsterdam: North-Holland Physics Publishing, 1985). See also Elizabeth Pennish, "Of Great God Cybernetics and His Fair-Haired Child," *The Scientist,* 14 November 1988, 5, 23.

247. William F. Allman, "Designing Computers That Think the Way We Do," *Technology Review,* May/June 1987, 59–65.

248. Restak, *Brain,* 226.

249. This is not a fanciful example. John J. Hopfield points out "the fact that the supernova in the crab nebula was extensively documented (described [in detail] over several months) in the Chinese literature of the 11th century. It produced a star visible in mid-day, so was clearly a singular event. The Chinese were interested in such events at the time. The European Christian culture was not. This singular event went unrecorded in western literature, although the star would have been readily visible in southern Europe in 1054" (John J. Hopfield, personal communication to author). For a detailed account of the event Hopfield is referring to, see Hans Breuer, *Columbus Was Chinese: Discoveries and Inventions of the Far East* (New York: Herder and Herder, 1972), 1–15.

250. As paraphrased in Gould, *Hen's Teeth and Horses' Toes,* 286.

Society as a Neural Net

251. Marvin Minsky, *The Society of Mind* (New York: Simon and Schuster, 1986).

252. Restak, *Mind,* 249; and "Innovation" 8 September 1986 (PBS Television).

253. "Decision making about a colony's food sources is not conducted by some small group of leader bees, but instead is a product of the intricately interwoven behaviors of thousands of individual bees" (Thomas D. Seeley, *Honeybee Ecology: A Study of Adaptation in Social Life* [Princeton, N.J.: Princeton University Press, 1985], 93).

254. Karl von Frisch, *Bees: Their Vision, Chemical Senses, and Language* (Ithaca, N.Y.: Cornell University Press, 1950), 53–96.

255. Thomas D. Seeley and Royce A. Levien, "A Colony of Mind: The Beehive as Thinking Machine," *The Sciences*, July/August 1987, 39–42.

256. Jeremy Campbell, *Winston Churchill's Afternoon Nap: A Wide-Awake Inquiry into the Human Nature of Time* (New York: Simon and Schuster, 1986), 237; and Carole Douglis, "The Beat Goes On," *Psychology Today*, November 1987, 37.

257. For more on this constant interplay of signals, see Michael Argyle, "Innate and Cultural Aspects of Human Non-verbal Communication," in Blakemore and Greenfield, *Mindwaves*, 55–74.

258. Salisbury, *Black Night, White Snow*, 310, 366, 380–81.

259. Salisbury, *Black Night, White Snow*, 360.

260. Shakespeare regularly drew material from Roman authors like Plutarch, Plautus, Lucius Annaeus Seneca, and Livy (A. L. Rowse, *Shakespeare, the Man* [New York: Harper & Row, 1973]; and Lawrence Danson, "Shakespeare," in *Academic American Encyclopedia* 17:237).

261. Peter Gay, *Freud: A Life for Our Time* (New York: W. W. Norton, 1988).

262. For a brilliant evocation of this aspect of life, see "Shedding Life: On the Mysteries of Dying, Cell by Cell," by Czechoslovakian research immunologist and poet Miroslav Holub, *Science 86*, April 1986, 51–53. See also Wicken, "Thermodynamics, Evolution and Emergence," in Weber, Depew, and J. D. Smith, *Entropy, Information, and Evolution*, 166.

The Expendability of Males

263. A. L. Kroeber, *The Nature of Culture* (Chicago: University of Chicago Press, 1952), 313; Charles Winick, *Dictionary of Anthropology* (New York: Philosophical Library, 1956), 19, 67–68, 265; Torrey E. Fuller, *Witchdoctors and Psychiatrists* (New York: Harper & Row, Perennial Library, 1986), 51; and Benedict, *Patterns of Culture*, 243.

264. David Lamb, *The Africans: Encounters from the Sudan to the Cape* (London: Methuen, 1986), 81.

265. Wilson, *Sociobiology*, 158.

266. Constance Holden, "Why Do Women Live Longer than Men?" *Science*, 9 October 1987, 158–60.

267. "Thabit: The Death of the Knight Rabia, Called Boy Longlocks," in *The Islamic World,* ed. William H. McNeill and Marilyn Robinson Waldman (Chicago: University of Chicago Press, 1983), 6–8.

268. Melvin Konner, "The Gender Option," *The Sciences,* November/December 1987, 3. The tendency of males to die violently at the hands of their fellows is by no means limited to humans. Among European moose and red deer, over 10 percent of the males die as a result of battles with their rivals (Morse, *Behavioral Mechanisms in Ecology,* 197).

269. As calculated by MIT's Richard Rhodes ("Epidemic of War Deaths," *Science News,* 20 August 1988, 124).

270. There have been exceptions to this rule. Some North American Indians, Aleuts, and ancient Irish were polygamous (H. R. Hays, *From Ape to Angel: An Informal History of Social Anthropology* [New York: Alfred A. Knopf, 1958], 167; and *Encyclopedia Americana* [Danbury, Conn.: Grolier, 1985], 22:365). But the bulk of polygamous societies have been located in more tropical regions. Of thirty-one polygamous cultures listed by William N. Stephens, twenty-seven are based in warm climates, and only four belong to the earth's chillier zones (William M. Stephens, *The Family in Cross-Cultural Perspective* [New York: Holt, Rinehart and Winston, 1964], 49–69. See also James Lowell Gibbs, Jr., "Polygamy," in *Academic American Encyclopedia* 15:419; and Gibbs, "Monogamy," in *Academic American Encyclopedia* 13:536).

271. For the relationship between the share of work handled by women and polygamy, see George Peter Murdock, *Social Structure* (New York: Macmillan 1949), 36. Murdock was professor of anthropology at Yale University.

272. Their name may sound silly, but mongongo nuts are serious groceries for the Kalahari Desert's !Kung bushmen. These little morsels supply 50 percent of the !Kung's vegetable diet. The average !Kung eats three hundred of them a day, yet gathering mongongo nuts is as easy as taking a stroll. The nuts strew the landscape in such abundance that thousands "rot on the ground each year for want of picking" (Richard Borshay Lee, "The Hunters: Scarce Resources in the Kalahari," in *Conformity and Conflict: Readings in Cultural Anthropology,* ed. James P. Spradley and David W. McCurdy [Boston: Little, Brown and Co., 1986], 195–96. See also A. W. Johnson and Earle, *Evolution of Human Societies,* 40–41).

273. Morse, *Behavioral Mechanisms in Ecology,* 170–71; and Wilson, *Sociobiology,* 165.

274. For more on the relationship between parental workloads, polygamy, and plumage, see Bonner, *The Evolution of Culture in Animals,* 156–58.

275. Morse, *Behavioral Mechanisms in Ecology,* 203–5; Lorenz, *On Aggression,* 40; and Wilson, *Sociobiology,* 68.

276. Morse, *Behavioral Mechanisms in Ecology,* 206.

277. John Naisbitt, *Megatrends: Ten New Directions Transforming Our Lives* (New York: Warner Books, 1984), 2–5. The Center for Popular Economics places the shift away from muscle-oriented jobs like farming, logging, and factory work even earlier. It points out that "as early as 1950, more than half the labor force worked in service jobs" (The Center for Popular Economics, Nancy Folbre, coord., *A Field guide to the U.S. Economy* (New York: Pantheon Books, 1987), 2.2.

278. U.S. Bureau of the Census, *Statistical Abstracts of the United States: 1988,* 108th ed. (Washington, D.C.: U.S. Government Printing Office, 1987), 428.

279. An additional factor may have contributed to the well-documented physiological demasculinization of men during the last half of the twentieth century: estrogen-aping environmental chemicals. For a thorough review of the research on this topic, see Janet Raloff, "That Feminine Touch: Are Men Suffering from Prenatal or Childhood Exposures to 'Hormonal' Toxicants?" *Science News,* 22 January 1994, 56–58.

How Men Are Society's Dice

280. Wilson, *Insect Societies,* 236–37.

281. Ordish, *Year of the Ant,* 84.

282. Ordish, *Year of the Ant,* 84–90.

283. For descriptions of Arab raiding, see Polk and Mares, *Passing Brave,* 104, 151, 153. For its role as one of the most highly valued activities of bedouin life until as late as the 1940s, see: Sir John Glubb, *A Short History of the Arab Peoples* (New York: Stein and Day, 1969), 25; and Philip K. Hitti, *The Arabs: A Short History* (South Bend, Ind.: Gateway Editions, 1970), 10–18. Hitti is professor emeritus of Semitic literature at Princeton University.

284. David Holden and Richard Jones, *The House of Saud* (London: Pan Books, 1982), 2–174.

Is Pitching a Genetically Acquired Skill?

285. C. T. Dourish, W. Rycroft, and S. D. Iversen, "Postponement of Satiety by Blockade of Brain Cholecystokinin (CCK-B) Receptors," *Science,* September 1989, 1509–11.

286. Anderson, *Food of China,* 9. Don't be misled by the title of this book. It is a breathtaking excursion through the material underpinnings of Chinese society, rich with insights into the hidden machinery of cultural development.

287. Goodall, *In the Shadow of Man,* 34, 171, 200–02, 205–7; and N. M. Tanner, *On Becoming Human,* 79–80.

288. Despite their poor aim, an impressive number of primates throw objects at intruders penetrating their territory. These include gorillas, orangutans, and patas mon-

keys. Others, like gibbons, howlers, red spider monkeys, and cebuses, drop branches and nuts on the invaders. Baboons and macaques roll stones down hills to discourage interlopers (K. R. L. Hall, "Tool-Using Performances as Indicators of Behavioral Adaptability," in Jay, *Primates*, 136–37).

289. Authority accrues to the successful hunter even in the most aggressively egalitarian primitive groups. For example, the Hadza of Tanzania seem staunchly opposed to anyone attempting to act superior or trying to be a boss. Men and women have nearly equal status, and no group has a formal leader. Yet the Hadza encampment eventually takes on the best hunter's name (Ernestine Friedl, "Society and Sex Roles," in Spradley and McCurdy, *Conformity and Conflict*, 162–63). The !Kung of the Kalahari declare that they will allow no one member of the tribe to climb above the others. Yet they generally name their watering holes after some outstanding member of the clan—a superior hunter, orator, or healer. And when they go hunting, the !Kung defer to the judgement of the man with the best track record at bringing home prey. Nunamiut Eskimos praise men who deny the wish "to place themselves above the heads of others." Yet they obediently follow the lead of the best hunter when it comes time to go after caribou (A. W. Johnson and Earle, *Evolution of Human Societies*, 52, 133–37).

290. Napoleon Chagnon, "Life Histories, Blood Revenge, and Warfare in a Tribal Population," *Science*, February 1988, 988–89.

291. The practice persists in primitive hunter-gatherer bands. Among the Ache of eastern Paraguay, top hunters "exchange game for sexual access to women," and "the children of productive hunters are treated better by band members" (Raymond Hames, "Time Allocation," in E. A. Smith and Winterhalder, *Evolutionary Ecology and Human Behavior*, 214).

292. William H. Calvin, *The Throwing Madonna: Essays on the Brain* (New York: McGraw-Hill Book Co., 1983), 28–42.

293. Fossey, *Gorillas in the Mist*, 150, 190, 193.

Oliver Cromwell—the Rodent Instincts Don a Disguise

294. William James, *The Varieties of Religious Experience* (New York: Collier Books, 1961), 215–16. The ancient Germans, the Celts of the Scottish highlands, the Lapps of northern Europe, and the Tlingit of Alaska found the sport of war so delightful that they imagined their dead war heroes had been granted the pleasure of the ultimate entertainment—a warfare that never ends (Canetti, *Crowds and Power*, 43–44).

295. Lorenz, *On Aggression*, 158–59.

296. Antonia Fraser, *Cromwell* (New York: Donald I. Fine, 1973), 17–18.

297. Fraser, *Cromwell*, 72–5. By the way, one of the Englishmen who spread these stories was John Milton, author of *Paradise Lost*, who would eventually become an official propagandist for Cromwell (Fraser, *Cromwell*, 304).

298. Fraser, *Cromwell,* 80–81.

299. Fraser, *Cromwell,* 326–57.

300. Fraser, *Cromwell,* 497–502.

The Invisible World as a Weapon

301. For a sense of what Mecca was like in those days, see the Editors of Time-Life Books, *The March of Islam* (Alexandria, Va.: Time-Life Books, 1988), 22–23.

302. This account of the life of Mohammed and the development of Islam is based on the following sources: Sarwat Saulat, *The Life of the Prophet* (Lahore: Islamic Publications, 1983); McNeill and Waldman, *Islamic World;* Wells, *Outline of History;* J. M. Roberts, *Pelican History of the World;* Gibbon, *Roman Empire* (Penguin Classics), 652.

303. Saint Paul's epiphany on the road to Damascus has also been attributed to epilepsy.

304. Tim Newark, *The Barbarians: Warriors & Wars of the Dark Ages* (London: Blandford Press, 1985), 86.

305. The sieges of Vienna began in 1529 under a second wave of Islamic empire builders, the Turks. See Stanford J. Shaw, *History of the Ottoman Empire and Modern Turkey* (Cambridge, England: Cambridge University Press, 1976), vol. 1, *Empire of the Gazis; The Rise and Decline of The Ottoman Empire 1280–1808,* 94.

The True Route to Utopia

306. Bainton, *Christianity,* 17–27.

307. Sources for this interpretation of Christianity's beginnings include: *The New English Bible: New Testament* (Oxford, England: Oxford University Press; Cambridge, England: Cambridge University Press, 1961); Charles Guignebert, *Jesus* (New York: Alfred A. Knopf, 1935); Joseph Klausner, *The Messianic Idea in Israel,* trans. W. F. Stinespring (New York: Macmillan, 1955); Edgar J. Goodspeed, *Introduction to the New Testament* (Chicago: University of Chicago Press, 1937); and Bainton, *Christianity.*

308. Gibbon, *Roman Empire* (Penguin Classics) 276.

309. Bainton, *Christianity,* 39, 47; and R. L. Fox, *Pagans and Christians,* 266–67.

310. E. Pagels, *Gnostic Gospels,* 28, 118, 127, 139, 141. For Saint Paul's invention of the phrase, see R. L. Fox, *Pagans and Christians,* 370.

311. Gibbon, *Roman Empire* (Penguin Classics), 309–15.

312. The story of Constantine's vision of a cross in the sun goes back to Bishop Eusebius of Caesarea. Twenty-five years after the battle at the Milvian Bridge, Eusebius,

a contemporary of Constantine's, declared that the emperor had mentioned seeing the vision of the cross in the sky. On that cross, Eusebius reported, were the words "By this conquer" (Bainton, *Christianity*, 90). Eusebius was in a privileged position to gain such knowledge: he had dinner with Constantine during the Council of Nicaea, delivered the eulogy to Constantine that opened the official deliberations, and sat at the emperor's right hand during the council's sessions (Boorstin, *Discoverers*, 572; and Bainton, *Christianity*, 96). On the other hand, Gibbon, in his *Decline and Fall of the Roman Empire*, claimed that the story of Constantine's vision was a fairy tale (Gibbon, *Roman Empire* [Penguin Classics], 383). Gibbon felt Constantine's conversion to Christianity was a more gradual process. But whether the emperor had visions or not, Constantine not only made Christianity the empire's official religion, he took to presiding over doctrinal councils and made himself the virtual head of the Christian church. (See also: George Ostrogorsky, *History of the Byzantine State,* trans. Joan Hussey [New Brunswick, N.J.: Rutgers University Press, 1969], 46–48; and R. L. Fox, *Pagans and Christians*, 613–22.)

313. Wells, *Outline of History*, 337.

314. Abû 'Ali al-Muhassin al-Tanûkhî, "Ruminations and Reminiscences," excerpt in McNeill and Waldman, *Islamic World*, 102.

Why Men Embrace Ideas—and Why Ideas Embrace Men

315. Szulc, *Fidel*, 80–86.

316. Even the august third-century Christian theologian Origen, who felt the notion of a flaming furnace beneath the earth was a fantasy, had to admit that hellfire was a devilishly convenient tool for keeping the faithful in line (R. L. Fox, *Pagans and Christians*, 327).

317. These insights into blind faith and hellfire originated with Richard Dawkins in *The Selfish Gene,* 212.

318. For a detailed, archaeologically based description of the Anglo-Saxon conquest, see Michael Wood, *In Search of the Dark Ages* (New York: Facts on File Publications, 1987), 1–60.

319. For a chilling firsthand account of how the Marxist meme was pounded into the brains of Bulgarians, see Georgi Markov, *The Truth That Killed* (New York: Ticknor & Fields, 1984). Markov was an award-winning writer in Communist Bulgaria and a member of the regime's intellectual elite. In 1969, disheartened by an appalling lack of freedom and by the corruption of party members, Markov defected to London and became a broadcaster, beaming his opinions back to the land he had left behind. Bulgarian President Todor Zhivkov was not pleased with Markov's open dissent. On 7 September 1978, the radio commentator was killed on a London street with a poisoned pellet shot from a James Bond–style umbrella.

Righteous Indignation=Greed for Real Estate

320. Paraphrased by Depew and Weber, *Entropy, Information, and Evolution,* 335–36.

321. Tuchman, *Proud Tower,* 177–78.

322. Quoted in William L. Shirer, *20th Century Journey: A Memoir of a Life and the Times* (New York: Simon and Schuster, 1976), vol. 1, *The Start, 1904–1930,* 68.

323. Herman Harvey, audiotaped interview with Hans Morgenthau, *Sum and Substance* (Newport Beach, Calif., Books on Tape, 1986).

324. Atiya, *Crusade, Commerce and Culture,* 18.

Shiites

325. John Reed, *Ten Days That Shook the World* (Harmondsworth, Middlesex: Penguin Books, 1977), 37–41. John Reed, the highest-paid American reporter of his day, was a witness to the critical events of the Russian Revolution. His account was highly sympathetic to the Bolshevik faction. In fact, it was a passionate exposition of the Bolshevik point of view. The introduction to Reed's book would eventually be written by none other than Vladimir Ilyich Lenin. See also Salisbury, *Black Night, White Snow,* 334–35.

326. J. Reed, *Ten Days That Shook The World,* 10, 35.

327. "In 1921 Russian pig-iron production was about one-fifth of its 1913 level, that of coal a tiny 3 per cent or so" (J. M. Roberts, *Pelican History of the World,* 842–43).

328. These figures come from the official Central Statistical Bureau of the U.S.S.R., cited in Heller and Nekrich, *Utopia in Power,* 120. See also Bruce W. Lincoln, *Red Victory: A History of the Russian Civil War* (New York: Simon and Schuster, 1989).

329. Yevgeny Yevtushenko, "Civic Timidity Is Killing Perestroika," *World Press Review,* July 1988, 27. (First published in *Literaturnaya Gazeta.*)

330. Heller and Nekrich, *Utopia in Power,* 235. The figure of fifteen million deaths comes from Iosif G. Dyadkin, *Unnatural Deaths in the USSR, 1928–1954* (New Brunswick, N. J.: Transaction Books), 25. Dyadkin was professor of geophysics at the All-Union Geophysical Research Institute in the Soviet town of Kalinin until he was arrested in 1980 for writing this book.

331. J. R. Tanner, C. W. Previte-Orton, and Z. N. Brooke, eds., *Cambridge Medieval History* (Cambridge, England: Cambridge University Press, 1968), 285; Norman Cohn, *The Pursuit of the Millennium* (New York: Oxford University Press, 1974), 136–40; and M. Harris, *Cows, Pigs, Wars and Witches,* 194–97.

332. For an extremely good account of the struggle that split the early Moslem world between Shiites and Sunnis, see Mohammad Heikal, *The Return of the Ayatollah* (1981; reprint, London: Andre Deutsch, 1983), 75–80. Heikal was the longtime editor of Egypt's leading newspaper, *Al Ahram,* and was a close confidant of Egyptian President Gamal Abdel Nasser. His analysis provides the foundation on which my narrative is based. For the emotional dimension of the Shiite faith, see Canetti, *Crowds and Power,* 146–54. See also E. L. Danie, "Abbasid Dynasty," in Embree, *Encyclopedia of Asian History* 1:3.

333. Saulat, *Life of the Prophet,* 17.

334. Wells, *Outline of History,* 375; and Saulat, *Life of the Prophet,* 39–40.

335. The followers of Ali wouldn't begin to formally define themselves as Shiites ("Shi'at Ali," the party of Ali) until after Ali's death (Heikal, *Return of the Ayatollah,* 79). For the sake of simplicity, I've taken the liberty of referring to them as Shiites from the beginning.

336. This is a figure of speech. Mohammed's revelations were not formally assembled into the Koran until twenty years after the Prophet's death.

337. Wells, *Outline of History,* 374–80.

338. Wells, *Outline of History,* 382; J. M. Roberts, *Pelican History of the World,* 323; and Ronald Grigor Suny, "Armenia," in *Academic American Encyclopedia* 2:172.

339. Fazlur Rahman, *Islam* (Chicago: University of Chicago Press, 1979), 171.

340. Wells, *Outline of History,* 384.

341. P. M. Holt, Ann K. S. Lambton, and Bernard Lewis, eds., *The Cambridge History of Islam* (Cambridge: Cambridge University Press, 1970), vol. 1, *The Central Islamic Lands,* 72; J. M. Roberts, *Pelican History of the World,* 326.

342. Polk and Mares, *Passing Brave,* 103. Polk is director of Middle Eastern studies at the University of Chicago and was a staffer in the State Department's Policy Planning Council under the Kennedy administration. Mares is a former reporter for the *Chicago Sun-Times.* In 1971, the pair mounted an expedition to cross the great sand barrier of Northern Arabia by camel. It was a deliberate effort to recapture the way of life of the bedouins who had founded Arabic culture.

Poetry and the Lust for Power

343. Harris L. Coulter, *Divided Legacy: The Conflict Between Homeopathy and the American Medical Association—Science and Ethics in American Medicine 1800–1910* (Berkeley: North Atlantic Books, 1982), 6, 22–23; Ansley, *Columbia Encyclopedia in One Volume,* 779, 842; and Peter L. Petrakis, "Homeopathy," in *Academic American Encyclopedia* 10:- 212.

344. Coulter, *Divided Legacy,* 328–31.

345. The assertions of the clinical ecologists have been supported by studies published in a wide variety of journals, including *Annals of Allergy, Allergy in Otolaryngologic Practice, Journal of the International Academy of Metabology, Proceedings Third World Congress of Psychiatry,* and Britain's prestigious *Lancet.* For a clear and impressive overview of the maverick field of clinical ecology, see Marshall Mandell, M.D., and Lynne Waller Scanlon, *Dr. Mandell's 5-Day Allergy Relief System* (New York: Pocket Books, 1980), 46–117. Despite the frivolous title of this book, the American Academy of Environmental Medicine calls it "remarkable" and Bernard Rimland, director of the Institute for Child Behavior Research and founder of the National Society for Autistic Children, calls it "excellent." See also: *What Is Clinical Ecology?* (Denver, Colo.: American Academy of Environmental Medicine); and William H. Philpott, M.D., and Dwight K. Kalita, *Brain Allergies* (New Canaan, Conn.: Keats Publishing, 1987), 7, 231. In 1991, a media attack was mounted to discredit clinical ecology as "bogus science." Indirect evidence suggested that the publicity assault may have been orchestrated by the insurance industry to discredit doctors who testified in damage suits against polluters.

346. In 1896, there were 110 homeopathic hospitals, "145 dispensaries, 62 orphan asylums and old people's homes, over thirty nursing homes and sanatoria, and 16 insane asylums." A count of homeopathic medical schools in 1900 put the total number at 22 (Coulter, *Divided Legacy,* 304, 442, 450).

347. Coulter, *Divided Legacy,* 298–302.

348. Coulter, *Divided Legacy,* 17, 59–60.

349. Coulter, *Divided Legacy,* 179–84.

350. Coulter, *Divided Legacy,* 140–450.

351. Coulter, *Divided Legacy,* 432–33.

352. Leo Braudy, *The Frenzy of Renown: Fame and Its History* (New York: Oxford University Press, 1987), 61, 65–66, 82; Gibbon, *Roman Empire* (Penguin Classics), 91–92; and Bradford, *Hannibal,* 73.

353. Braudy, *Frenzy of Renown,* 129–34.

When Memes Collide—the Pecking Order of Nations

354. Margulis and D. Sagan, *Microcosmos,* 64.

355. John Sparks, *The Discovery of Animal Behaviour* (Boston, Little, Brown and Co., 1982), 226–30; and Joseph Altman, *Organic Foundations of Animal Behavior* (New York: Holt Rinehart and Winston, 1966), 454.

356. David McFarland, ed., *The Oxford Companion to Animal Behavior* (New York: Oxford University Press, 1982), 139–40. For a description of pecking order among

grackles, see Burtt, *Psychology of Birds,* 23. See also Grier, *Biology of Animal Behavior,* 568–69.

357. Wilson, *Sociobiology,* 141. See also Robert Burton, *Bird Behavior* (New York: Alfred Knopf, 1985), 133.

358. Wilson, *Sociobiology,* 141.

359. Barash, *Whisperings Within,* 179–80; and Wilson, *Sociobiology,* 97, 118, 137, 139.

360. Lydia Tomoshok, Craig Van Dyke, M.D., and Leonard S. Zegans, M.D., *Emotions in Health and Illness: Theoretical and Research Foundations* (London: Grune & Stratton, 1983), 76; Maser and Seligman, *Psychopathology,* 287–88; Bruce Bower, "Chronic Hypertension May Shrink Brain," *Science News,* 12 September 1992, 166; Raven and Rubin, *Social Psychology,* 294; Ornstein and Sobel, *Healing Brain,* 161–72; Carol Tavris, *Anger: The Misunderstood Emotion* (New York: Simon and Schuster, 1982), 112; and "Hypertension, a Mental Handicap," *Brain Mind Bulletin,* August 1992, 1 (summary of research by Shari Waldstein and Steven Manuck, first published in *Psychological Bulletin* 110:451–68).

361. Konner, *Tangled Wing,* 119.

362. Robert M. Sapolsky, "Lessons of the Serengeti," *The Sciences,* May/June 1988, 42.

363. In one study, psychologist John Paul Scott was actually able to take advantage of this principle to produce mice who consistently won battles even against opponents who towered over them. As David P. Barash describes it in his book *The Hare and the Tortoise,* Scott exposed the little contenders "to a graded series of fights which were all 'fixed' to insure its [the privileged rat's] victory." After a nonstop string of wins, the rodent wrestler's sense of confidence made him unbeatable (Barash, *Hare and the Tortoise,* 154). See also Wilson, *Sociobiology,* 123.

364. "It is to the advantage of a chicken to live in a stable hierarchy. Members of flocks kept in disorder by experimental replacements eat less food, lose more weight when their diet is restricted, and lay fewer eggs" (Wilson, *Sociobiology,* 139). See also McFarland, *Oxford Companion to Animal Behavior,* 12.

365. See the section on "Intergroup Dominance" in Wilson, *Sociobiology,* 144.

366. Pecking order relationships between langur troops in the wild can be as complex as the relations between modern nations. For a description of "dominance structure among troops" of langurs, see Suzanne Ripley, "Intertroop Encounters among Ceylon Gray Langurs (*Presbytis entellus*)," in *Social Communication among Primates,* ed. Stuart A. Altmann (Chicago: University of Chicago Press, 1967), 237–54, especially 248.

367. Wilson, *Sociobiology,* 120.

368. Caesar, *The Conquest of Gaul,* trans. S. A. Handford (Harmondsworth, Middlesex: Penguin Books, 1982), 28–42.

Superior Chickens Make Friends

369. Bradford, *Hannibal,* 21.

370. Bradford, *Hannibal,* 23. See also James Mitchell, *The Illustrated Reference Book of Classical History* (Leicester: Windward, W. H. Smith & Son, 1982), 35.

371. For a map of the key centers of civilization in the days when Carthage's trading powers were at their peak, see Mitchell, *The Illustrated Reference Book of Classical History,* 40.

372. Bradford, *Hannibal,* 33.

373. Bradford, *Hannibal,* 25, 40.

374. Honor Frost, "How Carthage Lost the Sea," *Natural History,* December 1987, 58–67.

375. Bradford, *Hannibal,* 27–28.

376. Bradford, *Hannibal,* 210.

377. Bradford, *Hannibal,* 29–32.

378. Bradford, *Hannibal,* 34–36.

379. Bradford, *Hannibal,* 28.

380. Bradford, *Hannibal,* 36–42.

381. Historians' estimates of the number of men and animals Hannibal took with him vary widely. Ernle Bradford analyzed the classical sources and concluded that Hannibal had reached the Alps with 59,000 foot soldiers, 9,000 cavalrymen and their horses, plus 37 elephants (Bradford, *Hannibal,* 47–48).

382. J. M. Roberts, *Pelican History of the World,* 235.

383. Bradford, *Hannibal,* 115–16.

384. Bradford, *Hannibal,* 87–88.

385. Polybius, *The History of Polybius,* translated by Evelyn S. Shuckburgh from the text of F. Hultsch (Bloomington, Ind.: Indiana University Press, 1962), 2:64.

386. Bradford, *Hannibal,* 87, 182.

387. Bradford, *Hannibal,* 157–59.

388. Bradford, *Hannibal,* 161.

389. Bradford, *Hannibal,* 187–90.

390. Bradford, *Hannibal,* 187–204.

391. Bradford, *Hannibal,* 17–19, 207–9. Additional details for this narrative of Carthage and Hannibal come from Allen M. Ward, "Carthage," in *Academic American Encyclopedia* 4:173–74; and Ward, "Hannibal," in *Academic American Encyclopedia* 10:38.

392. Though Gadir (Cadiz) had been founded by the Phoenicians, it became the key base for Carthage in the development of her commercial relations with Spain. Later it was the center from which Hamilcar Barca carried out his subjugation of the Spanish tribes. Other important Carthaginian outposts in Spanish territory included Ebesus (Ibiza) and Carthago Nova (Carthagena) (Mitchell, *Classical History,* 35).

393. Manchester, *Glory and the Dream,* 789.

394. Szulc, *Fidel,* 509.

395. Napoleon Chagnon describes the troubles of those who move down in the intergroup pecking order in grim terms. For example, as a tribe called the Patanowa-teri slid to the bottom, they fell into "rather desperate straits. Their old enemies . . . began raiding them with even greater frequency. . . . A few additional villages began raiding the Patanowa-teri to settle old grudges, realizing that the Patanowa-teri had so many enemies that they could not possibly retaliate against all of them. The Panatowa-teri then began moving from one location to another, hoping to avoid and confuse their enemies. . . . Each group that raided them passed the word to other villages concerning the location of the Patanowa-teri. . . . The raids were frequent and took a heavy toll. . . . The Patanowa-teri were raided at least twenty-five times while I conducted my fieldwork" (Chagnon, *Yanomamo,* 127). For a full description of the Patanowa-teri's plight, which gives a graphic sense of how pecking order slippage can make life unbearable among the Yanomamo, see Chagnon, *Yanomamo,* 124–37.

Worldviews as the Welding Torch of the Hierarchical Chain

396. Romila Thapar, *A History of India* (1966; reprint, Harmondsworth, Middlesex: Penguin Books, 1985), 1:29–35; D. D. Kosambi, *Ancient India: A History of Its Culture and Civilization* (New York: Pantheon Books, 1965), 72–83; Patricia Bahree, *The Hindu World* (Morristown, N.J.: Silver Burdett Co., 1985), 10; and Chester G. Starr, *A History of the Ancient World* (New York: Oxford University Press, 1974), 166. For a dissenting view on the Aryan invasion, see Franklin Southworthy, "The Reconstruction of Prehistoric South Asian Language Contact," in *The Uses of Linguistics,* ed. Edward Bendix, New York Academy of Sciences, vol. 583 (New York, 1990), 207.

397. Thapar, *History of India,* 1:37–38; and al-Biruni, *Albiruni's India,* 100–01. Al-Biruni—who has shown up in these footnotes before—was a famed Arab mathematician, astronomer, and historian of the eleventh century who learned Sanskrit and traveled in India for thirteen years as a guest of Sultan Mahmud, ruler of territories in

Afghanistan and Iran. Al-Biruni's highly sympathetic account of the Hindu world of nearly one thousand years ago is extraordinary. See also Kosambi, *Ancient India,* 86.

398. Georg Wilhelm Friedrich Hegel, *The Philosophy of History* (New York: Willey Book Co., 1900), 146, 152, 153. Hegel cites "Manu's Code" as his source. For more on the privileges of the Brahmans and the penalties inflicted on the lower castes, see Al-Biruni, *Albiruni's India,* 136, 162, 163.

399. Al-Biruni, *Albiruni's India,* 50–58, 103.

400. Thapar, *History of India,* 1:39–40; and Al-Biruni, *Albiruni's India,* 100–01.

401. For a detailed review of the archaeological evidence on these invasions, see Wood, *In Search of the Dark Ages.*

402. Reischauer, *Japan Past and Present,* 11–12.

The Barbarian Principle

403. For the reality of barbarian lifestyles (they wore beards, dressed in furs, couldn't read, and drank like fiends), see Edward Gibbon, *The Decline and Fall of the Roman Empire* (New York: Modern Library), vol. 1; 190–93; Justine Davis Randers-Pehrson, *Barbarians and Romans* (Norman, Okla.: University of Oklahoma Press, 1983), 39; and Newark, *Barbarians,* 7. For the Roman view (barbarians were dirty, dressed in the tattered skins of mice, drank blood, ate raw food, slept on their horses, and couldn't comprehend even the most infantile technology), see Michael Rouche, "The Early Middle Ages in the West," in *A History of Private Life: From Pagan Rome to Byzantium,* ed. Paul Veyne, trans. Arthur Goldhammer (Cambridge, Mass.: Harvard University Press, 1987), 419, 421; Randers-Pehrson, *Barbarians and Romans,* 41; and Philip Dixon, *The Making of the Past: Barbarian Europe* (Oxford: Phaidon Press, 1976), 13.

404. For the latest archaeological evidence on Egyptian life shortly before Menes united Egypt, see Michael Hoffman, "Before the Pharaohs: How Egypt Became the World's First Nation-State," *The Sciences,* January/February 1988, 40–47.

405. For a firsthand sense of the opulence of early Egyptian life, see the Egyptian collection of the British Museum. See also the Editors of Time-Life Books, *The Age of the God Kings* (Alexandria, Va.: Time-Life Books, 1987), 67; and the diagram of a nobleman's home in B. W. B. Garthoff, "Egyptian Art and Architecture," in *Academic American Encyclopedia* 7:86. The living room of a wealthy Egyptian was an impressively high-ceilinged central hall.

406. As early as the fourth millennium B.C., for example, pottery was already mass produced in a large section of town set aside for this sort of organized enterprise. The industrialists who spearheaded the process shipped the finished goods to distant markets on paddle- and sail-driven riverboats, and used their profits to acquire political power. In the days before the pharaohs, some of these early titans of industry probably became

local kings. It is even possible that the first pharaoh came from among their ranks (Hoffman, "Before the Pharaohs," 44–47).

407. The Hyksos came up with a host of military innovations—among them, chain armor, the battle-ax, the composite bow, and the chariot (Editors of Time-Life Books, *Barbarian Tides: Time Frame 1500–600 B.C.* [Alexandria, Va.: Time-Life Books, 1987], 31).

408. Every Babylonian commercial transaction of any consequence was written down. In addition, Babylonians were prodigious letter writers. Most, however, could not read and write for themselves. To handle their correspondence and business needs, they turned to the local scribe. By the way, the one people who were running neck and neck with the Babylonians in the race for literacy were the Egyptians. For a marvelous account of the origins of Sumerian (and hence Babylonian) script, see Denise Schmandt-Besserat, "Oneness, Twoness, Threeness: How Ancient Accountants Invented Numbers," *The Sciences,* July/August 1987, 44–48. See also: H. W. F. Saggs, *Everyday Life in Babylonia & Assyria* (New York: Dorset Press, 1965), 80–81; Samuel Noah Kramer, *The Sumerians: Their History, Culture and Character* (Chicago: University of Chicago Press, 1963), 23; Time-Life Books, *Age of God Kings,* 16–21, 37–44; and *Encyclopedia Americana* 8:325. For a brief description of how the Babylonians simplified the Sumerian cuneiform, see Albertine Gaur, *A History of Writing* (New York: Charles Scribner's Sons, 1984), 17, 66. For a sense of the parallel development of hieratic script in Egypt, see Morris Bierbrier, *The Tomb Builders of the Pharaohs* (New York: Charles Scribner's Sons, 1982), 78.

409. Jaynes, *Origin of Consciousness in the Breakdown of the Bicameral Mind,* 208.

410. For a map of the Babylonian empire at the time of Nebuchadnezzar, see Wells, *Outline of History,* 184.

411. Jer. 52:24–29.

412. For example, Daniel—of lion's-den fame—made out very well as a consultant to King Nebuchadnezzar, interpreting dreams and giving input on public policy. Eventually, the king "made him ruler over the whole province of Babylon." Other Jews—among them Shadrach, Meshach, and Abednego—also ascended to high-level administrative posts (Dan. 1–2).

413. The Hebrews knew whereof they spoke. Assyria had smashed the ten tribes of the northern Jewish kingdom of Israel in 722 B.C. and deported the survivors. These were the famous lost ten tribes (J. M. Roberts, *Pelican History of the World,* 126). That left only the two tribes of the southern Hebrew state, Judah. These, too, would be carted off when the Babylonians destroyed Jerusalem in 587 B.C.

414. Herodotus, *The Histories,* trans. Aubrey de Selincourt (Harmondsworth, Middlesex: Penguin Books, 1972), 115.

415. "Prowess in fighting," says Herodotus, was the Persians' "chief proof of manliness" (Herodotus, *Histories,* 98). I base the statement that in the days before their conquests, the Persians were unlettered on Herodotus' claim—paraphrased by many modern scholars—that the Persians taught their sons "three things only: to ride, to use the bow, and to speak the truth" (Herodotus, *Histories,* 98). Reading, writing, and arithmetic are conspicuously absent from this list. Other sources for this portrait of the Persians include: G. B. Gray and M. Carey, "The Reign of Darius," in *Cambridge Ancient History* (Cambridge: Cambridge University Press, 1969), vol. 4, *The Persian Empire and the West,* ed. J. B. Bury, S. A. Cook, and F. E. Adcock, 189–91; G. Buchanan, "The Foundation and Extension of the Persian Empire," in *Cambridge Ancient History* 4:3–4; Edward Farmer et al., *Comparative History of Civilizations in Asia* (Boulder, Colo.: Westview Press, 1986), vol. 1, *10,000 B.C. to 1850,* 136; and Starr, *Ancient World,* 277–80.

416. For an account of how "the Medes, who had been masters of Asia beyond the Halys for a hundred and twenty-eight years . . . were forced to bow before the power of Persia," see Herodotus, *Histories,* 95–96.

417. In an unforeseen twist of fate, Queen Nitocris' massive defense apparatus helped undo the Babylonians. The Persians diverted the Euphrates River into the Queen's massive lake, which had degenerated into a swamp. This lowered the river's water to such a level that the Persian army was able to march across the normally mighty stream, clamber up the banks, and rush through Babylon's back gates, catching the inhabitants by surprise (Herodotus, *Histories,* 117–18).

418. A wealthy nobleman once came to the Persian ruler Cyrus the Great to petition that Persia's common folk be allowed to move from their barren homes in the mountains down into the rich valleys they and their leaders had conquered. Cyrus replied contemptuously that "soft countries breed soft men. It is not the property of any one soil to produce fine fruits and good soldiers too." In other words, the Persian people were ordered to stay on their rocky hillsides (Ernle Bradford, *The Battle for the West: Thermopylae* [New York: McGraw-Hill Book Co., 1980], 43).

419. Wells, *Outline of History,* 188.

420. Herodotus, *Histories,* 379–81; J. M. Roberts, *Pelican History of the World,* 191; and Wells, *Outline of History,* 191.

421. Herodotus, *Histories,* 423–30.

422. Herodotus, *Histories,* 472.

423. Herodotus, *Histories,* 465. Modern scholars feel the Persian army may have been far smaller than Herodotus thought. Some are convinced that it comprised closer to 100,000 men (J. M. Roberts, *Pelican History of the World,* 191). Others feel it was closer to 250,000 (Bradford, *Battle for the West,* 34). In 480 B.C., however, even this would have been an immense force.

424. Herodotus, *Histories,* 466–70.

425. Bradford, *Battle for the West.*

426. Herodotus, *Histories,* 452.

427. Herodotus, *Histories,* 453.

428. Kenneth Dover, *The Greeks* (Austin, Tex.: University of Texas Press, 1980), 13.

429. Manchester, *Arms of Krupp,* 125–37, 143–48; and Barraclough, *Origins of Modern Germany,* 422.

430. For a very different analysis of the barbarian menace through history, one which nonetheless supports the conclusions of this chapter, see Bennett Bronson, "The Role of Barbarians in the Fall of States," in Yoffee and Cowgill, *Collapse of Ancient States,* 196–218.

Are There Killer Cultures?

431. Heikal, *The Return of the Ayatollah,* 39–40. The Shah eagerly accepted Stalin's deal . . . until he found out that it entailed hosting a small army of Soviet advisers and technicians.

432. *The New Encyclopaedia Britannica* 1:640 gives the figure of ten thousand. British author David Pryce-Jones says the death toll was actually "several tens of thousands" (Pryce-Jones, "Self-Determination, Arab Style," 43). Twenty thousand is the most commonly used estimate.

433. "The Tale of the Recalcitrant Imam," *New York Times,* July 25, 1982, 12. In early 1989, shortly after he gained a new reputation as a man of peace, Arafat made it clear that he still felt disputes between Mohammedans should be settled with bullets. In an interview with Radio Monte Carlo about the Arab uprising in Israel, Arafat said, "Whoever thinks of stopping the *intifada* before it achieves its goals, I will give him ten bullets in the chest" ("Arafat Unmasked—by His Own Words," *New York Post,* January 23, 1989, 22).

434. Said Mohammed, "I have bequeathed to you that which will always be a guide to you, if you will take hold of it; the Book of God and the practices of my life." This saying and its implications have been critical in the development of Islam (Heikal, *Return of the Ayatollah,* 80).

435. "I will instil terror into the hearts of the unbelievers: smite their necks and smite all their finger-tips off them." *The Qu'ran,* downloaded from America Online's Islam Library, original upload 1996, Al-Anfal, sura 12.

"And slay them wherever ye catch them," *The Qu'ran,* Al-Baqara, sura 191.

In a petition sent to Atlantic Monthly Press objecting to this chapter, the American

Arab Anti-Discrimination Committee points out that many Moslems feel that these are the words of God, not of Mohammed, who was merely Allah's conduit. The true speaker, they wish to point out, is "God Almighty—not Prophet Muhammad, peace be upon him" ("Human Rights Petition For: Civil Liberties of Nasser Ahmed. In Camera," Washington, D.C., Arab American Anti-Discrimination Committee, 1996).

"Twenty-seven military campaigns," Sarwat Saulat, *Life of The Prophet,* 100.

"He personally led nine of them," D. S. Roberts, *Islam,* 42.

436. The idea of permanent warfare between the Moslem world and the non-Moslem world is so deeply ingrained that the phrase for the non-Moslem regions of the planet—Dar al-Harb—literally *means* "The Home of War" (Davidson, *Africa in History,* 219).

437. D. S. Roberts, *Islam,* 42–43. The interpretation cited as nearly universal by Roberts is probably derived from verses 191–93 of the Koranic chapter Al-Baqara, whose passages say, "slay them wherever ye catch them and turn them out from where they have turned you out; for tumult and oppression are worse than slaughter. . . . Such is the reward of those who suppress faith. . . . And fight them on until there is no more tumult or oppression and there prevail justice and faith in Allah." "Justice," in the eyes of many ancient and modern Moslems, means the imposition of Koranic law. "Faith" is understood to be faith in Islam. Hence the passage, in the eyes of many, instructs the pious to use slaughter to impose Koranic law and Islam, since slaughter is preferable to the continuance of non-Islamic codes and beliefs ("tumult and oppression").

438. Canetti, *Crowds and Power,* 141–42.

439. *The Qu'ran,* Al-Baqara, sura 25.

440. *The Qu'ran,* Al-Baqara, sura 42.

441. *The Qu'ran,* Al-Baqara, sura 83.

442. A Moslem cleric rises through five grades of rank in his career. He starts out as a *talib ilm*—a student. He moves on to *mujitahid*—a person capable of arriving at an opinion. Then a *mubelleg al-risala*—"a carrier of the message"; a *hojat al-Islam*—an "authority on Islam"; and *ayatollah*—a "sign of God." The final, and ultimate level of authority is *ayatollah al-uzma*—"great sign of God" (Heikal, *Return of the Ayatollah,* 83).

443. Khomeini, *Sayings of the Ayatollah Khomeini,* 4. For statements similar to this and to those that follow, see Ruhollah Khomeini, *Islam and Revolution,* 34, 48, 286, 287, 327. See also Shaul Bakash, *The Reign of the Ayatollahs,* 234.

444. Khomeini, *Sayings of the Ayatollah Khomeini,* 26.

445. Khomeini, *Sayings of the Ayatollah Khomeini,* 51. Some Western Islamic scholars are intent on "correcting" what they perceive as an undeserved negative image of Islam. They claim that the "believers in God's partners" so often accursed by Mohammed are "pagans" and not Christians. These savants point out quite validly that the

Koran says, "Those who believe [in the Koran] and those who follow the Jewish [Scriptures] and the Christians and the Sabians and who believe in Allah and the last day and work righteousness shall have their reward with their Lord; on them shall be no fear nor shall they grieve" (Al-Baqara, 62). The apologists then ignore literally dozens of passages like the following: "The Jews call Uzair a son of Allah and the Christians call Christ the son of Allah. That is a saying from their mouths; [in this] they but imitate what the unbelievers of old used to say. Allah's curse be on them" (At-Tauba, 30); "O ye who believe! take not the Jews and the Christians for your friends and protectors. . . . Verily Allah guideth not a people unjust" (Al-Maida, 51). Most important, the Koran is explicit in its pronouncement that Christians are, indeed, "those who believe in God's partners": "They do blaspheme who say: Allah is Christ the son of Mary. But said Christ: O children of Israel! worship Allah my Lord and your Lord. Whoever joins other gods with Allah Allah will forbid him the garden and the Fire will be his abode. There will for the wrong-doers be no one to help" (Al-Maida, 72). Also: "They do blaspheme who say: Allah is one of three in a Trinity: for there is no god except One Allah. If they desist not from their word [of blasphemy] verily a grievous penalty will befall the blasphemers among them" (Al-Maida, 73).

446. Bakash, *Reign of the Ayatollahs,* 233. As R. K. Ramazani, the Harry Flood Byrd, Jr., professor of government and foreign affairs at the University of Virginia, put it, "Khomeini believes that the export of revolution is obligatory" in the interests of "an overarching concept of Islamic world order." Ramazani pointed out that Khomeini "rejected . . . the very idea of the [secular, non-Islamic] nation-state. . . . In other words, in Khomeini's ideal Islamic world order there would be no room for the modern secular . . . international system." Khomeini, in Ramazani's words, felt "it is Iran that is uniquely qualified as a nation to pave the way for the ultimate founding of world government. . . . In Khomeini's words, '. . . the Iranian nation must grow in power and resolution until it has vouchsafed Islam to the entire world' " (Ramazani, *Revolutionary Iran,* 20–24).

447. Khomeini, *Islam and Revolution,* 35, 219. Khomeini's view of Christ may become a little clearer if you realize that standard Islamic doctrine asserts the Old and New Testament are insidious corruptions of God's word, and that these perversions were later "corrected" by the Koran.

448. Indyk, "Watershed in the Middle East," 70. For additional information on the progress of Islamic fundamentalism, see "Islam Resumes Its March," by the editors of the *Economist,* reprinted in the *National Times,* May 1992, 9.

449. Esposito, *The Islamic Threat,* 8.

450. Platlea, "Islamic Fever," 34; and "The Fight for African Souls."

451. Danziger, Himelfarb, and Weisenberg, "Schwarz 'Optimistic' on South Africa's Prospects."

452. Robert R. McMillan, publisher of *Long Island Economic Times,* "Do You Have a Stamp of Israel in Your Passport?" *Caucus Current,* May 1992, 28.

453. Andrew Giarelli, "Regional Reports: Asia/Pacific." The Sinkiang region commands fully a sixth of China's territory and is the source of most of the nation's oil and precious metals. As recently as the nineteenth century Moslem uprisings devastated China's Sinkiang and Yunnan sections. The Sinkiang revolt of 1875 temporarily tore the province from Chinese control. The eighteen-year-long Yunnan rising resulted in the deaths of a million people and the near depopulation of the territory's main cities (Eberhard, *History of China,* 301, 304; and Anderson, *Food of China,* 131). In the late 1980s, despite the repressive policies of the Chinese Communist government, Islamic religious leaders in Sinkiang—inspired by Teheran—were once again challenging "the authority of the secular state" (Delfs, "China's Unruly Minorities," 40). And in the 1990s, things had gotten worse.

454. Sumit Ganguly, "Avoiding War in Kashmir."

455. Esposito, *Islamic Threat,* 11, 12, 23, 203, 206; and "Spread of Islamic Rules."

456. Shelby, "Secessions."

457. Holmes, "Iran's Shadow."

458. According to the United Nations International Labor Organization, heavily armed Islamic militias ship black Sudanese men to agricultural forced labor camps, then send desirable women and children north—tied to each other by ropes around their necks—to become involuntary housekeepers and concubines. Meanwhile, Sudan keeps these activities quiet in the American media through the high-priced efforts of the Washington lobbying firm Pagonis & Donnelly (Ward Johnson, "Sudanese Government Wars with Populace," *New York Times,* April 3, 1993, 22).

459. Makram Muhammed Ahmed, "Algeria at the Brink"; Jacques Girardon, "A Veiled Future for Algeria;" and "Will Algeria Become a Second Iran?"

460. Stanley Reed, "Jordan and the Gulf Crisis," 28.

461. Marr, "The Islamic Revival," 37.

462. Kraven, "The Real Face of Kuwait."

463. McKenna, "The Subcontinental Blues."

464. Onaran, "Islamic Revival in Central Asia."

465. Alan Riding, "France, Reversing Course, Fights Immigrants Refusal to Be French," *New York Times,* 5 December 1993, 1, 14. "Alienated young men, who feel particular resentment that they have not found a place in French society, are the principle targets of recruitment by Islamic fundamentalists. 'I'm worried about the fundamen-

talists because police don't go into suburbs rampant with crime and drugs,' Amina [a Tunisian-born French singer] said.''

466. "Islamic fundamentalism in the Balkans . . . constitutes a direct threat to European peace'' (Varitsiotes, "Security in the Mediterranean and the Balkans''). Varitsiotes is defense minister of Greece. For the historical background of Islam in the Balkans, see Dvornik, *The Slavs in European History and Civilization,* and Zivojinovic, "Islam in the Balkans.'' Dragoljub R. Zivojinovic, a professor of history at the University of Belgrade, claims that "the ambitions of Islamic movements in the Balkans, notably in Bosnia-Hercegovina [are] threatening European and even global stability.'' Zivojinovic attempts to demonstrate that two books by Bosnian president Alija Izetbegovic—*Islamska deklaracia* and *Islam between East and West*—are "an exposition of the political essence of fundamentalism and its outlook on the world . . . , an invitation to Muslims around the world to awaken and stand up in order to accomplish a historical duty for which they are predestined.'' Zivojinovic feels Izetbovic's expansionist pronouncements come "close to the ideas of the Ayatollah Khomeini.''

467. For more on the rising threat of Islam, see Krauthammer, "The Unipolar Moment,'' and Tim Weiner, "Blowback from the Afghan Battlefield,'' *New York Times Magazine,* 13 March 1994, 53–55. And for the Islamic world's almost universal embrace of a fundamentalism tainted heavily by hatred of the U.S. after the 1991 Iraqi-American War, see Ahmad, "A Tug of War for Muslim's Allegiance.''

468. Islam is the fastest-expanding religion in the African-American community, with over a million black American adherents (Goldman, "Mainstream Islam Rapidly Embraced by Black Americans''). Funds from countries like Iran, Libya, and Saudi Arabia have made much of this expansion possible. In 1977, for example, three Saudi princes decided to funnel fifty million dollars into American black neighborhoods, but there would be a price. When Saudi Arabia sponsored a Black American Business Conference at Los Angeles's Century Plaza Hotel in 1979, Gerald E. Gray, head of the Pan American Steel Corporation, gave the six hundred African-American entrepreneurs assembled for the event the following advice. To get more Arab money, he said, blacks need to "establish some non-economic relationships [with Islamic interests]. . . . When Arabs attempted to boycott companies, we didn't say anything in their support. When Arabs were accused of creating inflation by raising the price of oil, we had a chance to articulate their position. . . . [W]e're going to have to be their voice in this country if we expect them to participate in business with us'' (Emerson, *The American House of Saud,* 73–74). For an indication of the manner in which Islamic groups have overcome the barriers erected between church and state to run "Islamic cultural programs'' in inner city public schools, see Michael Daly, "Pal Saw the Route of All Evil in Sheik,'' *New York Daily News,* 23 March 1993, 8, 18.

469. Barsky, *Al-Fuqra: Holy Warriors of Terrorism,* 1.

470. Michel, "Allah's G.I.s.''

471. Darlow, *Sword of Islam.*

472. Olmert, *Islam.*

473. For attempted Iranian inroads into the Central Asian Republics, see Olcott, "Central Asia's Catapult to Independence," 108; Robin Wright, "Islam, Democracy and the West"; LeCompte, "Communism Confronts Islam"; Rumer and Rumer, "Who Will Be the Next Yugoslavia?" 37; Mortimer, "New Ism in the East," 50; and Siddiqui, "The Scramble for Central Asia."

474. Curtin, *Cross-Cultural Trade in World History,* 107.

475. Telhami, "Arab Public Opinion and the Gulf War," 443.

476. Mackenzie, "Pitfalls in Policy on the Path to Kabul," 11.

477. Draper, "Visions of Turkey."

478. Darlow, *Sword of Islam.*

479. Olmert, *Islam.*

480. Olmert, *Islam.*

481. Kanan Makiya, *Cruelty and Silence.* Note also the following statement from Hisham Sharabi, professor of European intellectual history and Omar al-Mukhtar Professor of Arab Culture at Georgetown University, one of the Arab-American community's most influential secular intellectual leaders: "[T]he secularists' opposition to Islamic fundamentalism does not make them, as some Western observers seem to think, potentially objective allies of the West in its fight against Islamic fundamentalism. In the secularists' view Western hostility to Islamic fundamentalism, like its hostility to Arab nationalism, stems . . . from imperialist interests and hegemonic goals which the secularist intellectuals alongside the Muslim fundamentalists are unconditionally committed to oppose" (Sharabi, "Modernity and Islamic Revival").

482. Darlow, *Sword of Islam.* This extraordinary documentary, one of the few to probe the hostile world of Islamic fundamentalism, was the result of an eighteen-month investigation in Beirut, Cairo, and Iran.

483. Aziz Said, "Islamic Fundamentalism and the West."

484. Esposito, *Islamic Threat,* 173, 181.

485. Edward W. Said, "The Phony Islamic Threat," 62.

486. Esposito, *Islamic Threat,* 171.

487. Marr, "The Islamic Revival," 37, 43–44.

488. Hamdani, "Islamic Fundamentalism," 38, 44.

489. Khomeini, *Sayings of the Ayatollah Khomeini,* 3–7, 27–28, 31.

490. As long ago as 1983, the Chinese had sold Moslem Pakistan the technology for building atomic bombs the size of soccer balls. Pakistan, in turn, had built facilities for mass-producing these weapons and was fully equipped with the ballistic missiles to deliver them (John Dikkenburg, " 'Supermarket' in the Pacific''). By 1993, there were active nuclear weapons development programs in Iraq, Iran, Libya, and several other Islamic states. According to Harvard University's Samuel P. Huntington, "a top Iranian official has declared that all Muslim states should acquire nuclear weapons'' (Huntington, "The Clash of Civilizations?'').

Violence in South America and Africa

491. Georg Wilhelm Friedrich Hegel, *Lectures on the Philosophy of World History*, trans. H. B. Nisbet (London: Cambridge University Press, 1975), 166.

492. Quoted in Tom Buckley, *Violent Neighbors: El Salvador, Central America and the United States* (New York: Times Books, 1984), 39.

493. Though the United States fought Mexico in 1846, our "imperialist" involvements in Latin America wouldn't blossom until the end of the nineteenth century. The prime symbol of American interests in South America was the United Fruit Company, which became famous for its meddling in "banana republic" politics. As late as 1890, United Fruit (then called the Boston Fruit Company) was still only a fledgling firm sailing a handful of trading schooners to Jamaica. In 1899, J. P. Morgan finally amalgamated Boston Fruit with some smaller firms, cleared and drained a million acres of Central American land, and started the economic behemoth Latin Americans would soon love to hate (Buckley, *Violent Neighbors*, 226–27).

494. For an effort to explain the historical origins of Latin America's culture of violence, see Lawrence E. Harrison, *Underdevelopment Is a State of Mind: The Latin American Case,* Center for International Affairs, Harvard University (Lanham, Md.: Madison Books, 1988).

495. Ken C. Kotecha with Robert W. Adams, *The Corruption of Power: African Politics* (Washington, D.C.: University Press of America, 1981).

496. For a brilliantly detailed picture of modern Africa, its violence, and its political and economic turmoil, see David Lamb, *The Africans: Encounters from the Sudan to the Cape.*

497. "Huge Death Toll Feared in Burundi," *New York Times,* 28 November 1993.

498. Mooney's statement appeared in a report on the Ghost Dance Movement, a sect that briefly defied the Indian tradition of violence and rejected war (James Mooney, *The Ghost-Dance Religion and the Sioux Outbreak of 1890* [Chicago: University of Chicago Press, 1965], 25 [Originally published as part of the *Fourteenth Annual Report of the Bureau of Ethnology to the Secretary of the Smithsonian Institution, 1892–93*]). Thomas Jefferson wrote a spirited defense of Indians based on his own firsthand and extremely methodical

observations (among Jefferson's many accomplishments was a detailed analysis of the structural relationships between Native American languages). One of the criticisms Jefferson addressed was the charge that Native Americans "have no ardor for their females." Here is the founding father's reply: "It's true they [Native Americans] do not indulge those excesses, nor discover that fondness which is customary in Europe; but this is not owing to a defect in nature but to manners. Their soul is wholly bent upon war" (from Thomas Jefferson, *Notes on Virginia,* quoted in Daniel J. Boorstin, *Hidden History* [New York: Harper & Row, Cornelia and Michael Bessie Book, 1987], 117). More recently, anthropologists studying the Kwakiutl Indians of the Pacific Northwest have discovered that these coast dwellers engaged in sophisticated wars designed to exterminate or enslave rival clans. Only the coming of the white man forced them to stop (A. W. Johnson and Earle, *Evolution of Human Societies,* 164). For a disturbing description of the joy that Plains Indians took in killing, see Benedict, *Patterns of Culture,* 106.

The Importance of Hugging

499. William James, *Will, Emotion Instinct and Life's Ideals,* Halvorson Dixit Recording (Newport Beach, Calif.: Books on Tape).

500. Judith Hooper and Dick Teresi, "Sex and Violence," *Penthouse,* February 1987, 42. The classic anecdotal example of this principle is Margaret Mead's contrast between the Arapesh and the Mundugamor of New Guinea. See Margaret Mead, *Male and Female: A Study of the Sexes in a Changing World* (1949; New York: Dell Publishing, 1968), 76–77, 86–88, 117, 134–35. See also the summary of Mead's findings in Hays, *From Ape to Angel,* 347.

501. Halim Barakat, "The Arab Family and the Challenge of Social Transformation," in *Women and Family in the Middle East: New Voices of Change,* ed. Elizabeth Warnock Fernea (Austin, Tex.: University of Texas Press, 1985), 27, 31, 32, 37, 44.

502. Juliette Minces, *The House of Obedience: Women in Arab Society,* trans. Michael Pallis (London: Zed Press, 1982), 33; and Soraya Altorki, *Women in Saudi Arabia: Ideology and Behavior among the Elite* (New York: Columbia University Press, 1986), 31.

503. Ibn Ishaq, *Biography of the Messenger of God,* excerpted in McNeill and Waldman, *The Islamic World,* 16–17.

504. Minces, *House of Obedience,* 33–34. For a blood-chilling account of the treatment of women in Islamic societies, see Jan Goodwin, *Price of Honor: Muslim Women Lift the Veil of Silence on the Islamic World* (Boston: Little, Brown and Company, 1994). Goodwin lived four years in the Moslem world, compiling this book.

505. Cairo University professor of psychology Dr. Yousry Abdel Mohsen says that a similar coldness in the relationship between city men and women lay behind a rash of Egyptian murders during the late eighties in which wives did away with their husbands, stabbing them as many as twenty times, or cutting them in small pieces "for easy dis-

posal'' (Alan Cowell, "Egypt's Pain: Wives Killing Husbands," *New York Times,* 23 September 1989, 4).

506. Lila Abu-Lughod, "Bedouin Blues," *Natural History* 96, no. 7 (July 1987): 24–33.

507. Uris spent years in travel and research preparing for *The Haj.* He employed a research associate on the project—Diane Eagle—whom he credits with "pulling a thousand and one brilliant reports." His goal was to place his fiction in a thoroughly authentic, factual setting. (See the acknowledgments and introduction to Leon Uris, *The Haj* [New York: Bantam Books, 1985]. Uris has also discussed the thoroughness of his research in personal communication with the author.)

508. Halim Barakat cites Shirabi's "important study of the Arab family," which concludes that "the most repressed elements of Arab society are the . . . women, and the children" (Barakat, "Arab Family," in Fernea, *Women and Family,* 27, 31, 32, 37, 44.

509. Minces, *House of Obedience,* 29, 35.

510. Barakat, "Arab Family," in Fernea, *Women and Family,* 27, 31, 32, 37, 44.

511. E. Pagels, *Gnostic Gospels,* xi–xii.

512. Charles Lyall, *Ancient Arabian Poetry,* London, 1930, p. xxiii, quoted in Polk and Mares, *Passing Brave,* 37.

513. Lawrence Stone, *The Family, Sex and Marriage in England, 1500–1800* (New York: Harper & Row, 1977), 161–68; John Cleverley and D. C. Phillips, *Visions of Childhood: Influential Models from Locke to Spock* (New York: Columbia University, Teachers College Press, 1986), 28–29; L. A. Sagan, "Family Ties," 28.

514. Elizabethan children entertained themselves in a variety of other ways. They "caught birds and put their eyes out, tied bottles or tin cans to the tails of dogs, killed toads by putting them on one end of a lever and hurling them into the air by striking the other end, dropped cats from great heights to see whether they would land on their feet, cut off pigs' tails as trophies . . . , inflated the bodies of live frogs by blowing into them with a straw," and stoned dogs to death or drowned them (Keith Thomas, *Man and the Natural World: A History of the Modern Sensibility* [New York: Pantheon Books, 1983], 147).

515. K. Thomas, *Man and the Natural World,* 144.

516. Stone, *England, 1500–1800,* 433; and K. Thomas, *Man and the Natural World,* 45, 186.

The Puzzle of Complacency

517. Eberhard, *History of China,* 63; and Boorstin, *Discoverers,* 74.

518. Eberhard, *History of China,* 198; Dennis Bloodworth and Ching Ping Bloodworth, *The Chinese Machiavelli: 3,000 Years of Chinese Statecraft* (New York: Farrar, Straus and Giroux, 1976), 80; and Boorstin, *Discoverers,* 141.

519. Eberhard, *History of China,* 272.

520. D. Bloodworth and C. P. Bloodworth, *Chinese Machiavelli,* 87.

521. Eberhard, *History of China,* 116–24.

522. Lai Po Kan, *The Ancient Chinese* (1980; reprint, Morristown, N.J.: Silver Burdett Co., 1985), 53.

523. Eberhard, *History of China,* 225.

524. Though Constantine experienced his vision of the cross in A.D. 312 and ordered his soldiers to carry banners and shields inscribed with the symbol of the cross, he did not formally declare himself a Christian until A.D. 324 (J. M. Roberts, *Pelican History of the World,* 283).

525. Ostrogorsky, *History of the Byzantine State,* 44–46; and Ira M. Shiskin, "Istanbul," in *Academic American Encyclopedia* 11:307.

526. Gibbon, *The Decline and Fall of the Roman Empire* (Penguin Classics), 331–32; and J. M. Roberts, *Pelican History of the World,* 284.

527. Gibbon, *The Decline and Fall of the Roman Empire* (Modern Library) 2:163–70; Ostrogorsky, *History of the Byzantine State,* 55; and Boorstin, *Discoverers,* 568. Rome continued to have its own emperor until 476. But from 410 on, the emperor's territory broke down into a collection of "independent estates ruled by [barbarian] Germanic warlords" (Newark, *Barbarians,* 50). As historian J. M. Roberts puts it, Rome's "independence of action was gone" (see J. M. Roberts, *Pelican History of the World,* 287–89; and Wells, *Outline of History,* 341). Meanwhile, the Byzantines "settled down to enjoy a considerable breathing space" (Ostrogorsky, *History of the Byzantine State,* 55).

528. See *Hammond's Historical Atlas* (New York: C. S. Hammond, 1948), map H-9. Though barbarian tribes had bitten off parts of the empire, even these acknowledged Byzantium's preeminence. George Ostrogorsky says, "The lands which had once belonged to the Roman Empire were held to belong to her inalienably and in perpetuity, even though they were under the actual control of Germanic kings" (Ostrogorsky, *History of the Byzantine State,* 69). At the head of the "Roman" Empire was Byzantium.

529. Ostrogorsky, *History of the Byzantine State,* 80.

530. Ostrogorsky, *History of the Byzantine State,* 66–67.

531. Gibbon, *Roman Empire* (Penguin Classics), 640–46.

532. Ostrogorsky, *History of the Byzantine State,* 83–85.

533. The Caliph Omar first entered Byzantine territory in 634. In 636, he trounced the Byzantine army at the Battle of Jarmuk and went on to sweep up Syria, besiege Jerusalem, and humble the Persians. By 640, the Mohammedans had taken Mesopotamia and Armenia, and had begun their conquest of Egypt. By 646, they had taken Alexandria—and hence Egypt—for good (Ostrogorsky, *History of the Byzantine State,* 113–15).

Poverty with Prestige Is Better Than Affluent Disgrace

534. In the Naval Treaty of 1922, the United States, Britain, and Japan agreed to a strict, ten-year limitation on new warship construction and to the abandonment of two million tons of planned or actual military vessels. American Secretary of State Charles Evans Hughes said that as a result of the treaty, "preparation for naval warfare will stop now" (Frederick Lewis Allen, *Only Yesterday: An Informal History of the 1920's* [New York: Harper & Row, Perennial Library, 1964], 109–10 [originally published 1931]). Then, in 1925, negotiators signed the Pact of Locarno, "in which the Western powers guaranteed their mutual frontiers and promised never to go to war over them again." Everyone thought that Locarno was a virtual guarantee of peace . . . except the Germans. German Foreign Minister Gustav Stresemann, who won a Nobel Peace Prize for negotiating the treaty, told his confidants that the document simply bought Germany time in which to rearm (Shirer, *20th Century Journey* 1:250, 415–16).

But this was not the first modern effort to end war by rational means. Almost twenty years earlier, at the Hague Peace Conference of 1907, the Great Powers had made an equally futile effort to eliminate war via negotiation. The agreement hammered out at the Hague prohibited the launch of explosives from balloons, guaranteed the safety of neutral territory, outlawed surprise attack, and limited the use of naval mines. Unfortunately, these well-intentioned resolutions were unable to slow the approach of World War I (Tuchman, *Proud Tower,* 335–36).

535. M. Harris, *Cannibals and Kings,* 104–8; and Marshall D. Sahlins, "Poor Man, Rich Man, Big-Man, Chief," in Spradley and McCurdy, *Conformity and Conflict,* 308.

536. Benedict, *Patterns of Culture,* 178; A. W. Johnson and Earle, *Evolution of Human Societies,* 168–69; and M. Harris, *Cows, Pigs, Wars and Witches,* 94–98.

537. James Burke, *Connections* (Boston: Little, Brown and Co., 1978), 87.

538. Goodall, *In the Shadow of Man,* 34, 171, 200–202, 205–7; and N. M. Tanner, *On Becoming Human,* 79–80.

539. The power of generosity to arouse resentment in the recipient has been confirmed by social psychology experiments performed in the United States, Sweden, and Japan. Among the conclusions: "the subjects who received the help were more likely to appreciate and like the donor if the donor was relatively poor rather than wealthy"; and "recipients . . . are . . . likely to reject the helper if they see the help as diminishing their own self-esteem" (Raven and Rubin, *Social Psychology,* 337–44).

540. "One of the most widely spread traits of human beings, manifest under the most diverse types of social order, is the desire for prestige" (Herskovits, *Economic Anthropology*, 38).

541. Claude Brown, "Manchild in Harlem," *New York Times Magazine*, 16 September 1984, 36–41.

542. Only three of the thousand Masada defenders remained alive.

543. One of the Iranian officers of the Cossack Brigade would eventually make himself shah, and his son would be Shah Mohammad Reza Pahlavi.

544. Heikal, *Return of the Ayatollah*, 27–30.

545. Heikal, *Return of the Ayatollah*, 37.

546. Christopher T. Rand, *Making Democracy Safe for Oil: Oilmen and the Islamic East* (Boston: Little, Brown and Co., 1975), 145.

547. Fifteen thousand Iranian officers went to the United States for between two and three years of training (Heikal, *Return of the Ayatollah*, 68).

548. See Heikal, *Return of the Ayatollah*, 67, for the seven-point plan that the Americans handed to the shah, instructing him on how to boost his popularity.

549. Heikal, *Return of the Ayatollah*, 165.

550. Heikal, *Return of the Ayatollah*, 56; and Daniel Yergin, *The Prize: The Epic Quest for Oil, Money and Power* (New York: Simon and Schuster, 1991).

551. Heikal, *Return of the Ayatollah*, 58–59.

552. Anthony Sampson, *The Seven Sisters: The Great Oil Companies and the World They Shaped* (New York: Bantam Books, 1976), 140–51; and Heikal, *Return of the Ayatollah*, 63.

553. Heikal, *Return of the Ayatollah*, 61–62; I. G. Edmonds, *Allah's Oil: Mideast Petroleum* (New York: Thomas Nelson, 1977), 114–16; and Rand, *Making Democracy Safe for Oil*, 133–39.

554. Heikal, *Return of the Ayatollah*, 62–63; Paul Johnson, *Modern Times: The World from the Twenties to the Eighties* (New York: Harper & Row, Harper Colophon Books, 1985), 491; and Edmonds, *Allah's Oil*, 116–18; and Ramazani, *Revolutionary Iran*, 201.

555. Bakash, *Reign of the Ayatollahs*, 12–13.

556. Attending the festivities were "the Kings of Norway and Sweden, of Thailand and Denmark, of Belgium and Greece. Prince Philip and Princess Anne came from Britain, the Emperor Haile Selassie and President Senghor of Senegal from Africa, Vice-President Agnew from the United States and President Podgorny from the Soviet Union; King Hussein and Presidents Franjieh of Lebanon and Bourguiba of Tunisia . . . not to

mention the Prime ministers of France, Italy and Portugal'' (Heikal, *Return of the Ayatollah*, 94).

557. The official caterer was Maxim's of Paris. But if you didn't care for Maxim's cuisine, the shah graciously flew in the chef of your choice.

558. Heikal, *Return of the Ayatollah*, 94–97.

559. Khomeini, *Sayings of the Ayatollah Khomeini*, 5–27.

Why Prosperity Will Not Bring Peace

560. David Blundy and Andrew Lycett, *Quaddafi and the Libyan Revolution* (London: Weidenfeld and Nicolson, 1987), 107.

561. Blundy and Lycett, *Quaddafi and the Libyan Revolution*, 105, 108, 111.

562. In 1973, for example, OPEC's oil embargo produced a dizzying rise in oil prices—and in income for Libyans. But during the year that followed, the Libyan murder rate shot up 55 percent (Blundy and Lycett, *Quaddafi and the Libyan Revolution*, 111).

563. Lunde, *Murder and Madness*, 32. Lunde is a criminal psychiatrist at Stanford University.

564. Robert L. O'Connell, *Of Arms and Men: A History of War, Weapons, and Aggression* (New York: Oxford University Press, 1989), 10.

565. Marco Polo, *The Travels of Marco Polo* (New York: Dorset Press, 1987), 128–29; David Morgan, *The Mongols* (New York: Basil Blackwell, 1986), 32; R. P. Lister, *Genghis Khan* (New York: Stein and Day, 1969), 54–56, 128, 213; J. J. Saunders, *The History of the Mongol Conquests* (London: Routledge & Kegan Paul, 1971), 63–64; and Chambers, *Devil's Horsemen*, 56–59.

566. Anderson, *The Food of China*, 58; and Harrison E. Salisbury, *War between Russia and China* (New York: W. W. Norton & Co., 1969), 18.

567. Morgan, *Mongols*, 5; J. M. Roberts, *Pelican History of the World*, 364–66; and Boorstin, *Discoverers*, 126.

568. Glubb, *A Short History of the Arab People*, 226. Tamerlane made a regular habit of erecting pyramids of skulls in the towns he had conquered. He had no interest in being remembered as a man of mercy (B. Spuler, ''The Disintegration of the Caliphates in the East,'' in Holt, Lambton, and Lewis, *Cambridge History of Islam* 1:170). Like his Mongol progenitors two hundred years before, Tamerlane was spurred on to his conquests by a burst of good weather that covered the steppes with a bounty of grass and brought the wealth of water to the oases of the deserts (*New Encyclopaedia Britannica* 11:785).

569. For the manner in which wealth brings about warfare among the Maring of New Guinea, see Reader, *Man on Earth*, 44.

570. Dean Archer and Rosemary Gartner, *Violence and Crime in Cross-National Perspective* (New Haven, Conn.: Yale University Press, 1984), 86. Archer's information is based on a ten-year study that drew from the statistics of 110 nations.

571. Wilson, *Sociobiology*, 144.

572. Konner, *Tangled Wing*, 119, 193–94, 472.

573. James M. Dabbs, Jr., and Robin Morris, "Testosterone, Social Class, and Antisocial Behavior in a Sample of 4,462 Men," *Psychological Science*, May 1990, 209–11. For a dissenting view of testosterone's effects, see Marvin Harris, *Our Kind: Who We Are, Where We Came from, Where We Are Going* (New York: Harper & Row, 1989), 264–66.

574. Eleanor Grant, "Of Muscles and Mania," *Psychology Today*, September 1987, 12. For citations of studies showing that steroids heighten aggressive behavior in laboratory animals, see Bruce Svare, "Steroid Use and Aggressive Behavior," *Science*, 2 December 1988, 1227.

575. McFarland, *Oxford Companion to Animal Behavior*, 10.

576. From an article in the *Journal of Neuro-Science*, abstracted in *Brain / Mind Bulletin*, March 1992, 7.

577. Wilson, *Sociobiology*, 124. The same phenomenon has been observed in female ring doves, rats, and a variety of other species. (J. Altman, *Organic Foundations of Animal Behavior*, 453.)

578. Wilson, *Sociobiology*, 139.

579. David Attenborough, *The Living Planet: A Portrait of the Earth* (Boston: Little, Brown and Company), 1984, 156–59; Lon L. McClanahan, Rodolfo Ruibal, and Vaughan H. Shoemaker, "Frogs and Toads in Deserts," *Scientific American*, March 1994, 82–88.

580. Anne Scott Beller, *Fat & Thin: A Natural History of Obesity* (New York: Farrar, Straus and Giroux, 1977), 251.

581. David Holzman, "How Gray Matter Can Mend Itself," *Insight*, 6 February 1989, 51.

582. Metabolism increases by as much as 30 percent after the intake of food. Oxygen consumption and body temperature both go up. The phenomenon is called specific dynamic action (Saul Balagura, *Hunger: A Biophysical Analysis* [New York: Basic Books, 1973], 94; Beller, *Fat & Thin*, 157). In rats, body temperature rises before the first bite of food is even swallowed (Alfred J. Rampone and Myron E. Shirasu, "Temperature Changes in the Rat in Response to Feeding," *Science*, 17 April 1964, 317–19).

583. It takes roughly four hours for the stomach to completely process a single meal.

584. This phenomenon was demonstrated most dramatically in the siege of the Warsaw Ghetto. As the Nazis tried to starve the Jews to death, Jewish physicians carefully studied the impact of food deprivation on themselves and on their fellow ghetto residents. They discovered that the lack of nourishment resulted in a radical decrease of metabolic activity. Since then, the same mechanism has been shown at work in those attempting to lose weight by lowering their food intake (Dr. Julian Fliederbaum et al. "Metabolic Changes in Hunger Disease," in *Hunger Disease: Studies by the Jewish Physicians in the Warsaw Ghetto*, ed. Myron Winick, M.D., trans. Martha Osnos [New York: John Wiley & Sons, 1979], 84–121; and Beller, *Fat & Thin*, 225–26, 251). The strategy of reducing metabolism to cope with shortages is universal. It occurs even in a microscopic organism like daphnia, the water flea (Evelyn Morholt and Paul F. Brandwein, *A Sourcebook for the Biological Sciences* [San Diego: Harcourt Brace Jovanovich, 1986], 267).

585. J. A. Deutsch and D. Deutsch, *Physiological Psychology* (Homewood, Ill.: Dorsey Press, 1966), 7–11, 26–29.

586. J. A. Deutsch and D. Deutsch, *Physiological Psychology*, 186–92; and Restak, *Brain*, 132.

587. The flow of profits from oil picked up dramatically in the years immediately preceding the embargo of 1973. For example, Libya's oil revenues went "from $1.3 billion in 1970 to $2.3 billion in 1973," an increase of 77 percent. (Blundy and Lycett, *Quaddafi and the Libyan Revolution*, 108). It is significant that one of the first major acts of international Arab terrorism performed in the West—the PLO murder of athletes at the Munich Olympics—took place in 1972.

The Secret Meaning of "Freedom," "Peace," and "Justice"

588. According to Caesar, the Gauls used the Greek alphabet for most of their normal purposes, such as keeping public and private accounts. Only the Gallic priests—the Druids—refused to commit their lore to writing, preferring to pass their extensive canon of verses from generation to generation as a strictly oral tradition (Caesar, *Conquest of Gaul*, 141).

589. Favorite topics of debate among the Druids, for example, included "the heavenly bodies and their movements, the size of the universe and of the earth, the physical constitution of the world, and the power and properties of the gods" (Caesar, *Conquest of Gaul*, 141).

590. Caesar, *Conquest of Gaul*, 157.

591. Heikal, *Return of the Ayatollah*, 121.

592. The politician was Mossadeq—the premier who had nationalized the oil industry. The story comes from his son. Reportedly, the result was Mossadeq's complete paralysis (Heikal, *Return of the Ayatollah*, 65).

593. Bakash, *Reign of the Ayatollahs,* 12; and Heikal, *Return of the Ayatollah,* 121.

594. Khomeini first showed signs of political potential in 1941 when he wrote his book *Kashf ol-Asrar.* In it, the cleric attacked the shah's father as a usurper, belittled the legitimacy of the parliament, called the governmental ministries corrupt, and declared the police unspeakably cruel; but the future ayatollah kept relatively quiet for the next twenty years. Then, in 1962, Khomeini erupted when the government passed a new law that, among other things, made it possible for women to vote. This, in the Ayatollah's opinion, was an outrage against Islam (Bakash, *Reign of the Ayatollahs,* 24; and Heikal, *Return of the Ayatollah,* 86).

595. *Taghutis*—the emotionally charged, Koranic term for "tyrants"—was the ayatollah's favorite term for his opponents (Heikal, *Return of the Ayatollah,* 88).

596. Bakash, *Reign of the Ayatollahs,* 86–88.

597. Bakash, *Reign of the Ayatollahs,* 4. See also Shaul Bakash, "The Islamic Republic of Iran, 1979–1989," *Wilson Quarterly,* Autumn 1989, 54–62.

598. Bakash, *Reign of the Ayatollahs,* 62.

599. Bakash, *Reign of the Ayatollahs,* 111–12.

600. Bakash, *Reign of the Ayatollahs,* 111.

601. Bakash, *Reign of the Ayatollahs,* 79.

602. Bakash, *Reign of the Ayatollahs,* 232.

603. "News of the Week in Review," 29 July 1988 (PBS). See also the London monthly *South,* cited in Sterett Pope, "Reconstruction Race," *World Press Review,* November 1988, 44.

604. Bakash, *Reign of the Ayatollahs,* 235; Peter Scholl-Latour, *Adventures in the East: Travels in the Land of Islam,* trans. Ruth Hein (Stuttgart, 1983; New York: Bantam Books, 1988), 151. Peter Scholl-Latour is former editor-in-chief and publisher of the leading German magazine *Der Stern.*

605. Bakash, *Reign of the Ayatollahs,* 237. The Iranian attempts to spread fundamentalism to the USSR had a powerful impact. In 1988, when Soviet Azerbaidzhanis rioted in the streets, they carried pictures of the ayatollah (Marshall I. Goldman, "The USSR's New Class Struggle," *World Monitor,* February 1989, 49).

606. Saburo Eguchi and Vince Sherry (producers), *Asia Now,* 11 September 1993 (Seattle: KCTS; Hawaii Public Television; and NHK Tokyo); Leonard Davis, *The Philippines: People, Poverty and Politics* (New York: St. Martin's Press, 1987), 131. Leonard Davis is a lecturer at the Department of Social Administration, City Polytechnic, Hong Kong. See also: Bakash, *Reign of the Ayatollahs,* 235; and R. J. May, "Muslim and Tribal Filipinos," in *The Philippines after Marcos,* by Ronald James May and Francisco Nemenzo

(New York: St. Martin's Press, 1985), 120. The Moro National Liberation Front has several factions. Some lean toward Islamic fundamentalism. Others seem to favor the more secular Islamic revolutionary approach of Libya's Qaddafi or Iraq's Saddam Hussein. The Moro National Liberation Front has received heavy support from a wide range of Islamic states, including Libya and Saudi Arabia. For a detailed portrait of the Moros, see May, "Muslim and Tribal Filipinos," 110–29. For an example of Iranian-sponsored violence against Islamic regimes closer to home, see Ihsan A. Hijazi, "Pro-Iranian Terror Groups Targeting Saudi Envoys," *New York Times,* 6 January 1989, sec. 1, 15.

607. Thierry Lalevee, "Tehran's New Allies in Africa," *World Press Review,* September 1993, 20–21. (First published in the Arab-oriented publication *Arabies* [Paris].)

The Victorian Decline and the Fall of America

608. Tuchman, *Proud Tower,* 63.

609. The source of the vast majority of the statistics in this chapter is Paul Kennedy, *The Rise and Fall of the Great Powers: Economic Change and Military Conflict from 1500 to 2000* (New York: Random House, 1987).

610. Paul Kennedy, "The (Relative) Decline of America," *Atlantic,* August 1987, 34.

611. Actually, the Chinese and Indians perfected cotton cloth far before the British, who then figured out how to mass produce what the Asians had made by hand.

612. Curtin, *Cross-Cultural Trade In World History,* 252.

613. Rosalind Williams, "Reindustrialization Past and Present," *Technology Review,* November/December 1982, 449–50.

614. The first commercially operated railroad engine in America, the "Stourbridge Lion," was made in Britain. So was the steam engine that powered American industrial hero Robert Fulton's famous steamship the *Clermont* (Harry Edward Neal, *From Spinning Wheel to Spacecraft: The Story of the Industrial Revolution* [New York: Julian Messner, 1965], 59, 73; Bryan Morgan, *Early Trains* [London: Camden House Books, n.d.], 19; H. Philip Spratt, "The Marine Steam-Engine," in *A History of Technology,* ed. Charles Singer et al. [Oxford: Oxford University Press, 1958], vol. 5, *The Late Nineteenth Century, c. 1850 to c. 1900,* 143). British firms built entire railways in Canada, France, Jamaica, Guiana, Argentina, Kenya, Uganda, and numerous other countries (B. Morgan, *Early Trains,* 39; Edward A. Haine, *Seven Railroads* [Cranbury, N.J.: A. S. Barnes & Co., 1979], 24–27, 81, 83, 147–69; and S. Nock, *The Dawn of World Railways, 1800–1850* [New York: Macmillan, 1972], 2–5, 121–25). They supervised the building of the first railways in Japan (Beasley, *Meiji Restoration,* 356). British engineers even played a critical role in the development of new railroad technology for the Germans and Austrians (C. Hamilton Ellis, "The Development of Railway Engineering," in Singer et al., *History of Technology* 5:328).

615. India imported a million yards of cotton fabric in 1814. By 1870, that figure was up to a startling 995 million. Many of India's native weavers simply couldn't compete with the higher quality, cheaper mass-produced cloth and quietly went out of business (Kennedy, *Great Powers,* 148).

616. For further information on the role of steam technology in the British dominance of ocean transport, see Curtin, *Cross-Cultural Trade in World History,* 252.

617. Neal, *From Spinning Wheel to Spacecraft,* 36.

618. MIT's System Dynamics National Model, a computerized approach to macro-economics, gives a vivid sense of how the country that rides the crest of a new technological wave surfs its way to power. For an account of the system and a hint at its historical implications, see Nathaniel J. Mass and Peter M. Senge, "Reindustrialization: Aiming at the Right Targets," *Technology Review,* August/September 1981, 56–65.

619. For a vivid sense of the implacable barrier malaria posed to Europeans, see Sanche de Gramont, *The Strong Brown God: The Story of the Niger River* (Boston: Houghton Mifflin Co., 1976), 161–73, 195–200. See also Curtin, *Cross-Cultural Trade in World History,* 15, 57.

620. Burke, *Connections,* 204–7.

621. Burke, *Connections,* 78; James Burke, *The Day the Universe Changed* (Boston: Little, Brown and Co., 1986), 282–89; and Boorstin, *Discoverers,* 679–84.

622. Burke, *Day the Universe Changed,* 277.

623. It's not that England was totally devoid of entrepreneurs and inventors. Swan, for example, came up with an incandescent lightbulb very early in the game, and Henry Wilde was producing dynamos at a respectable clip. But with their patrician British rhythms, these men of the old order didn't stand a chance. Swan, for example, felt it would be unsporting to patent his inventions. He also made sure that each bulb he produced was individually graded according to its filament's characteristics before it left the plant. The Americans and Germans were not slowed down by such niceties. Edison, for example, patented almost everything he—or his employees—could possibly think up; and he managed to sell eighty thousand lightbulbs in just their first fifteen months on the market.

624. Neal, *From Spinning Wheel to Spacecraft,* 92–93.

625. Sources for this history of the commercial development of electricity include: C. Mackechnie Jarvis, "The Generation of Electricity," in Singer et al., *History of Technology* 5:184–99; C. Mackechnie Jarvis, "The Distribution and Utilization of Electricity," in Singer et al., *History of Technology* 5:211–16; A. Stowers, "The Stationary Steam Engine—1830–1900," in Singer et al., *History of Technology* 5:133–34.

626. Williams, "Reindustrialization Past and Present," 50.

627. Williams, "Reindustrialization Past and Present," 50.

628. Tuchman, *Proud Tower,* 356.

629. Williams, "Reindustrialization Past and Present," 54.

630. Kennedy, *The Rise and Fall of the Great Powers,* 211.

631. Steven Greenhouse, "Germany=#1 Exporter," *New York Times,* 6 October 1988, sec. D.

632. Kennedy, "The (Relative) Decline of America," 30.

633. Needless to say, our addiction to debt had knocked us out of our old position as the world's biggest lending power. By 1988, the country that had moved to the top of the lending pyramid was Japan (Swaminathan S. Anklesaria Aiyar, "The Power Passes," *World Press Review,* October 1988, 55 [First published in *Indian Express* {New Delhi}]).

634. Mayo Mohs, "I.Q.: New Research Shows That the Japanese Outperform All Others in Intelligence Tests. Are They Really Smarter?" *Discover,* September 1982, 22.

635. Adam Smith, "Adam Smith's Money World," 2 November 1987, no. 408 (Educational Broadcasting Cos.).

636. According to the *New York Times,* most American companies were "doing only basic research, waiting for Government research grants or just monitoring the field. . . . [C]ompanies are reluctant to engage in long-term efforts that turn new technology into products." Meanwhile, the Japanese had put together a consortium of forty-five companies and plunged enthusiastically into new product development. However, the United States was held back because company heads were uninterested in projects that didn't promise immediate profits. Their easiest way to make a quick buck off superconductors was to land a fat, superconductor-related defense contract. But, as John A. Alic, who headed an Office of Technology Assessment review of superconductor commercialization, said, "It takes time to move technology from the military to the commercial sector, and we don't have that time anymore" (Andrew Pollack, "U.S. Reported Trailing Japan in the Superconductor Race," *New York Times,* 16 October 1988).

637. Ira C. Magaziner, *The Silent War* (New York: Random House, 1989), 201–30. For Japanese progress on commercializing solar technology, see Tatsuya Anzai, "Will the Market for Solar Cells Ever Heat Up?" *Tokyo Business,* October 1993, 48–50.

638. John Naisbitt, in *Megatrends,* makes a similar point. He notes that America's growth rate has been lagging behind that of Japan for many years. Over the long term, he explains, this sluggishness in the economic race can be deadly. Naisbitt says, "The United Kingdom, by growing just 1 percent less than France, Germany, and the United States, managed in a couple of generations to transform itself from the wealthiest society

on Earth to a relatively poor member of the Common Market" (Naisbitt, *Megatrends*, 58–59).

Scapegoats and Sexual Hysteria

639. Williams, "Reindustrialization Past and Present," 50.

640. Tuchman, *Proud Tower,* 37–38.

641. Theodore H. White, *America in Search of Itself: The Making of the President, 1956–1980* (New York: Harper & Row, Cornelia and Michael Bessie Book, 1982), 258.

642. The figure is from Caddell's memo (White, *America in Search of Itself,* 258).

643. Louis Harris, *Inside America* (New York: Vintage Books, 1987), 33.

644. Allan Bloom, *The Closing of the American Mind* (New York: Simon and Schuster, 1987), 74.

645. A. Bloom, *Closing of the American Mind,* 79.

646. "The Culture of Apathy," *New Republic,* 8 February 1988, 7–8.

647. The campaign against the record industry is one that I chronicle from personal experience. As co-founder of a group called Music in Action, I led a battle against music censorship, keeping detailed files on developments in this area, since I was called on frequently to give interviews on the subject to newspaper, radio, or television journalists.

648. Moira McCormick, "VSDA Applauds As Ill. Gov. Amends Antiobscenity Bill," *Billboard,* 20 February 1988, 43.

649. The author was a leader of the fight against this rather insidious legislation, which is now the law of the land.

650. Bradford, *Hannibal,* 94.

651. The unfortunate victims from out of town included two Greeks and two Gauls (Bradford, *Hannibal,* 123–24).

Laboratory Rats and the Oil Crisis

652. For information on a vast variety of similar experiments, see: Berkowitz, *Aggression,* 4, 7, 8, 19, 22, 34, 41, 42; Hilgard, *Psychology in America,* 371–72; Wilson, *Sociobiology,* 123; and Ulrich and Azrin, "Reflexive Fighting In Response to Aversive Stimulation," 518.

653. Ulrich and Azrin, "Reflexive Fighting" 511–20, especially 516.

654. Goodall, "Life and Death at Gombe," 598–99. Goodall says that scapegoating also occurs among rhesus macaques, baboons, vervets, and langurs (Van Lawick-

Goodall, "A Preliminary Report on Expressive Movements and Communication in the Gombe Stream Chimpanzees," 332).

655. Dollard et al., *Frustration and Aggression,* 31; and Altemeyer, "Marching in Step," 35; and Raven and Rubin, *Social Psychology,* 271–73.

656. Salisbury, *War between Russia and China,* 184.

657. The eight countries—Albania, Bulgaria, Czechoslovakia, East Germany, Hungary, Poland, Romania, and Yugoslavia—all fell under the control of Communist governments between 1944 and 1949.

658. Manchester, *Glory and the Dream,* 412.

659. P. Johnson, *Modern Times;* and Kirk Gentry, *J. Edgar Hoover: The Man and the Secrets* (New York: W. W. Norton & Co., 1991).

660. Manchester, *Glory and the Dream,* 520–30, 700–718.

661. Manchester, *Glory and the Dream,* 406–10, 531.

662. Manchester, *Glory and the Dream,* 957.

663. Ira M. Sheskin, "Suez Canal," in *Academic American Encyclopedia* 18:324.

664. Yergin, *Prize;* and P. Johnson, *Modern Times,* 491–93.

665. White, *America in Search of Itself,* 86–93.

666. President Dwight D. Eisenhower restricted his response to providing food and medicine for the Hungarians, and to sending protests to Soviet premier Bulganin (Manchester, *Glory and the Dream,* 765).

667. Manchester, *Glory and the Dream,* 764–65.

668. Within months of the Suez incident, Nasser was exploiting this prestige to plot the overthrow of governments in Libya, Saudi Arabia, and Iraq (White, *America in Search of Itself,* 93).

669. Shig Fujita, "Japan's CD Imports Top Exports," *Billboard,* 29 October 1988, 86.

670. From a speech by Michael Eisner, chairman of the board and chief executive officer of the Walt Disney Company, delivered in May 1988 before the World Council of Affairs. Eisner pointed out that entertainment produced a far greater trade surplus than even the computer business. Entertainment was our second biggest export, computers our thirteenth.

Why Nations Pretend to Be Blind

671. De Waal, *Chimpanzee Politics,* 23–26.

672. De Waal, *Chimpanzee Politics,* 121.

673. De Waal, *Chimpanzee Politics,* 91–108, 116–21.

674. Manchester, *Glory and the Dream,* 6.

675. Manchester, *Glory and the Dream,* 173–76.

676. P. Johnson, *Modern Times,* 394; Manchester, *Glory and the Dream,* 251; Louis L. Snyder, "Pearl Harbor," in *Academic American Encyclopedia* 15:126; and Ashley Brown, *Modern Warfare: From 1939 to the Present Day* (New York: Crescent Books, 1986), 49.

677. For some of the biological reasons, see the earlier chapter titled "Why Prosperity Will Not Bring Peace."

678. Ladislas Farago and Andrew Sinclair, *Royal Web: The Story of Princess Victoria and Frederick of Prussia* (New York: McGraw-Hill Book Co., 1982), 22. Farago and Sinclair's account of nineteenth-century Prussia, my major source for the story of Bismarck, is based largely on the private papers of the central participants in the epoch's events, papers in the possession of the prince of Hesse, the marquess of Salisbury, and of the British royal family.

679. Barraclough, *Origins of Modern Germany,* 421; and H. W. Koch, *A History of Prussia* (New York: Dorset Press, 1978), 241. The new German prosperity was to Bismarck what water is to a fish. This son of a nobleman first appeared on the pan-German stage in 1851 as Prussian minister to the German Confederation. The Prussian eruption of wealth had begun to flow only the year before. Prussian prosperity rose steadily until the Europe-wide depression of 1873. In those twenty-two years, Bismarck ascended to the Prussian premiership, firmly seized the reins of state, defeated France, and unified Germany.

680. Farago and Sinclair, *Royal Web,* 85.

681. Farago and Sinclair, *Royal Web,* 87–88, 120.

682. Manchester, *Arms of Krupp,* 131.

683. Farago and Sinclair, *Royal Web,* 156.

684. Farago and Sinclair, *Royal Web,* 87, 100–101, 137–38, 182.

685. Farago and Sinclair, *Royal Web,* 171–77.

686. For a catalog of the military "advantages" that convinced the world of France's invincibility, see Kennedy, *The Rise and Fall of the Great Powers,* 186.

687. France's Louis Napoléon fought with an army of only 104,000 men. Meanwhile, the Germans had mobilized 1,183,000, and had sent 400,000 to the French front (Manchester, *Arms of Krupp,* 127, 136).

688. Kennedy, *The Rise and Fall of the Great Powers,* 186; and Farago and Sinclair, *Royal Web,* 175.

689. Manchester, *Arms of Krupp,* 131.

690. Manchester, *Arms of Krupp,* 125–37, 143–48; and Barraclough, *Origins of Modern Germany,* 422.

691. William Shawcross, *The Quality of Mercy: Cambodia, Holocaust and Modern Conscience* (New York: Simon and Schuster, 1984), 18, 20, 21, 26–27, 32, 40–43, 51, 52.

692. Shawcross, *Quality of Mercy,* 110.

693. Shawcross, *Quality of Mercy,* 53.

694. Don't console yourself with the pleasant thought that Pol Pot and the Khmer Rouge have melted away. As of 1994, they were mounting new military attacks in an attempt to regain power. During the years since they'd been ousted by the Vietnamese, the followers of Pol Pot had continued to control refugee camps on the Thai border. There, they kept tens of thousands of camp residents as their virtual prisoners. Though international relief agencies provided the food and medicine that kept inmates of these camps alive, the Khmer Rouge often refused to allow relief agency officials in for inspection. With good reason. The Khmer Rouge used the camp inmates as virtual slave labor, and shot or punished refugees under their control who expressed any objection. Meanwhile, the Khmer Rouge continued to hold a seat at the United Nations, operated a governmental cabinet manned by the same figures who had run the Cambodian holocaust, and waited for the Vietnamese to withdraw their troops. After the pullout, the Khmer Rouge intended to step back into power. What's more, they anticipated that they would have the support of China, the United States, and most of the Southeast Asian countries when they did so ("The Second Coming of Pol Pot: Fears of a Return to the Killing Fields," *World Press Review,* October 1988, 25–28 [First published in *Asiaweek* (Hong Kong)]; Keith Richburg, "Back to Vietnam," *Foreign Affairs,* Fall 1991, 111–32; and *Asia Now,* KCTS, Seattle, Hawaii Public Television, and NHK Tokyo, 7 May 1994).

How the Pecking Order Reshapes the Mind

695. Morse, *Behavioral Mechanisms in Ecology,* 78.

696. Morse, *Behavioral Mechanisms in Ecology,* 87.

697. Anthropologist E. N. Anderson sees this as a fundamental mechanism underlying human creativity. He says, "We know from modern experience that people living on the margin of real want do not experiment: they cannot afford to. Much more innovation takes place among the rich than among the poor. . . . Necessity . . . is no mother of invention." Anderson goes on to demonstrate how periods of abundance may have led the Chinese in 6,000 B.C. to the experimentation that produced a shift from hunting and gathering to agriculture (Anderson, *Food of China,* 13–14). The swing to conserva-

Notes

tive behavior in humans may actually be triggered by a shifting balance of internal chemicals. According to a study conducted by Marvin Zuckerman, a psychologist from the University of Delaware in Newark, Delaware, humans who court adventure have low levels of monoamine oxidase and DBH, and high levels of gonadal hormones (among the gonadal hormones is testosterone). The study implies that humans who avoid the new are carting around the opposite chemical complement (Rick Weiss, "How Dare We: Scientists Seek the Sources of Risk-taking Behavior," *Science News,* 25 July 1987, 57–59).

698. Kennedy, *The Rise and Fall of the Great Powers,* 12.

699. As a publicity adviser to numerous rock stars in the early and mid-seventies, I gave exactly this advice. I repeated the message as a speaker at several national record-industry symposia.

700. Elizabeth Stark, "Mom and Dad: The Great American Heroes," *Psychology Today,* May 1986, 12–13.

701. Birnbach, author of *The Preppie Handbook,* conveyed her astonished impressions to me in a phone call just after she'd come back from her field trip. Her observations were supported by the University of California Cooperative Institutional Research Program's annual survey of 300,000 college freshmen. According to this survey, from 1976 to 1986, the number of students who said that one primary reason for attending college was "to be very well off financially" had climbed by over 30 percent. The number of students majoring in business had doubled from its 1966 level. And students were steering clear of the arts and sciences in their quest for the major that would win them the highest income with the lowest number of years in school (Paul Chance, "The One Who Has the Most Toys When He Dies, Wins," *Psychology Today,* May 1987, 54).

702. For a portrait of fifties' college students that is startlingly similar to what Birnbach saw on the college campuses of the eighties, see Manchester, *Glory and the Dream,* 576–79.

703. This figure is given in an analysis by Northern Illinois University sociologist Robert W. Suchner, who examined the results of national surveys involving over 5,400 undergraduate students (Richard Camer, "Science and Religion: Divided We Stand?" *Psychology Today,* June 1987, 61).

704. Conservatives outnumbered liberals among American voters by almost two and a half to one. While 39 percent called themselves conservative, a meager 16 percent were willing to assume the label of a liberal (L. Harris, *Inside America,* 297).

705. In 1986, 53 percent of Americans under thirty were Republicans and only 47 percent were Democrats (L. Harris, *Inside America,* 298).

706. Fred Barnes, "The GOP Lives: Right Back," *New Republic,* 5 July 1993, 19.

707. By 1977, fully 40 percent of mothers with children under six were holding down a job (Folbre, *U.S. Economy,* chart 3.1).

708. U.S. Bureau of the Census, *Statistical Abstract of the United States: 1988,* 108th ed., 166; Michael E. Porter, "Why U.S. Business Is Falling Behind," *Fortune,* 28 April 1986, 255. Michael E. Porter is a Harvard professor and a former member of President Reagan's Commission on Industrial Competitiveness.

709. For a short history of the U.S. trade deficit, see Thomas Ferguson and Joel Rogers, "The Reagan Victory: Corporate Coalitions in the 1980 Campaign," in *The Hidden Election: Politics and Economics in the 1980 Presidential Campaign,* ed. Thomas Ferguson and Joel Rogers (New York: Pantheon Books, 1981), 10.

710. Porter, "U.S. Business Is Falling Behind," 258; and U.S. Bureau of the Census, *Statistical Abstract of the United States: 1988,* 108th ed., chart 647. For the further decline of wages from 1980 to 1989, see Peter Passell, "America's Position in the Economic Race: What the Numbers Show and Conceal," *New York Times,* 4 March 1990, sec. E4–5.

711. The youth suicide rate in America went up sharply in the 1970s and peaked in 1980. See Constance Holden, "Youth Suicide: New Research Focuses on a Growing Social Problem," *Science,* 22 August 1986, 839. See also David Gelman et al., "Depression," *Newsweek,* 4 May 1987, 48.

712. Porter, "Why U.S. Business Is Falling Behind," 258.

713. Alfred E. Eckes, "Trading American Interests," *Foreign Affairs,* Fall 1992, 152.

Perceptual Shutdown and the Future of America

714. To be specific, the average American was working 40.6 hours in 1973. By 1985, that figure was up to 48.8 (L. Harris, *Inside America,* 17–18, 122).

715. Lefcourt, *Locus of Control,* 8–18; Miller, Rosellini, and Seligman, "Learned Helplessness and Depression," 104–30; Franklin, *Molecules of the Mind,* 131–33; L. A. Sagan, "Family Ties," 28; Goleman, *Vital Lies, Simple Truths,* 38; Restak, *Brain,* 167–69; and Restak, *Mind,* 152.

716. Goleman, *Vital Lies, Simple Truths,* 34–36.

717. Sabine Delanglade and Renaud Belleville, "Competitive Does Not Mean Cheap," *World Press Review,* October 1988, 31. (First published in *L'Express.*) This is an interview with Sony's Akio Morita.

718. Azby Brown, "Japan's Moonhouses," *Omni,* July 1989, 17.

719. Francis Narin and J. Davidson Frame, "The Growth of Japanese Science and Technology," *Science,* 11 August 1989, 600–605.

720. *The 1987 Information Please Almanac* (Boston: Houghton Mifflin Co., 1987), 134.

721. *1987 Information Please Almanac,* 144–284.

722. Anderson, *Food of China,* 69–70.

723. Anderson, *Food of China,* 65, 70, 71.

724. Eberhard, *History of China,* 255.

725. These naval behemoths were powered by two hundred men operating a treadmill. They were seventy-two feet high, extraordinarily maneuverable, and often had rams on their prows (Robert Temple, *The Genius of China: 3,000 Years of Science, Discovery, and Invention* [New York: Simon and Schuster, 1987], 192–94).

726. Boorstin, *Discoverers,* 56–64; D. Bloodworth and C. P. Bloodworth, *Chinese Machiavelli,* 263–64.

727. Breuer, *Columbus Was Chinese,* 166; and Eberhard, *History of China,* 272. Robert Temple says, "In China . . . perfectly cast iron cannons were being produced before Europe even learned how to make cast iron" (Temple, *Genius of China,* 246).

728. Eberhard, *History of China,* 267; Boorstin, *Discoverers,* 191; Kennedy, *The Rise and Fall of the Great Powers,* 6–7; D. Bloodworth and C. P. Bloodworth, *Chinese Machiavelli,* 263–64; and Breuer, *Columbus Was Chinese,* 51.

729. Curtin, *Cross-Cultural Trade in World History,* 125–27; Boorstin, *Discoverers,* 199; and Kennedy, *The Rise and Fall of the Great Powers,* 7.

730. One of the sixteenth-century Chinese critics of the misbegotten policy, Chang Han, wrote, "Those who are in charge of state economic matters . . . [are] ignoring the benefits of the sea trade. How can they be so blind" (Anderson, *Food of China,* 87).

731. D. Bloodworth and C. P. Bloodworth, *Chinese Machiavelli,* 261.

732. Beasley, *Meiji Restoration,* 75.

733. Even humiliation in war did not shatter the Chinese sense of superiority. The Manchus invaded Peking in 1644, and by 1683, they had subdued all China. These barbarians installed their own dynasty on the imperial throne. Inadvertently reinforcing the Chinese notion of cultural and technological supremacy, the Manchus adopted the Chinese system of government and numerous other Chinese practices. They ruled through the old Chinese bureaucratic structure. They secured the cooperation of the Chinese elite. In many ways, they became more Chinese than their new subjects. This shouldn't have been surprising, for even in the days of their initial conquests, the Manchus had tapped the wisdom of Chinese advisers (D. Bloodworth and C. P. Bloodworth, *Chinese Machiavelli,* 269–71; and Eberhard, *History of China,* 278–81).

734. Eberhard, *History of China,* 299–300; and Curtin, *Cross-Cultural Trade,* 243–44.

735. William J. Duiker, *Cultures in Collision: The Boxer Rebellion* (San Rafael, Calif.: Presidio Press, 1978), 19–25; Immanuel D. Y. Hsu, *The Rise of Modern China* (New

York: Oxford University Press, 1975), 426–28; and Reischauer, *Japan, Past and Present,* 138.

736. Duiker, *Cultures in Collision,* 150; and Hsu, *Rise of Modern China,* 490.

The Myth of Stress

737. See James H. Humphrey and Joy N. Humphrey, "Stress in Childhood," in Selye, *Selye's Guide to Stress Research* 3:136–63, for a typical example of educators who fear competition and stress in the classroom.

738. Pelletier, "Stress," in Selye, *Selye's Guide to Stress Research* 3:49.

739. Edward A. Kravitz, "Hormonal Control of Behavior: Amines and the Biasing of Behavioral Output in Lobsters," *Science,* 30 September 1988, 1779.

740. Sharon Begley with Louise Lief, "The Way We Were," *Newsweek,* 10 November 1986, 62–72.

741. Ornstein and Sobel, *Healing Brain,* 164–65.

742. Sapolsky, "Lessons of the Serengeti," 38–42.

743. Lefcourt, *Locus of Control,* 108.

744. Hans Selye, *Stress without Distress* (New York: New American Library, 1975), 19–20. Other writers also acknowledge that some stress is good. Langley Neuropsychiatric Institute's Kenneth R. Pelletier says, "It is abundantly clear that stress is not inherently destructive and is, in fact, often highly beneficial" (Pelletier, "Stress," in Selye, *Selye's Guide to Stress Research* 3:48–49). Jack C. Horn and Jeff Meer, two editors at *Psychology Today,* put it succinctly when they said, "People rust out faster from disuse than they wear out from overuse" (Jack C. Horn and Jeff Meer, "The Vintage Years," *Psychology Today,* May 1987, 83). Richard Restak, M.D., writer of two highly acclaimed PBS television series on the brain and mind, acknowledges that "some people actually thrive on stress" (Restak, *Brain,* 168).

745. Osteoporosis can even be caused by the lack of exercise of skeletal tissue in prolonged space flight. The syndrome is called disuse osteoporosis (Peter L. Petrakis, "Osteoporosis," in *Academic American Encyclopedia* 14:457).

746. William T. Greenough and Fred R. Volkmar, "Pattern of Dendritic Branching in Occipital Cortex of Rats Reared in Complex Environments," *Experimental Neurology,* August 1973, 491–504; Mark R. Rosenzweig, "Environmental Complexity, Cerebral Change, and Behavior," *American Psychologist* 21 (1966): 321–42; Mark R. Rosenzweig, Edward L. Bennett, and Cleeves Diamond, "Brain Changes in Response to Experience," *Scientific American,* February 1972, 22–29; Marion C. Diamond, "Enrichment Response of the Brain," in *Encyclopedia of Neuroscience* 1:396–97; and Grier, *Biology of Animal Behavior,* 568–69. For information on how sensory deprivation in young mon-

keys damages their visual systems, see Austin H. Riesen, "Plasticity of Behavior: Psychological Aspects," in *Biological and Biochemical Bases of Behavior,* ed. Harry F. Harlow and Clinton N. Woolsey (Madison, Wis.: University of Wisconsin Press, 1965), 425–50.

747. M. C. Diamond, "Enrichment Response of the Brain," in *Encyclopedia of Neuroscience* 1:396–97; Altman, *Organic Foundations of Animal Behavior,* 372–73, 376–77; and Restak, *Mind,* 76–77.

748. Harvey D. Goldstein, *Ceremony of Innocence* Lecture series (Broadcast and Media Services, University of Southern California). 1986. Early Christian tradition identified the author of Ecclesiastes as King Solomon. Robin Lane Fox says modern scholars disagree, ascribing the work to an unknown Jewish author of the third century B.C. (R. L. Fox, *Pagans and Christians,* 322).

749. G. K. Chesterton, *What I Saw in America* (1922; New York: Da Capo Press, 1968), 105.

750. Reischauer, *Japanese,* 154–55.

751. Peter Tasker, *The Japanese: A Major Exploration of Modern Japan* (New York: E. P. Dutton & Co., Truman Talley Books, 1988), 88–92; and Shotaro Ishinomori, *Japan, Inc.: An Introduction to Japanese Economics,* trans. Betsey Scheiner (Berkeley: University of California Press, 1988), 284–86. For insights into the philosophy behind this zeal, see Michio Morishima, *Why Has Japan 'Succeeded'? Western Technology and the Japanese Ethos* (New York: Cambridge University Press, 1982), 117.

752. L. A. Sagan, "Family Ties," 22.

Tennis Time and the Mental Clock

753. McFarland, *Oxford Companion to Animal Behavior,* 479–80; and J. Altman, *Organic Foundations of Animal Behavior,* 425.

754. Mihaly Csikszentmihalyi, "Memes Vs. Genes: Notes from the Culture Wars," in Brockman, *Reality Club,* 117. For the fraction of a second of light an eye can discern as a discrete flicker, see J. A. Deutch and D. Deutch, *Physiological Psychology,* 350.

755. A few rare athletes can actually glean a message from a mere hundredth of a second of a ball's trajectory. Baseball player Ted Williams at the age of fifty demonstrated that he could register exactly where the seams were on the ball as it smacked into his bat at eighty miles an hour. Williams smeared pine tar on the bat's barrel and called out the part of the ball that he'd hit. A sample call: "one quarter of an inch above the seam." When the ball was checked to see where the tar had left its mark, Williams was right five out of seven times (Arthur Seiderman and Steven Schneider, *The Athletic Eye* (New York: Hearst Books, 1983), 17–18, 91).

756. "In these coffee houses you could borrow money, lend it, invest it, or spend it." In fact, one coffeehouse owner actually began selling insurance to his merchant-capi-

talist clientele. In time, the insurance venture proved more lucrative than serving cups of Java. The coffeehouse proprietor was Edward Lloyd, as in Lloyd's of London (Burke, *Connections,* 193–94. See also Fernand Braudel, *The Structures of Everyday Life: Civilization & Capitalism, 15th–18th Century,* trans. Sian Reynolds [New York: Harper & Row, Perennial Library, 1981], 1:254–60; and Mitchell Stephens, *A History of the News: From the Drum to the Satellite* [New York: Viking, 1988], 41–43).

757. Eberhard, *History of China,* 169, 196; Anderson, *Food of China,* 55–56; and Curtin, *Cross-Cultural Trade,* 104–5.

758. Robert Christopher, *The Japanese Mind* (New York: Fawcett Columbine, 1983), 163.

759. Frederick Jackson Turner first presented his thesis, "The Significance of the Frontier in American History," in 1893. The concept wasn't published in book form until almost thirty years later when *The Frontier in American History* appeared in print (New York: Henry Holt, 1920). For a modern variation on the frontier hypothesis, see Boorstin, *Hidden History,* ix–xxv.

760. For the Hundred Years' War, see Tuchman, *Distant Mirror,* 48–594; and G. M. Trevelyan, *A Shortened History of England* (1942; Harmondsworth, Middlesex: Penguin Books, 1959), 181–88. For the loss of Calais in 1558, see James A. Williamson, *The Evolution of England: A Commentary on the Facts* (Oxford: Oxford University Press, 1944), 179; and *New Encyclopaedia Britannica* 2:731.

761. A. L. Rowse, *The Expansion of Elizabethan England* (Newport Beach, Calif.: Books on Tape). Cambridge University historian Eric Walker agrees with Rowse's assessment (Eric A. Walker, *The British Empire: Its Structure and Spirit, 1497–1953* [Cambridge, Mass.: Harvard University Press, 1956], 2). For the futile campaigns against the French with which the English king nearly bankrupted his government, see J. J. Scarisbrick, *Henry VIII* (Berkeley, Calif.: University of California Press, 1968), 33–35, 453–56, and virtually the entire rest of the book; J. D. Mackie, *The Oxford History of England: The Earlier Tudors, 1485–1558* (London: Oxford University Press, 1962), 410–12; Kenneth O. Morgan, ed., *The Oxford Illustrated History of Britain* (New York: Oxford University Press, 1984), 256; and *New Encyclopedia Britannica* 5:840–41.

762. K. Thomas, *Man and the Natural World,* 116–17.

763. When granted anything they wished, the heroes of these tales chose such items as "a bun, a sausage, and as much wine as he can drink," "white bread and chicken," or "crude wine and a bowl of potatoes in milk" (Robert Darnton, *The Great Cat Massacre and Other Episodes in French Cultural History* [New York: Vintage Books, 1985], 22, 24–34).

764. Boorstin, *Discoverers,* 236–44.

765. On 12 March 1519, Cortés landed at Tabasco and overwhelmed an Aztec army that outnumbered his tiny force three hundred to one (Hammond Innes, *The Conquistadors* [New York: Alfred A. Knopf, 1969], 42–52).

766. Trevelyan, *Shortened History of England,* 206.

767. See Steven Levy, *Artificial Life: A Report from the Frontier Where Computers Meet Biology* (New York: Vintage Books, 1993).

768. For the feasibility of space colonies, see Gerard K. O'Neill, *2081: A Hopeful View of the Human Future* (New York: Simon and Schuster, Touchstone Book, 1981), 61–75. When he wrote this book, O'Neill was president of the Space Studies Institute.

The Lucifer Principle

769. Paul G. Hattersley, M.D., "Blood," in *Academic American Encyclopedia* 3:335.

770. Quoted in Canetti, *Crowds and Power,* 230–31.

771. Caesar bragged of wanting "to punish" the Eburones' "heinous crime with total annihilation" (Caesar, *Conquest of Gaul,* 149). When his troops overwhelmed the leading Gallic city of Avaricum—a town Caesar himself called "almost the finest in Gaul"—his troops killed nearly forty thousand inhabitants, including all the women, children, and elderly (Caesar, *Conquest of Gaul,* 169).

772. Gibbon, *The Decline and Fall of the Roman Empire* (Modern Library), 1:25–26.

773. Burke, *Connections,* 84.

774. Burke, *Connections,* 82.

Epilogue

775. David Layzer, "Growth of Order in the Universe," in Weber, Depew, and J. D. Smith, *Entropy, Information and Evolution,* 23–24.

776. *New Encyclopaedia Britannica* 11:702.

777. Daniel R. Brooks, D. David Cumming, and Paul LeBlond, "Dollo's Law and the Second Law of Thermodynamics: Analogy or Extension?" in Weber, Depew, and J. D. Smith, *Entropy, Information and Evolution,* 190–91; and Layzer, "Growth of Order in the Universe," 28.

778. Ornstein and Sobel, *Healing Brain,* 159.

779. Hull, *Science as a Process.*

780. Bruce Russett, "Peace among Democracies," *Scientific American,* November 1993, 120.

781. Jesse H. Ausubel, "2020 Vision," *The Sciences,* November/December 1993, 16–18. For further information on the low rate of war between democracies, see Carol R. Ember, Bruce Russett, and Melvin Ember, "Political Participation and Peace: Cross-Cultural Codes," *Cross-Cultural Research,* February/May 1993, 97–145.

782. Canetti, *Crowds and Power,* 99–103.

BIBLIOGRAPHY

≫≪

Abbott, Edwin A. *Flatland: A Romance of Many Dimensions*. New York: Barnes & Noble Books, 1983.

Abelson, Philip H. "Soviet Science." *Science,* 26 February 1988.

Abu-Lughod, Lila. "Bedouin Blues." *Natural History* 96, no. 7 (July 1987): 24–33.

Academic American Encyclopedia. Danbury, Conn.: Grolier, 1985.

Adelman, George, ed. *Encyclopedia of Neuroscience*. Boston: Birkhauser, 1987.

Ahmad, Eqbal. "A Tug of War for Muslim's Allegiance: Fundamentalist Currents Vie for Ascendancy." *World Press Review,* November 1991, 24–25. (First published in *New Statesman and Society.*)

Ahmed, Makram Muhammed. "Algeria at the Brink." *World Press Review,* September 1991, 34. (First published in *Al-Musawar* [Cairo].)

Aiyar, Swaminathan S. Anklesaria. "The Power Passes." *World Press Review,* October 1988, 55. (First published in *Indian Express* [New Delhi].)

Alexander, Franz G., and Sheldon T. Selesnick. *The History of Psychiatry: An Evaluation of Psychiatric Thought and Practice from Prehistoric Times to the Present*. New York: Harper & Row, 1966.

Allen, Frederick Lewis. *Only Yesterday: An Informal History of the 1920's*. 1931. New York: Harper & Row, Perennial Library, 1964.

Allman, William F. "Designing Computers That Think the Way We Do." *Technology Review,* May/June 1987, 59–65.

———. "Mindworks." *Science 86,* May 1986.

Alper, Joseph. "Depression at an Early Age." *Science 86,* May 1986.

Altemeyer, Bob. "Marching in Step: A Psychological Explanation of State Terror." *The Sciences,* March/April 1988, 30–38.

Altman, Joseph. *Organic Foundations of Animal Behavior.* New York: Holt Rinehart and Winston, 1966.

Altmann, Stuart A., ed. *Social Communication among Primates.* Chicago: University of Chicago Press, 1967.

Altorki, Soraya. *Women in Saudi Arabia: Ideology and Behavior among the Elite.* New York: Columbia University Press, 1986.

Anderson, E. N. *The Food of China.* New Haven, Conn.: Yale University Press, 1988.

Ansley, Clarke F., ed. *The Columbia Encyclopedia in One Volume.* New York: Columbia University Press, 1940.

Anzai, Tatsuya. "Will the Market for Solar Cells Ever Heat Up?" *Tokyo Business,* October 1993, 48–50.

Archer, Dean, and Rosemary Gartner. *Violence and Crime in Cross-National Perspective.* New Haven, Conn.: Yale University Press, 1984.

Argyle, Michael. "Innate and Cultural Aspects of Human Non-verbal Communication." In *Mindwaves: Thoughts on Intelligence, Identity and Consciousness,* edited by Colin Blakemore and Susan Greenfield. Oxford: Basil Blackwell, 1989.

"Ariane 4 Launch Considered Crucial to Satellite Programs." *Aviation Week & Space Technology,* 20 June 1988, 18.

"Arianespace to Solicit Payloads for Low-Cost Launches on Ariane 5." *Aviation Week & Space Technology,* 20 June 1988, 18.

Aristotle. *Ethica Nicomachea.* Translated by W. D. Ross. In *Introduction to Aristotle,* by Richard McKeon. Chicago: University of Chicago Press, 1973.

Arjomand, Said Amir. *The Turban for the Crown: The Islamic Revolution in Iran.* New York: Oxford University Press, 1988.

Atiya, Aziz. *Crusade, Commerce and Culture.* Bloomington, Ind.: Indiana University Press, 1962.

Attenborough, David. *The Living Planet: A Portrait of the Earth.* Boston: Little, Brown and Co., 1984.

Austin, M. M., and P. Vidal-Naquet. *Economic and Social History of Ancient Greece: An Introduction.* Berkeley: University of California Press, 1977.

Ausubel, Jesse H. "2020 Vision." *The Sciences,* November/December 1993, 16–18.

Aziz Said, Abdul. "Islamic Fundamentalism and the West." *Mediterranean Quarterly,* Fall 1992, 21–36.

Azrin, N. H., R. R. Hutchinson, and D. F. Drake. "Extinction Induced Aggression." In *Aggression: A Re-Examination of the Frustration-Aggression Hypothesis,* edited by Leonard Berkowitz. New York: Atherton Press, 1969.

Bahree, Patricia. *The Hindu World.* Morristown, N.J.: Silver Burdett Co., 1985.

Bainton, Roland H. *Christianity.* Boston: Houghton Mifflin Co. American Heritage Library, 1987.

Bakash, Shaul. "The Islamic Republic of Iran, 1979–1989." *Wilson Quarterly,* Autumn 1989, 54–62.

————. *The Reign of the Ayatollahs: Iran and the Islamic Revolution.* New York: Basic Books, 1984.

Balagura, Saul. *Hunger: A Biophysical Analysis.* New York: Basic Books, 1973.

Baldwin, Marshall W., ed. *A History of the Crusades: The First Hundred Years.* Philadelphia: University of Pennsylvania Press, 1955.

Barakat, Halim. "The Arab Family and the Challenge of Social Transformation." In *Women and Family in the Middle East: New Voices of Change,* edited by Elizabeth Warnock Fernea. Austin, Tex.: University of Texas Press, 1985.

Barash, David P. *The Hare and the Tortoise: Culture, Biology, and Human Nature.* New York: Penguin Books, 1987.

————. *Sociobiology and Behavior.* New York: Elsevier Scientific Publishing Co., 1977.

————. *The Whisperings Within: Evolution and the Origin of Human Nature.* New York: Penguin Books, 1979.

Barinaga, Marcia. "Cell Suicide: By Ice not Fire: a cascade of new research findings is giving researchers insights into the genes that cause programmed cell death," *Science,* 11 February 1994, 754–56.

Barnes, Fred. "The GOP Lives: Right Back." *New Republic,* 5 July 1993, 19.

Barraclough, Geoffrey. *The Origins of Modern Germany.* New York: W. W. Norton & Co., 1984.

Barsky, Yehudit. *Al-Fuqra: Holy Warriors of Terrorism.* New York: Anti-Defamation League, 1993.

Beasley, W. G. *The Meiji Restoration*. Stanford, Calif.: Stanford University Press.

Begley, Sharon, with Louise Lief. "The Way We Were." *Newsweek,* 10 November 1986, 62–72.

Beller, Anne Scott. *Fat & Thin: A Natural History of Obesity*. New York: Farrar, Straus and Giroux, 1977.

Benedict, Ruth. *Patterns of Culture*. 1934. New York: New American Library, Mentor Book, 1950.

Benet's Reader's Encyclopedia. 3d ed. New York: Harper & Row, 1987.

Bergland, Richard. *The Fabric of Mind*. Harmondsworth, Middlesex: Viking Penguin, 1986.

Berkowitz, Leonard, ed. *Aggression. See* Azrin.

Bierbrier, Morris. *The Tomb Builders of the Pharaohs*. New York: Charles Scribner's Sons, 1982.

The Biography of Thomas Edward Lawrence, Lawrence of Arabia, 1883–1935. Pasadena, Calif.: Cassette Book Co., 1983.

Birdsell, J. B. *Human Evolution: An Introduction to the New Physical Anthropology*. Chicago: Rand McNally & Co., 1972.

al-Biruni, Abu Raihan Muhammad ibn Ahmad. *Albiruni's India*. Translated by Edward Sachau. Edited by Ainslie T. Embree. New York: W. W. Norton & Co., 1971.

Black, J. B. *The Reign of Elizabeth: 1558–1603*. Oxford: Oxford University Press, 1959.

Blakemore, Colin, and Susan Greenfield, eds. *Mindwaves. See* Argyle.

Bloodworth, Dennis, and Ching Ping Bloodworth. *The Chinese Machiavelli: 3,000 Years of Chinese Statecraft*. New York: Farrar, Straus and Giroux, 1976.

Bloom, Allan. *The Closing of the American Mind*. New York: Simon and Schuster, 1987.

Bloom, Floyd E. "Endorphins." In *Encyclopedia of Neuroscience*. Vol. 1. *See* Adelman.

Blundy, David. "The U.S. in Space." Excerpted in *World Press Review,* November 1988, 10. (First published in *London Sunday Telegraph*.)

Blundy, David, and Andrew Lycett, *Quaddafi and the Libyan Revolution*. London: Weidenfeld and Nicolson, 1987.

Boesch, Christophe, and Hedwige Boesch-Acherman. "Dim Forest, Bright Chimps." *Natural History,* September 1991, 50.

Bonner, John Tyler. *The Evolution of Culture in Animals.* Princeton, N.J.: Princeton University Press, 1980.

Boorstin, Daniel J. *The Discoverers: A History of Man's Search to Know His World and Himself.* New York: Random House, Vintage Books, 1985.

————. *Hidden History.* New York: Harper & Row, Cornelia and Michael Bessie Book, 1987.

Bower, Bruce. "The Character of Cancer." *Science News,* 21 February 1987, 120–21.

————. "Chronic Hypertension May Shrink Brain." *Science News,* 12 September 1992, 166.

————. "Heart Attack Victims Show Fatal Depression." *Science News,* 23 October 1993, 263.

————. "Million Cell Memories." *Science News,* 15 November 1986, 313–15.

————. "Personality Linked to Immunity." *Science News,* 15 November 1986, 310.

————. "Taking Hopelessness to Heart." *Science News,* 31 July 1993, 79.

Bradford, Ernle. *The Battle for the West: Thermopylae.* New York: McGraw-Hill Book Co., 1980.

————. *Hannibal.* New York: McGraw-Hill Book Co., 1981.

————. *The Shield and the Sword: The Knights of St. John, Jerusalem, Rhodes and Malta.* New York: E. P. Dutton & Co., 1973.

————. *The Sword and the Scimitar: The Saga of the Crusades.* New York: G. P. Putnam's Sons, 1974.

Braudel, Fernand. *The Structures of Everyday Life: Civilization and Capitalism, 15th–18th Century.* Translated by Sian Reynolds. Vol. 1. New York: Harper & Row, Perennial Library, 1981.

Braudy, Leo. *The Frenzy of Renown: Fame and Its History.* New York: Oxford University Press, 1987.

Braun, Wernher von, and Frederick I. Ordway III. *Space Travel: A History,* an update of *History of Rocketry and Space Travel.* Revised in collaboration with Dave Dooling. New York: Harper & Row, 1985.

Breuer, Hans. *Columbus Was Chinese: Discoveries and Inventions of the Far East.* New York: Herder and Herder, 1972.

Brien, Alan. Letter. *New York Times Book Review,* 1 January 1989, 2.

Brokhin, Yuri. *Hustling on Gorky Street.* Newport Beach, Calif.: Books on Tape, 1989.

Bronson, Bennett. "The Role of Barbarians in the Fall of States." In *The Collapse of Ancient States and Civilizations,* edited by Norman Yoffee and George L. Cowgill. Tucson, Ariz.: University of Arizona Press, 1988.

Brooks, Daniel R., D. David Cumming, and Paul LeBlond. "Dollo's Law and the Second Law of Thermodynamics: Analogy or Extension?" In *Entropy, Information and Evolution: New Perspectives on Physical and Biological Evolution,* edited by Bruce H. Weber, David J. Depew, and James D. Smith. Cambridge: MIT Press, Bradford Book, 1988.

Brown, Ashley. *Modern Warfare: From 1939 to the Present Day.* New York: Crescent Books, 1986.

Brown, Azby. "Japan's Moonhouses." *Omni,* July 1989, 17.

Brown, Claude. "Manchild in Harlem." *New York Times Magazine,* 16 September 1984, 36–41.

Brown, Peter. *The Body and Society: Men, Women, and Sexual Renunciation in Early Christianity.* New York: Columbia University Press, 1988.

Bruner, Jerome S. *Beyond the Information Given: Studies in the Psychology of Knowing.* Edited by Jeremy M. Anglin. New York: W. W. Norton & Co., 1973.

Buchanan, G. "The Foundation and Extension of the Persian Empire." In *The Cambridge Ancient History.* Vol. 4, *The Persian Empire and the West,* edited by J. B. Bury, S. A. Cook, and F. E. Adcock. Cambridge: Cambridge University Press, 1969.

Buckley, Tom. *Violent Neighbors: El Salvador, Central America and the United States.* New York: Times Books, 1984.

Burke, James. *Connections.* Boston: Little, Brown and Co., 1978.

————. *The Day the Universe Changed.* Boston: Little, Brown and Co., 1986.

Burton, Robert. *Bird Behavior.* New York: Alfred A. Knopf, 1985.

Burtt, Harold E. *The Psychology of Birds: An Interpretation of Bird Behavior.* New York: Macmillan, 1967.

Buss, Leo W. *The Evolution of Individuality.* Princeton, N.J.: Princeton University Press, 1987.

Butler, R. A. "The Effect of Deprivation of Visual Incentives on Visual Exploration Motivation in Monkeys." *Journal of Comparative and Physiological Psychology* 50 (1957): 177–79.

Byrne, Donn. *An Introduction to Personality.* Englewood Cliffs, N.J.: Prentice-Hall, 1966.

Caesar, *The Conquest of Gaul.* Translated by S. A. Handford. Harmondsworth, Middlesex: Penguin Books, 1982.

Calvin, William H. *The Throwing Madonna: Essays on the Brain.* New York: McGraw-Hill Book Co., 1983.

Camer, Richard. "Science and Religion: Divided We Stand?" *Psychology Today,* June 1987, 61.

Cammer, Leonard. *Up from Depression.* New York: Simon and Schuster, 1969.

Campbell, Jeremy. *Winston Churchill's Afternoon Nap: A Wide-Awake Inquiry into the Human Nature of Time.* New York: Simon and Schuster, 1986.

Campbell, John H. "Evolution as Nonequilibrium Thermodynamics: Halfway There?" In *Entropy, Information, and Evolution. See* Brooks.

Canetti, Elias. *Crowds and Power.* Translated by Carol Stewart. New York: Farrar, Straus and Giroux, 1984.

Carpenter, Rhys. *Discontinuity in Greek Civilization.* Cambridge: Cambridge University Press, 1966.

Castro, Janice. "Blast-off for Profits: A New Roster of Space Racers Line up to Launch the World's Satellites." *Time,* 2 March 1987.

Chagnon, Napoleon "Life Histories, Blood Revenge, and Warfare in a Tribal Population." *Science,* February 26 1988, 985–92.

———. *Yanomamo: The Fierce People.* New York: Holt, Rinehart and Winston, 1968.

Chambers, James. *The Devil's Horsemen: The Mongol Invasion of Europe.* New York: Atheneum, 1979.

Chance, Paul. "The One Who Has the Most Toys When He Dies, Wins." *Psychology Today,* May 1987, 54.

Chesterton, G. K. *What I Saw in America.* 1922. New York: Da Capo Press, 1968.

Christopher, Robert. *The Japanese Mind.* New York: Fawcett Columbine, 1983.

Cialdini, Robert B. *Influence: How and Why People Agree to Things.* New York: William Morrow, 1984.

Clark, Matt, and David Gelman with Mariana Gosnell, Mary Hager, and Barbara Schuler. "A User's Guide to Hormones." *Newsweek,* 12 January 1987, 50–59.

Cleverley, John, and D. C. Phillips. *Visions of Childhood: Influential Models from Locke to Spock.* New York: Teachers College Press, Columbia University, 1986.

Clubb, O. Edmund. *20th Century China.* New York: Columbia University Press, 1978.

Cohen, Sheldon, J. R. Kaplan, Joan E. Kunick, Steven E. Manuck, and Bruce S. Rabin. "Chronic Social Stress Affiliation and Cellular Immune Response in Non-Human Primates." *Psychological Science,* September 1992, 301–04.

Cohn, Norman. *The Pursuit of the Millennium.* New York: Oxford University Press, 1974.

Coleman, Ray. *Lennon.* New York: McGraw-Hill Book Co., 1986.

Corsini, ed., Raymond J. and Bonnie D. Ozaki, assistant ed. *Encyclopedia of Psychology.* New York: John Wiley & Sons, 1984.

Cowell, Alan. "Egypt's Pain: Wives Killing Husbands," *New York Times,* 23 September 1989, 4.

Coulter, Harris L. *Divided Legacy: The Conflict between Homeopathy and the American Medical Association—Science and Ethics in American Medicine 1800–1910.* Berkeley: North Atlantic Books, 1982.

Cousins, Norman. *Human Options.* New York: W. W. Norton & Co., 1981.

Csikszentmihalyi, Mihaly. "Memes Vs. Genes: Notes from the Culture Wars." In *The Reality Club,* edited by John Brockman. New York: Lynx Books, 1988.

Culbert, T. Patrick. "The Collapse of Classic Maya Civilization." In *The Collapse of Ancient States and Civilizations. See* Bronson.

"The Culture of Apathy." *New Republic,* 8 February 1988, 7–8.

Curtin, Philip D. *Cross-Cultural Trade in World History.* New York: Cambridge University Press, 1984.

Dabbs, Jr., James M., and Robin Morris. "Testosterone, Social Class, and Antisocial Behavior in a Sample of 4,462 Men." *Psychological Science,* May 1990, 209–11.

Daly, Martin, and Margo Wilson. "Evolutionary Social Psychology and Family Homicide." *Science,* October 1988, 519–23.

Danie, E. L. "Abbasid Dynasty." In *Encyclopedia of Asian History,* edited by Ainslie Thomas Embree. New York: Charles Scribner's Sons, 1988.

Danziger, Raphael, Joel Himelfarb, and Mindy Weisenberg. "Schwarz 'Optimistic' on South Africa's Prospects." *Near East Report,* 3 August 1992, 146.

Darlow, David (producer and director). *The Sword of Islam*. Manchester: Granada TV, 1987.

Darnton, Robert. *The Great Cat Massacre and Other Episodes in French Cultural History*. New York: Vintage Books, 1985.

Darwin, Charles. *The Descent of Man and Selection in Relation to Sex*. London: John Murray, 1871.

————. *The Origin of Species by Means of Natural Selection or the Preservation of Favoured Races in the Struggle for Life*. Edited by J. W. Burrow. London: Penguin Books, 1968. (Originally published in 1859.)

Davidson, Basil. *Africa in History*. New York: Collier Books, 1974.

Davies, A. Powell. *The First Christian*. New York: Mentor Books, 1959.

Davies, Paul. *The Cosmic Blueprint: New Discoveries in Nature's Creative Ability to Order the Universe*. New York: Simon and Schuster, 1988.

Davis, James C. "Toward a Theory of Revolution." In *Roots of Aggression*. *See* Berkowitz.

Davis, Leonard. *The Philippines: People, Poverty, and Politics*. New York: St. Martin's Press, 1987.

Davis, Mark J. (writer, director, and editor). *The Private Lives of Dolphins*. Nova. Boston: WGBH, 1992.

Dawkins, Richard. *The Selfish Gene*. New York: Oxford University Press, 1976.

Delanglade, Sabine, and Renaud Belleville. "Competitive Does Not Mean Cheap." *World Press Review*, October 1988, 31–32. (First published in *L'Express*.)

Delfs, Robert. "China's Unruly Minorities." *World Press Review*, December 1988, 40. (First published in *Far Eastern Economic Review*.)

Denber, Herman C. B. "Depression: Pharmacological Treatment." In *International Encyclopedia of Psychiatry, Psychology, Psychoanalysis and Neurology*, edited by Benjamin B. Wolman. Vol. 4. New York: Van Nostrand Reinhold Co., 1977.

Deng, Francis M., "The Tragedy in Sudan Must End: A Personal Appeal to Compatriots and to Humanity," *Mediterranean Quarterly*, Winter 1994, 45–55.

Depew, David J., and Bruce H. Weber. "Consequences of Nonequilibrium Thermodynamics for Darwinism." In *Entropy, Information, and Evolution*. *See* Brooks.

Desborough, V. R. "The End of Mycenaean Civilization and the Dark Ages." In *The Cambridge Ancient History*. Vol. 2, part 2, *The History of the*

Middle East and the Aegean Region, 1380–1000 B.C., edited by I. Edwards, C. Gadd, N. Hammond, and E. Sollberger. *See* Buchanan.

Deutsch, J. A., and D. Deutsch. *Physiological Psychology*. Homewood, Ill.: Dorsey Press, 1966.

de Waal, Frans. *Chimpanzee Politics: Power & Sex among Apes*. New York: Harper Colophon Books, 1984.

Diamandopoulos, P. "Thales of Miletus." In *The Encyclopedia of Philosophy*, edited by Paul Edwards. Vol. 8. New York: Macmillan, 1967.

Diamond, Jared. "The Great Leap Forward." *Discover*, May 1989, 50–60.

Diamond, Marian C. "Enrichment Response of the Brain." In *Encyclopedia of Neuroscience*. Vol. 1. *See* Adelman.

Dikkenburg, John. " 'Supermarket' in the Pacific." *World Press Review*, September 1992, 14–16. (First published in *Asia Magazine* [Hong Kong].)

Dixon, Philip. *The Making of the Past: Barbarian Europe*. Oxford: Phaidon Press, 1976.

Dollard, John, Neal E. Miller, Leonard W. Doob, O. H. Mowrer, Robert R. Sears, Clellan S. Ford, Carl Iver Hovland, and Richard E. Sollenberger. *Frustration and Aggression*. New Haven, Conn.: Yale University Press, 1957.

Douglas, Mary. *Natural Symbols: Explorations in Cosmology*. New York: Pantheon Books, 1982.

Douglis, Carole. "The Beat Goes On." *Psychology Today*, November 1987, 37–42.

Dourish, C. T., W. Rycroft, and S. D. Iversen. "Postponement of Satiety by Blockade of Brain Cholecystokinin (CCK-B) Receptors." *Science*, September 1989, 1509–11.

Dover, Kenneth. *The Greeks*. Austin, Tex.: University of Texas Press, 1980.

Draper, Roger. "Visions of Turkey." *World Press Review*, May 1990, 44.

Duiker, William J. *Cultures in Collision: The Boxer Rebellion*. San Rafael, Calif.: Presidio Press, 1978.

Duncalf, Frederic. "The First Crusade: Clermont to Constantinople." In *A History of the Crusades. See* Baldwin.

Dunlop, D. M. *Arab Civilization to A.D. 1500*. New York: Praeger, 1971.

Durkheim, Emile. *Suicide: A Study in Sociology*. Translated by John A. Spaulding and George Simpson. New York: Free Press, 1951.

Dutt, Gargi, and V. P. Dutt. *China's Cultural Revolution.* Bombay: Asia Publishing House, 1970.

Dvornik, Francis. *The Slavs in European History and Civilization.* New Brunswick, N.J.: Rutgers University Press, 1962.

Dyadkin, Iosif G. *Unnatural Deaths in the USSR, 1928–1954.* New Brunswick, N.J.: Transaction Books, 1983.

Eagle, Morris, and David Wolitzky. "Perceptual Defense." In *International Encyclopedia of Psychiatry, Psychology, Psychoanalysis & Neurology.* Vol. 8. *See* Denber.

Eberhard, Wolfram. *A History of China.* London: Routledge & Kegan Paul, 1977.

Eckes, Alfred E. "Trading American Interests." *Foreign Affairs,* Fall 1992, 135–54.

Economist, Editors of. "Islam Resumes Its March." *National Times,* May 1992, 9. (First published in the *Economist.*)

Edmonds, I. G. *Allah's Oil: Mideast Petroleum.* New York: Thomas Nelson, 1977.

Eguchi, Saburo, and Vince Sherry (producers). *Asia Now.* 11 September 1993. Seattle: KCTS; Hawaii Public Television; and NHK Tokyo.

Einstein, Albert. *The Meaning of Relativity.* 5th ed. Princeton, N.J.: Princeton University Press, 1955.

Ember, Carol R., Bruce Russett, and Melvin Ember, "Political Participation and Peace: Cross Cultural Codes," *Cross-Cultural Research,* February/May 1993, 97–145.

Embree, Ainslie Thomas, ed. *Encyclopedia of Asian History. See* Danie.

Emerson, Steven. *The American House of Saud: The Secret Petrodollar Connection.* New York: Franklin Watts, 1985.

The Encyclopedia Americana. Danbury, Conn.: Grolier, 1985.

"Epidemic of War Deaths." *Science News,* 20 August 1988, 124.

Erickson, Marilyn T. *Child Psycho-pathology: Behavior Disorders and Developmental Disabilities.* Englewood Cliffs, N.J.: Prentice-Hall, 1982.

Erikson, Erik H. *Childhood and Society.* 2d ed. New York: W. W. Norton & Co., 1953.

Esposito, John L. *The Islamic Threat: Myth or Reality?* New York: Oxford University Press, 1992.

Farago, Ladislas, and Andrew Sinclair. *Royal Web: The Story of Princess Victoria and Frederick of Prussia.* New York: McGraw-Hill Book Co., 1982.

Farmer, Doyne, Alan Lapedes, Norman Packard, and Burton Wendroff, eds. *Evolution, Games and Learning: Models for Adaptation in Machines and Nature, Proceedings of the Fifth Annual International Conference of the Center for Nonlinear Studies, Los Alamos, NM 87545, USA, May 20–24, 1985.* Amsterdam: North-Holland Physics Publishing, 1985.

Farmer, Edward, Gavin Hambly, David Kopf, Byron Marshall, and Romeyn Taylor. *Comparative History of Civilizations in Asia.* Vol. 1, *10,000 B.C. to 1850.* Boulder, Colo.: Westview Press, 1986.

Fenton, John Y., Norvin Hein, Frank E. Reynolds, Alan L. Miller, and Niels C. Nielsen, Jr. *Religions of Asia.* New York: St. Martin's Press, 1983.

Ferguson, Thomas, and Joel Rogers, "The Reagan Victory: Corporate Coalitions in the 1980 Campaign." In *The Hidden Election: Politics and Economics in the 1980 Presidential Campaign,* edited by Thomas Ferguson and Joel Rogers. New York: Pantheon Books, 1981.

Festinger, Leon, Henry W. Riecken, and Stanley Schachter. *When Prophecy Fails: A Social and Psychological Study of a Modern Group That Predicted the Destruction of the World.* New York: Harper Torchbooks, 1966.

"The Fight for African Souls," *World Press Review,* June 1992, 48. (First published in *Der Spiegel.*)

Fliederbaum, Julian, Ari Heller, Kazimierz Zweibaum, Suzanne Szejnfinkel, Regina Elbinger, and Fajga Ferszt. "Metabolic Changes in Hunger Disease." In *Hunger Disease: Studies by the Jewish Physicians in the Warsaw Ghetto,* edited by Myron Winick, M.D. Translated by Martha Osnos. New York: John Wiley & Sons, 1979.

Foell, Earl W. "Making Sense of the World." *World Monitor,* October 1988, 28–29.

Folbre, Nancy, coord. The Center for Popular Economics, *A Field Guide to the U.S. Economy.* New York: Pantheon Books, 1987.

Foote, P. G., and D. M. Wilson. *The Viking Achievement.* London: Sidgwick & Jackson, 1980.

Fossey, Dian. *Gorillas in the Mist.* Boston: Houghton Mifflin Co., 1983.

Fox, Robin. *The Red Lamp of Incest: An Enquiry into the Origin of Mind and Society.* New York: E. P. Dutton & Co., 1981.

———. *The Search for Society: Quest for a Biosocial Science and Morality.* New Brunswick, N.J.: Rutgers University Press, 1989.

———, ed. *Biosocial Anthropology.* London: Malaby Press, 1975.

Fox, Robin Lane. *Pagans and Christians.* San Francisco: Harper & Row, 1986.

Frank, Jerome D., M.D. "Some Psychopathological and Sociopsychological Determinants of Bloodthirstiness," *Medicine and War,* January–March 1994, 36–49.

Franklin, Jon. *Molecules of the Mind: The Brave New Science of Molecular Psychology.* New York: Atheneum, 1987.

Fraser, Antonia. *Cromwell.* New York: Donald I. Fine, 1973.

————. *The Warrior Queens.* New York: Alfred A. Knopf, 1989.

Frautschi, Steven. "Entropy in an Expanding Universe." In *Entropy, Information and Evolution. See* Brooks.

Freedman, Daniel G. *Human Sociobiology: A Holistic Approach.* New York: Free Press, 1979.

Freedman, Daniel X. "Psychic Energizer." *McGraw-Hill Encyclopedia of Science and Technology.* Vol. 11. New York: McGraw-Hill Book Co., 1982.

Frend, W. H. C. *The Rise of Christianity.* Philadelphia: Fortress Press, 1984.

Freud, Sigmund. *The Future of an Illusion.* New York: W. W. Norton & Co., 1989.

Friedl, Ernestine. "Society and Sex Roles." In *Conformity and Conflict: Readings in Cultural Anthropology,* edited by James P. Spradley and David W. McCurdy. Boston: Little, Brown and Co., 1986.

Frisch, Karl von. *Bees: Their Vision, Chemical Senses, and Language.* Ithaca, N.Y.: Cornell University Press, 1950.

Fromm, Erich. *The Art of Loving.* 1956. New York: Harper & Row, 1974.

Frost, Honor. "How Carthage Lost the Sea." *Natural History,* December 1987, 58–67.

Fujita, Shig. "Japan's CD Imports Top Exports." *Billboard,* 29 October 1988, 86.

Fuller, Torrey E. *Witchdoctors and Psychiatrists.* New York: Harper & Row, Perennial Library, 1986.

Galbraith, John Kenneth. *The Great Crash: 1929.* 1954. Boston: Houghton Mifflin, Co., 1988.

Gallenkamp, Charles. *Maya: The Riddle and Rediscovery of a Lost Civilization.* New York: Penguin Books, 1976.

Gamow, George. *One, Two, Three—Infinity.* New York: Dover Publications, 1988.

Ganguly, Sumit. "Avoiding War in Kashmir." *Foreign Affairs,* Winter 1990/91, 59–73

Gaur, Albertine. *A History of Writing.* New York: Charles Scribner's Sons, 1984.

Gay, Peter. *Freud: A Life for Our Time.* New York: W. W. Norton & Co., 1988.

Gelman, David, Mary Hager, Shawn Doherty, Mariana Gosnell, George Raine, and Daniel Shapiro. "Depression." *Newsweek,* 4 May 1987.

Gentry, Kirk. *J. Edgar Hoover: The Man and the Secrets.* New York: W. W. Norton & Co., 1991.

Ghiglieri, Michael Patrick. *The Chimpanzees of Kibale Forest: A Field Study of Ecology and Social Structure.* New York: Columbia University Press, 1984.

————. *East of the Mountains of the Moon.* New York: Free Press, 1988.

————. "War among the Chimps." *Discover,* November 1987, 66–76.

Giarelli, Andrew. "Regional Reports: Asia/Pacific." *World Press Review,* June 1992, 34.

Gibbon, Edward. *The Decline and Fall of the Roman Empire* (unabridged). 3 vols. New York: Modern Library, n.d.

————. *The Decline and Fall of the Roman Empire* (abridged version). Edited by Dero Saunders. New York: Penguin Classics, 1985.

Gibbons, Ann. "Chimps, More Diverse Than a Barrel of Monkeys." *Science,* 17 January 1992, 287–88.

————. "Evolutionists Take the Long View on Sex and Violence: Warring over Women." *Science,* 20 August 1993, 987–88.

Gilgen, Albert R. *American Psychology since World War I: A Profile of the Discipline.* Westport, Conn., Greenwood Press, 1982.

Girardon, Jacques. "A Veiled Future for Algeria: Fundamentalist Power Gives Rise to Uncertainty." *World Press Review,* August 1990, 32–33. (First published in *L'Express.*)

Glubb, Sir John. *A Short History of the Arab Peoples.* New York: Stein and Day, 1969.

Goffman, Ervin. *The Presentation of Self in Everyday Life.* New York: Doubleday, Anchor Books, 1959.

Goldman, Albert. *The Lives of John Lennon.* New York: William Morrow, 1988.

Goldman, Ari L. "Mainstream Islam Rapidly Embraced by Black Americans," *The New York Times,* 21 February 1989, 1 and B4.

Goldman, Marshall I. "The USSR's New Class Struggle." *World Monitor,* February 1989, 46–50.

Goldstein, Harvey D. *Ceremony of Innocence.* Lecture series. University of Southern California, Broadcast and Media Services.

Goleman, Daniel. *Vital Lies, Simple Truths: The Psychology of Self-Deception.* New York: Simon and Schuster, 1985.

Goodall, Jane. *Among the Wild Chimpanzees.* Edited and produced by Barbara Jampel. Produced by the National Geographic Society and WQED/Pittsburgh. National Geographic Special. Stamford, Conn.: Vestron Video, 1987.

———. *In the Shadow of Man.* 1971. Boston: Houghton Mifflin Co., 1983.

———. "Life and Death at Gombe." *National Geographic Magazine,* May 1979, 592–621.

———. "A Preliminary Report on Expressive Movements and Communication in the Gombe Stream Chimpanzees." In *Primates: Studies in Adaptation and Variability,* edited by Phyllis C. Jay. New York: Holt, Rinehart and Winston, 1968.

Goodspeed, Edgar J. *Introduction to the New Testament.* Chicago: University of Chicago Press, 1937.

Goodwin, Jan. *Price of Honor: Muslim Women Lift the Veil of Silence on the Islamic World.* Boston: Little, Brown and Company, 1994.

Gould, Stephen Jay. *Hen's Teeth and Horses' Toes.* New York: W. W. Norton & Co., 1984.

Gramont, Sanche de. *The Strong Brown God: The Story of the Niger River.* Boston: Houghton Mifflin Co., 1976.

Grant, Eleanor. "Of Muscles and Mania." *Psychology Today,* September 1987, 12.

Grant, Michael. *The Rise of the Greeks.* New York: Charles Scribner's Sons, 1987.

Grant, Michael, and John Hazel. *Gods and Mortals: Classical Mythology, a Dictionary.* New York: Dorset Press, 1985.

Gratus, Jack. *The False Messiahs.* New York: Taplinger Publishing Co., 1975.

Graves, Robert. *The Greek Myths.* Vol. 2. New York: Penguin Books, 1960.

————. *I, Claudius.* New York: Vintage Books, 1934.

Gray, G. B., and M. Cary. "The Reign of Darius." In *The Cambridge Ancient History.* Vol. 4, *The Persian Empire and the West. See* Buchanan.

Greenberg, Daniel S. "A Hidden Cost of Military Research: Less National Security." *Discover,* January 1987, 94–101.

Greenough, William T., and Fred R. Volkmar. "Pattern of Dendritic Branching in Occipital Cortex of Rats Reared in Complex Environments." *Experimental Neurology,* August 1973, 491–504.

Grier. James W. *Biology of Animal Behavior.* St. Louis, Mo.: Times-Mirror, 1984.

Griffin, Donald R. *Animal Thinking.* Cambridge: Harvard University Press, 1984.

Grzimek, Bernhard. *Grzimek's Animal Life Encyclopedia.* New York: Van Nostrand Reinhold Co., 1972.

Guignebert, Charles. *Jesus.* New York: Alfred A. Knopf, 1935.

Guillen, Michael. *Bridges to Infinity: The Human Side of Mathematics.* Los Angeles: Jeremy P. Tarcher, 1983.

Gula, Robert J., and Thomas H. Carpenter. *Mythology, Greek and Roman.* Wellesley Hills, Mass.: Independent School Press, 1977.

Hacker, Peter. "Languages, Minds and Brains." In *Mindwaves. See* Argyle.

Hahn, Emily. *The Islands: America's Imperial Adventure in the Philippines.* New York: Coward, McCann & Geoghegan, 1981.

Haine, Edward A. *Seven Railroads.* Cranbury, N.J.: A. S. Barnes & Co., 1979.

Hall, Edward T. *Beyond Culture.* New York: Doubleday, Anchor Books, 1977.

Hall, K. R. L. "Aggression in Monkey and Ape Societies." In *Primates. See* Goodall, "Gombe Stream Chimpanzees."

————. "Tool-Using Performances as Indicators of Behavioral Adaptability." In *Primates. See* Goodall, "Gombe Stream Chimpanzees."

Hamdani, Abbas. "Islamic Fundamentalism." *Mediterranean Quarterly,* Fall 1993, 38–47.

Hames, Raymond. "Time Allocation." In *Evolutionary Ecology and Human Behavior,* edited by Eric Alden Smith and Bruce Winterhalder. New York: Aldine de Gruyter, 1992.

Hammond's Historical Atlas. New York: C. S. Hammond, 1948.

Harlow, Harry F. *Learning to Love.* New York: Jason Aronson, 1974.

Harlow, Harry F., and Gary Griffin. "Induced Mental and Social Deficits in Rhesus Monkeys." In *Biological Basis of Mental Retardation,* edited by Sonia F. Osler and Robert E. Cooke. Baltimore: Johns Hopkins Press, 1965.

Harlow, Harry F., and Margaret Kuenne Harlow. "Social Deprivation in Monkeys." *Scientific American,* November 1962, 136–46.

Harrer, Heinrich. *Seven Years in Tibet.* Translated by Richard Graves. Los Angeles: Jeremy P. Tarcher, 1982.

Harris, Louis. *Inside America.* New York: Vintage Books, 1987.

Harris, Marvin. *Cannibals and Kings: The Origins of Cultures.* New York: Vintage Books, 1977.

―――. *Cows, Pigs, Wars and Witches: The Riddles of Culture.* New York: Vintage Books, 1978.

―――. "India's Sacred Cow." In *Conformity and Conflict. See* Friedl.

―――. *Our Kind: Who We Are, Where We Came From, Where We Are Going.* New York: Harper & Row, 1989.

Harrison, Lawrence E. *Underdevelopment Is a State of Mind: The Latin American Case.* The Center for International Affairs, Harvard University. Lanham, Md.: Madison Books, 1988.

Hartcher, Peter. "Guess Who's Carrying a Bigger Stick?" *World Press Review,* July 1988, 20–22. (First published in *Sydney Morning Herald.*)

Harvey, Herman. *Sum and Substance.* Newport Beach, Calif.: Books on Tape, 1985.

Hastings, James, ed. *Encyclopaedia of Religion and Ethics.* New York: Charles Scribner's Sons, 1908–1927.

Hatano, Giyoo, and Kayoko Inagaki. "Sharing Cognition through Collective Comprehension Activity." In *Perspectives on Socially Shared Cognition,* edited by Lauren B. Resnick, John M. Levine, and Stephanie D. Teasley. Washington, D.C.: American Psychological Association, 1991.

Hays, H. R. *From Ape to Angel: An Informal History of Social Anthropology.* New York: Alfred A. Knopf, 1958.

"Heart Disease and Type A Behavior." *Sources Digest: Psychology Research and Social Trends Forecasting,* December 1988, 3.

Hegel, Georg Wilhelm Friedrich. *Lectures on the Philosophy of World History.* Translated by H. B. Nisbet. London: Cambridge University Press, 1975.

————. *The Philosophy of History.* New York: Wiley Book Co., 1900.

Heikal, Mohamed. *Autumn of Fury: The Assassination of Sadat.* London: Corgi Books, 1983.

————. *The Return of the Ayatollah.* 1981. Reprint. London: Andre Deutsch, 1983.

Heller, Mikhail, and Aleksandr M. Nekrich. *Utopia in Power: The History of the Soviet Union from 1917 to the Present.* Translated by Phyllis B. Carlos. New York: Summit Books, 1986.

Herodotus. *The Histories.* Translated by Aubrey de Selincourt. Harmondsworth, Middlesex: Penguin Books, 1972.

Herskovits, Melville J. *Economic Anthropology: The Economic Life of Primitive Peoples.* 1940. New York: W. W. Norton & Co., 1965.

Hilgard, Ernest R. *Psychology in America: A Historical Survey.* San Diego: Harcourt Brace Jovanovich, 1987.

Hill, Rosalind, ed. *Gesta Francorum et Aliorum Hierosolimitanorum—Deeds of the Franks and Other Pilgrims to Jerusalem.* London: Thomas Nelson and Sons, 1962.

Hitti, Philip K. *The Arabs: A Short History.* South Bend, Ind.: Gateway Editions, 1970.

Hoffman, Michael. "Before the Pharaohs: How Egypt Became the World's First Nation-State." *The Sciences,* January/February 1988, 40–47.

Hofstadter, Douglas R., and Daniel C. Dennet. *The Mind's I: Fantasies and Reflections on Self and Soul.* New York: Bantam Books, 1981.

Holden, Constance. "Why Do Women Live Longer Than Men?" *Science,* 9 October 1987, 158–60.

————. "Youth Suicide: New Research Focuses on a Growing Social Problem." *Science,* 22 August 1986,

Holden, David, and Richard Jones. *The House of Saud.* London: Pan Books, 1982.

Holmes, Steven A. "Iran's Shadow: Fundamentalism Alters the Mideast's Power Relationships," *The New York Times,* 22 August 1993, Section 4, 1.

Holt, P. M., Ann K. S. Lambton, and Bernard Lewis, eds. *The Cambridge History of Islam.* Vol. 1, *The Central Islamic Lands.* Cambridge: Cambridge University Press, 1970.

Holub, Miroslav. "Shedding Life: On the Mysteries of Dying, Cell by Cell." *Science 86,* April 1986, 51–53.

Holy Bible, New King James Version. Nashville, Tenn.: Thomas Nelson Publishers, 1982.

Holzman, David. "How Gray Matter Can Mend Itself." *Insight*, 6 February 1989, 50–51.

————. "Medicine Minus a Cost Tourniquet." *Insight*, 8 August 1988, 9–16.

Homer. *The Iliad.* Translated by Richmond Lattimore. Chicago: University of Chicago Press, 1961.

Hooper, Judith, and Dick Teresi. "Sex and Violence." *Penthouse*, February 1987, 40–96.

Hopkirk, Peter. *Setting the East Ablaze.* Newport Beach, Calif.: Books on Tape, 1984.

House, James S., Karl R. Landis, and Debra Umberson. "Social Relationships and Health." *Science*, July 1988.

Howarth, Stephen. *The Knights Templar.* New York: Atheneum, 1982.

Hsu, Immanuel D. Y. *The Rise of Modern China.* New York: Oxford University Press, 1975.

Hull, David L. *Science As a Process: An Evolutionary Account of the Social and Conceptual Development of Science.* Chicago: University of Chicago Press, 1988.

Humphrey, James H., and Joy N. Humphrey. "Stress in Childhood." In *Selye's Guide to Stress Research,* edited by Hans Selye. Vol. 3. New York: Van Nostrand Reinhold Co., Scientific and Academic Editions, 1983.

Huntington, Samuel P. "The Clash of Civilizations?" *Foreign Affairs*, Summer 1993, 46.

"Hypertension a Mental Handicap." *Brain Mind Bulletin*, August 1992, 1.

Indyk, Martin. "Watershed in the Middle East." *Foreign Affairs, America and the World*, 1991/1992, 70–93.

Innes, Hammond. *The Conquistadors.* New York: Alfred A. Knopf, 1969.

"Innovation." 18 February 1986. PBS Television.

Ishaq, Ibn. *Biography of the Messenger of God.* Excerpted in *The Islamic World*, edited by William H. McNeill and Marilyn Robinson Waldman. Chicago: University of Chicago Press, 1983.

Ishinomori, Shotaro. *Japan, Inc.: An Introduction to Japanese Economics.* Translated by Betsey Scheiner. Berkeley: University of California Press, 1988.

James, William. *The Varieties of Religious Experience.* New York: Collier Books, 1961.

———. *Will, Emotion, Instinct* and *Life's Ideals.* Newport Beach, Calif.: Books on Tape, Halvorson Dixit Recording.

Jammer, Max. *The History of Theories of Space in Physics.* Cambridge: Harvard University Press, 1954.

Janson, Charles. "Evolutionary Ecology of Primate Social Structure." In *Evolutionary Ecology and Human Behavior. See* Hames.

Jarvis, C. Mackechnie. "The Distribution and Utilization of Electricity." In *A History of Technology,* edited by Charles Singer, A. R. Hall, and Trevor I. Williams. Vol. 5, *The Late Nineteenth Century, c. 1850 to c. 1900.* Oxford: Oxford University Press, 1958.

———. "The Generation of Electricity." In *A History of Technology.* Vol. 5, *The Late Nineteenth Century, c. 1850 to c. 1900. See* Jarvis, "Distribution and Utilization of Electricity."

Jastrow, Robert. *The Enchanted Loom: Mind in the Universe.* New York: Simon and Schuster, Touchstone Book, 1983.

Jay, Phyllis C., ed. *Primates. See* Goodall, "A Preliminary Report on Expressive Movements and Communication in the Gombe Stream Chimpanzees."

Jaynes, Julian. *The Origin of Consciousness in the Breakdown of the Bicameral Mind.* Boston: Houghton Mifflin Co., 1976.

"Jerry Falwell; Circuit Rider to Controversy." *U.S. News and World Report,* 2 September 1985.

Johnson, Allen W., and Timothy Earle. *The Evolution of Human Societies: From Foraging Group to Agrarian State.* Stanford, Calif.: Stanford University Press, 1987.

Johnson, Paul. *Modern Times: The World from the Twenties to the Eighties.* New York: Harper & Row, Harper Colophon Books, 1985.

Johnston, Alan. *The Emergence of Greece.* Oxford: Elsevier-Phaidon, 1976.

Joint Staff. *United States Military Posture for FY 1988.* Washington, D.C.: U.S. Government Printing Office, 1987.

Kaiser, Robert G. *Russia: The People and the Power.* New York: Washington Square Press, 1984.

Kan, Lai Po. *The Ancient Chinese.* 1980. Reprint. Morristown, N.J.: Silver Burdett Co., 1985.

Kanner, Leo, M.D. *Child Psychiatry.* 4th ed. Springfield, Ill.: Charles C. Thomas Publisher, 1972.

Kantor, MacKinlay. "Then Came the Legions." In *75 Short Story Master-*

pieces: Stories from the World's Literature, edited by Roger B. Goodman. New York: Bantam Books, 1961.

Kaplan, Jay R., Stephen B. Manuck, Thomas B. Clarkson, Frances M. Lusso, David M. Taub, and Eric W. Miller. "Social Stress and Atherosclerosis in Normocholesterolemic Monkeys," *Science,* 13 May 1983.

Karol, K. S. *The Second Chinese Revolution.* Translated by Mervyn Jones. New York: Hill and Wang, 1974.

Keating, Robert. "Live Aid: The Terrible Truth." *Spin Magazine,* July 1986, 75–80.

Keegan, John. *The Mask of Command.* New York: Viking, Elisabeth Sifton Books, 1987.

Kelly, Thomas. *A History of Argos to 500 B.C.* Minneapolis: University of Minnesota Press, 1976.

Kennedy, Paul. "The (Relative) Decline of America." *Atlantic Monthly,* August, 1987, 29–38.

———. *The Rise and Fall of the Great Powers: Economic Change and Military Conflict from 1500 to 2000.* New York: Random House, 1987.

Khaldun, Ibn. *The Muqaddimah: An Introduction to History.* Edited by N. J. Dawood. Translated by Franz Rosenthal. Princeton, N.J.: Princeton University Press, 1967.

Khomeini, Ayatollah Sayyed Ruhollah Mousavi. *A Clarification of Questions: An Unabridged Translation of Resaleh Towzih al-Masael.* Translated by J. Borujerdi. Boulder, Colo.: Westview Press, 1984.

———. *Islam and Revolution: Writings and Declarations of Imam Khomeini.* Translated by Hamid Algar. Berkeley: Mizan Press, 1981.

———. *Sayings of the Ayatollah Khomeini: Political, Philosophical, Social, and Religious.* New York: Bantam Books, 1980.

Kirk, G. S. "The Homeric Poems As History." In *The Cambridge Ancient History—Vol. 2, part 2 The History of the Middle East And the Aegean Region, 1380–1000 B.C.,* edited by I. Edwards, C. Gadd, N. Hammond, and E. Sollberger. *See* Buchanan.

Klausner, Joseph. *From Jesus to Paul.* Translated by W. F. Stinespring. New York: Macmillan, 1943.

———. *The Messianic Idea in Israel.* Translated by W. F. Stinespring. New York: Macmillan, 1955.

Knightley, Philip, and Colin Simpson. *The Secret Life of Lawrence of Arabia.* New York: McGraw-Hill Book Co., 1969.

Koch, H. W. *A History of Prussia.* New York: Dorset Press, 1978.

Konner, Melvin. "False Idylls." *The Sciences,* September/October 1987, 8–10.

——. "The Gender Option." *The Sciences.* November/December 1987, 2–4.

——. *The Tangled Wing: Biological Constraints on the Human Spirit.* New York: Holt, Rinehart and Winston, 1982.

Kosambi, D. D. *Ancient India: A History of Its Culture and Civilization.* New York: Pantheon Books, 1965.

Kotecha, Ken C., with Robert W. Adams. *The Corruption of Power: African Politics.* Washington, D.C.: University Press of America, 1981.

Kramer, Michael. "Are You Running with Me Jesus? Televangelist Pat Robertson Goes for the White House." *New York* magazine, 18 August 1986.

Kramer, Samuel Noah. *The Sumerians: Their History, Culture and Character.* Chicago: University of Chicago Press, 1963.

Krauthammer, Charles. "The Unipolar Moment." *Foreign Affairs: America and the World,* Winter 1990/91, 23–33.

Kraven, Ken. "The Real Face of Kuwait." *National Times,* November 1992, 2. (First published in the *Washington Post.*)

Kravitz, Edward A. "Hormonal Control of Behavior: Amines and the Biasing of Behavioral Output in Lobsters." *Science,* 30 September 1988.

Kroeber, A. L. *The Nature of Culture.* Chicago: University of Chicago Press, 1952.

LaHaye, Tim. *Has the Church Been Deceived?* Washington, D.C.: American Coalition for Traditional Values, n.d.

Lalevee, Thierry. "Tehran's New Allies in Africa." *World Press Review,* September 1993, 20–21. (First published in *Arabies* [Paris].)

Lamb, David. *The Africans: Encounters from the Sudan to the Cape.* London: Methuen, 1986.

Lansing, J. Stephen (writer). *The Three Worlds of Bali.* Produced and directed by Ira R. Abrams. Odyssey television series. Co-produced by Public Broadcasting Associates and the University of Southern California, 1981.

Layzer, David. "Growth of Order in the Universe." In *Entropy, Information and Evolution. See* Brooks.

Lawrence, T. E. *Seven Pillars of Wisdom*. 1926. New York: Dell Publishing, 1962.

Leakey, Richard E., and Richard Lewin. *People of the Lake: Mankind and Its Beginnings*. New York: Avon Books, 1983.

LeCompte, Bernard. "Communism Confronts Islam." *World Press Review*, July 1992, 10. (First published in *L'Express*.)

Lee, Ki-baik. *A New History of Korea*. Translated by Edward W. Wagner. Cambridge: Harvard University Press, 1984.

Lee, Richard Borshay. "The Hunters: Scarce Resources in the Kalahari." In *Conformity and Conflict*. See Friedl.

Lefcourt, Herbert M. *Locus of Control: Current Trends in Theory and Research*. 2d ed. Hillsdale, N.J.: Lawrence Erlbaum Associates, 1982.

Lenorovitz, Jeffrey M. "Europe Presses U.S. to Agree on Launch Competition Rules." *Aviation Week & Space Technology*, 27 June 1988, 36–37.

———. "Europe's First Ariane 4 Launched Successfully." *Aviation Week & Space Technology*. 20 June 1988, 16–17.

Levey, Judith S., and Agnes Greenhall, eds. *The Concise Columbia Encyclopedia*. New York: Avon Books, 1983.

Levy, Steven. *Artificial Life*. New York: Vintage Books, 1992.

Liebman, Andrew (writer, director, and producer). *The Secret of Life: Conquering Cancer*. London: BBC-TV; Boston: WGBH, 1993.

"Life in the Unpromised Land: East Germans Migrate to the West." *World Press Review*, November 1988, 17. (First published in *Der Spiegel*.)

Lincoln, Bruce W. *Red Victory: A History of the Russian Civil War*. New York: Simon and Schuster, 1989.

Lister, R. P. *Genghis Khan*. New York: Stein and Day, 1969.

Livermore, Beth. "At Least Take a Deep Breath." *Psychology Today*, September 1992, 44.

Loftus, Elizabeth. *Memory*. Reading, Mass.: Addison-Wesley Publishing Co., 1980.

Lorenz, Konrad. *On Aggression*. New York: Harcourt Brace Jovanovich, 1974.

Lunde, Donald T. *Murder and Madness*. Portable Stanford Series. San Francisco: San Francisco Book Co., 1976.

Maalouf, Amin. *The Crusades through Arab Eyes*. New York: Schocken Books, 1985.

McClanahan, Lon L. Rodolfo Ruibal, and Vaughan H. Shoemaker. "Frogs and Toads in Deserts." *Scientific American,* March 1994, 82–88.

McClelland, James L., David E. Rumelhart, and the PDP Research Group. *Parallel Distributed Processing: Explorations in the Microstructure of Cognition.* Vol. 2, *Psychological and Biological Models.* Cambridge: MIT Press, Bradford Book, 1986.

McCormick, Moira. "VSDA Applauds As Ill. Gov. Amends Antiobscenity Bill." *Billboard,* 20 February 1988.

McDermott, John F., Jr. "Child Psychiatry." In *International Encyclopedia of Psychiatry, Psychology, Psychoanalysis and Neurology.* Vol. 3. *See* Denber.

McFarland, David, ed. *The Oxford Companion to Animal Behavior.* New York: Oxford University Press, 1982.

McGraw-Hill Encyclopedia of Science and Technology. See Freedman, Daniel X.

Machlowitz, Marilyn. *Whiz Kids: Success at an Early Age.* Newport Beach, Calif.: Books on Tape.

McKendrick, Paul. *The Greek Stones Speak: The Story of Archaeology in Greek Lands.* New York: W. W. Norton & Co., 1981.

McKenna, Bruce C. "The Subcontinental Blues." *National Review,* 27 May 1991, 21–22.

Mackenzie, Richard. "Pitfalls in Policy on the Path to Kabul." *Insight,* 9 April 1990, 8–15.

Mackie, J. D. *Oxford History of England: The Earlier Tudors, 1485–1558.* London: Oxford University Press, 1962.

MacLean, Paul D. *A Triune Concept of the Brain and Behavior.* Toronto: University of Toronto Press, 1973.

McLellan, David. *Karl Marx: His Life and Thought.* New York: Harper Colophon Books, 1973.

McMillan, Robert R. "Do You Have a Stamp of Israel in Your Passport?" *Caucus Current,* May 1992, 28–29.

McNeill, William H., and Marilyn Robinson Waldman. *The Islamic World. See* Ishaq.

Magaziner, Ira C. *The Silent War.* New York: Random House, 1989.

Makiya, Kanan. *Cruelty and Silence: War, Tyranny, Uprising and the Islamic World.* New York: W. W. Norton & Co., 1993.

Manchester, William. *The Arms of Krupp: 1587–1968.* New York: Bantam Books, 1978.

————. *The Glory and the Dream: A Narrative History of America—1932–1972.* New York: Bantam Books, 1974.

Mandell, Marshall, and Lynne Waller Scanlon, Dr. *Mandell's 5-Day Allergy Relief System.* New York: Pocket Books, 1980.

Margulis, Lynn, and Dorion Sagan. *Microcosmos: Four Billion Years of Microbial Evolution.* New York: Summit Books, 1986.

Markov, Georgi. *The Truth That Killed.* New York: Ticknor & Fields, 1984.

Marr, Phebe. "The Islamic Revival: Security Issues." *Mediterranean Quarterly,* Fall 1992, 37–50.

Martin, Walter T. "Theories of Variation in the Suicide Rate." In *Suicide,* edited by Jack P. Gibbs. New York: Harper & Row, 1968.

Marx, Karl, and Friedrich Engels. *The Communist Manifesto.* London: Penguin Books, 1967.

Maser, Jack D., and Martin E. P. Seligman, eds. *Psychopathology: Experimental Models.* San Francisco: W. H. Freeman & Co., 1977.

Mass, Nathaniel J., and Peter M. Senge. "Reindustrialization: Aiming at the Right Targets." *Technology Review,* August/September 1981, 56–65.

Massie, Robert K. *Peter the Great.* New York: Ballantine Books, 1986.

Mattingly, Garrett. *The Armada.* Boston: Houghton Mifflin Co., 1959.

Mauss, Marcel. *Sociology and Psychology: Essays by Marcel Mauss.* Translated by Ben Brewster. London: Routledge & Kegan Paul, 1979.

May, R. J. "Muslim and Tribal Filipinos." In *The Philippines after Marcos,* by Ronald James May and Francisco Nemenzo. New York: St. Martin's Press, 1985.

Mead, Margaret. *Male and Female: A Study of the Sexes in a Changing World.* 1949. New York: Dell Publishing, 1968.

Michel, Olivier. "Allah's G.I.s." *World Press Review,* September 1992, 40–41. (First published in *Le Figaro.*)

Miller, William R., Robert A. Rosellini, and Martin E. P. Seligman. "Learned Helplessness and Depression." In *Psychopathology.* See Maser and Seligman.

Minces, Juliette. *The House of Obedience: Women in Arab Society.* Translated by Michael Pallis. London: Zed Press, 1982.

Minsky, Marvin. *The Society of Mind.* New York: Simon and Schuster, 1986.

Mitchell, James. *The Illustrated Reference Book of Classical History.* Leicester: Windward, W. H. Smith & Son, 1982.

Mohs, Mayo. "I.Q.: New Research Shows That the Japanese Outperform All Others in Intelligence Tests. Are They Really Smarter?" *Discover,* September 1982, 18–24.

Mooney, James. *The Ghost-Dance Religion and the Sioux Outbreak of 1890.* Chicago: University of Chicago Press, 1965. Originally published as part of the *Fourteenth Annual Report of the Bureau of Ethnology to the Secretary of the Smithsonian Institution, 1892–93.*

Moore, R. Laurence. *Religious Outsiders and the Making of America.* New York: Oxford University Press, 1986.

Moorehead, Alan. *Darwin and the Beagle.* Newport Beach, Calif.: Books on Tape, 1969.

———. *The Russian Revolution.* New York: Bantam Books, 1959.

Morell, Virginia. "Dian Fossey: Field Science and Death in Africa." *Science 86,* April 1986, 17–21.

Morgan, Bryan. *Early Trains.* London: Camden House Books, n.d.

Morgan, David. *The Mongols.* New York: Basil Blackwell, 1986.

Morgan, Kenneth O., ed. *The Oxford Illustrated History of Britain.* New York: Oxford University Press, 1984.

Morholt, Evelyn, and Paul F. Brandwein. *A Sourcebook for the Biological Sciences.* San Diego: Harcourt Brace Jovanovich, 1986.

Morishima, Michio. *Why Has Japan 'Succeeded'? Western Technology and the Japanese Ethos.* New York: Cambridge University Press, 1982.

Morris, Donald. *The Washing of the Spears.* New York: Simon and Schuster, Touchstone Book, 1965.

Morse, Douglass H. *Behavioral Mechanisms in Ecology.* Cambridge: Harvard University Press, 1980.

Mortimer, Edward. "New Ism in the East." *World Monitor,* September 1992, 50–52.

Moyers: God & Politics—the Battle for the Bible. Produced by Gail Pellett. 16 December 1987. New York: Public Affairs Television.

Mullen, Bryan. "Atrocity As a Function of Lynch Mob Composition." *Personality and Social Psychology Bulletin,* June 1986, 187–97.

Murdock, George Peter. *Social Structure.* New York: Macmillan, 1949.

"Muscles in Space Forfeit More Than Fibers." *Science News,* 29 October 1988, 277.

Naisbitt, John. *Megatrends: Ten New Directions Transforming Our Lives.* New York: Warner Books, 1984.

Narin, Francis, and J. Davidson Frame. "The Growth of Japanese Science and Technology." *Science,* 11 August 1989, 600–5.

Neal, Harry Edward. *From Spinning Wheel to Spacecraft: The Story of the Industrial Revolution.* New York: Julian Messner, 1965.

Neihardt, John G. *Black Elk Speaks: Being the Life Story of a Holy Man of the Oglala Sioux.* New York: Pocket Books, 1972.

Newark, Tim. *The Barbarians: Warriors & Wars of the Dark Ages.* London: Blandford Press, 1985.

The New Encyclopaedia Britannica. Chicago: Encyclopaedia Britannica, 1986.

The New English Bible: New Testament. Copub., England: Oxford University Press and Cambridge University Press, 1961.

"News of the Week in Review." 29 July 1988 (PBS).

The 1987 Information Please Almanac. Boston: Houghton Mifflin Co., 1987.

Nock, S. *The Dawn of World Railways, 1800–1850.* New York: Macmillan, 1972.

O'Connell, Robert L. *Of Arms and Men: A History of War, Weapons, and Aggression.* New York: Oxford University Press, 1989.

Oden, Michael Dee. "Military Spending Erodes Real National Security." *Bulletin of the Atomic Scientists,* June 1988, 38–42.

O'Hanlon, Redmond. *Into the Heart of Borneo.* Edinburgh: Salamander Press Edinburgh, 1984.

Olcott, Martha Brill. "Central Asia's Catapult to Independence." *Foreign Affairs,* Summer 1992, 108–30.

Olmert, Michael (writer). *Islam.* Produced and directed by Steve York. 22 July 1987. Smithsonian World, no. 305. Co-produced by the Smithsonian Institution and WETA, Washington, D.C.

Onaran, Yalman. "Islamic Revival in Central Asia." *Near East Report,* 31 August 1992, 166.

"104 National Groups, 70 Languages." *World Press Review,* May 1988, 21–22. (First published in *Europeo* [Milan].)

O'Neill, Gerard K. *2081: A Hopeful View of the Human Future.* New York: Simon and Schuster, Touchstone Book, 1981.

Ordish, George. *The Year of the Ant.* New York: Charles Scribner's Sons, 1978.

Ornstein, Robert, and David Sobel. *The Healing Brain.* New York: Simon and Schuster, 1987.

Osman, Tony. *Space History.* New York: St. Martin's Press, 1983.

Ostrogorsky, George. *History of the Byzantine State.* Translated by Joan Hussey. New Brunswick, N.J.: Rutgers University Press, 1969.

Padover, Saul K. *Karl Marx: An Intimate Biography.* New York: McGraw-Hill Book Co., 1978.

Pagels, Elaine. *The Gnostic Gospels.* New York: Vintage Books, 1981.

Pagels, Heinz. *The Dreams of Reason: The Computer and the Rise of the Sciences of Complexity.* New York: Simon and Schuster, 1988.

Pelletier, Kenneth R. "Stress: Etiology, Assessment, and Management in Holistic Medicine." In *Selye's Guide to Stress Research.* Vol. 3. *See* Humphrey.

Pennish, Elizabeth. "Of Great God Cybernetics and His Fair-Haired Child." *The Scientist,* 14 November 1988.

Pfeiffer, John. "Listening for Emotions: Videotapes Show That Many Doctors Aren't—and Patients Suffer." *Science 86,* June 1986, 14–16.

Phillips, David Atlee. *The Nightwatch.* New York: Ballantine Books, 1982.

Phillips, David P. "A Dip in Deaths before Ceremonial Occasions: Some New Relationships between Social Integration and Mortality." *American Journal of Sociology* 84 (1979): 1150–74.

Phillips, David P., and Judith Lu. "The Frequency of Suicides around Major Public Holidays: Some Surprising Findings." *Suicide and Life Threatening Behavior* (Spring 1980): 41–50.

Phillipson, O. T. "Endorphins." In *The Oxford Companion to the Mind,* edited by Richard L. Gregory. New York: Oxford University Press, 1987.

Philpott, William H., M.D., and Dwight K. Kalita. *Brain Allergies.* New Canaan, Conn.: Keats Publishing, 1987.

Platlea, Daniel. "Islamic Fever—Too Hot for Churches." *Insight,* 22 January 1990, 34.

Pletka, Danielle. "Hell-bent to Build a Nuclear Bomb." *Insight,* 30 April 1990, 35–36.

Polk, William R., and William J. Mares. *Passing Brave.* New York: Ballantine Books, 1973.

Pollack, Andrew. "U.S. Reported Trailing Japan in the Superconductor Race," *The New York Times,* 16 October 1988, 1, 12.

Polo, Marco. *The Travels of Marco Polo.* New York: Dorset Press, 1987.

Polybius. *The History of Polybius.* Translated from the text of F. Hultsch by Evelyn S. Shuckburgh. Bloomington, Ind.: Indiana University Press, 1962.

Pope, Sterett. "Kurdish Horror." *World Press Review,* November 1988, 44.

————. "Reconstruction Race." *World Press Review,* November 1988, 44.

Porter, Michael E. "Why U.S. Business Is Falling Behind." *Fortune,* 28 April 1986.

Prigogine, Ilya, and Isabelle Stengers. *Order out of Chaos: Man's New Dialogue with Nature.* New York: Bantam Books, 1984.

Proffer, Ellendea. "We Kill for Mankind." Review of *Lenin: The Novel,* by Alan Brien. *New York Times Book Review,* 16 November 1988, 53.

Pryce-Jones, David. "Self-Determination, Arab Style." *Commentary,* January 1989, 39–46.

Purves, William K., and Gordon H. Orians. *Life: The Science of Biology.* Sunderland, Mass.: Sinauer Associates, 1987.

Queller, David C., Joan E. Strassman, and Colin R. Hughes. "Genetic Relatedness in Colonies of Tropical Wasps with Multiple Queens." *Science,* November 1988, 1155–57.

Raff, Martin C., Barbara A. Barres, Julia F. Burne, Harriet S. Coles, Yasuki Ishizaki, and Michael D. Jacobson. "Programmed Cell Death and the Control of Cell Survival: Lessons from the Nervous System." *Science,* 29 October 1993, 695–99.

Rahman, Fazlur. *Islam.* Chicago: University of Chicago Press, 1979.

Raloff, Janet. "That Feminine Touch: Are Men Suffering from Prenatal or Childhood Exposures to 'Hormonal' Toxicants?" *Science News,* 22 January 1994, 56–58.

Ramazani, R. K. *Revolutionary Iran: Challenge and Respect in the Middle East.* Baltimore: Johns Hopkins University Press, 1986.

Ramet, Pedro ed., *Religion and Nationalism in Soviet and East European Politics.* Durham, N.C.: Duke University Press, 1984.

Rampone, Alfred J., and Myron E. Shirasu. "Temperature Changes in the Rat in Response to Feeding." *Science,* 17 April 1964, 317–19.

Rand, Christopher T. *Making Democracy Safe for Oil: Oilmen and the Islamic East.* Boston: Little, Brown and Co., 1975.

Randers-Pehrson, Justine Davis. *Barbarians and Romans.* Norman, Okla.: University of Oklahoma Press, 1983.

Rao, Radhakrishna. "China Joins the Space Race." *World Press Review,* May 1988, 51. (First published in *Compass News Features* [Luxembourg].)

Rasor, Dina. *The Pentagon Underground.* New York: Times Books, 1985.

Rathus, Spencer A. *Psychology.* New York: Holt, Rinehart and Winston, 1987.

Raven, Bertram H., and Jeffrey Z. Rubin. *Social Psychology.* New York: John Wiley & Sons, 1983.

Raytheon Company. "Backgrounder." News Release 3-1535. April 1988.

Reader, John. *Man on Earth.* Austin, Tex.: University of Texas Press, 1988.

Reed, John. *Ten Days That Shook the World.* Harmondsworth, Middlesex: Penguin Books, 1977.

Reed, Stanley. "Jordan and the Gulf Crisis." *Foreign Affairs,* Winter 1990/91, 21–35.

Reischauer, Edwin O. *The Japanese.* Cambridge: Harvard University Press, Belknap Press, 1981.

————. *Japan Past and Present.* 3d ed. Tokyo: Charles E. Tuttle Co., 1964.

Reiter, Carla. "Toy Universes." *Science 86,* June 1986, 54–59.

Restak, Richard. *The Brain.* New York: Bantam Books, 1984.

————. *The Mind.* New York: Bantam Books, 1988.

Restelli, Marcio. "China's Secret Holy War." *World Press Review,* May 1994, 43. (First published in *L'Europeo* [Milan].)

Richardson, H. E. *A Short History of Tibet.* New York: E. P. Dutton & Co., 1962.

Richburg, Keith. "Back to Vietnam." *Foreign Affairs,* Fall 1991, 111–32.

Richelson, Eliott. "Antidepressants." In *Encyclopedia of Neuroscience.* Vol. 1. *See* Adelman.

Ridley, Matt. "Swallows and Scorpionflies Find Symmetry Beautiful." *Science,* 17 July 1992, 327–28.

Riesen, Austin H. "Plasticity of Behavior: Psychological Aspects." In *Biological and Biochemical Bases of Behavior,* edited by Harry F. Harlow and Clinton N. Woolsey. Madison, Wis.: University of Wisconsin Press, 1965.

Ripley, Suzanne. "Intertroop Encounters among Ceylon Gray Langurs (*Presbytis entellus*)." In *Social Communication among Primates. See* Altmann.

Roberts, D. S. *Islam: A Concise Introduction.* New York: Harper & Row, 1981.

Roberts, J. M. *The Pelican History of the World.* Harmondsworth, Middlesex: Penguin Books, 1983.

Roberts, Marjory. "Patient Knows Best." *Psychology Today,* June 1987, 10.

Rosenberg, Tina. *Children of Cain: Violence and the Violent in South America.* New York: William Morrow, 1991.

Rosenzweig, Mark R. "Environmental Complexity, Cerebral Change, and Behavior." *American Psychologist* 21 (1966): 321–42.

Rosenzweig, Mark R., Edward L. Bennett, and Marian Cleeves Diamond. "Brain Changes in Response to Experience." *Scientific American,* February 1972, 22–29.

Ross, Ishbel. *The General's Wife: The Life of Mrs. Ulysses S. Grant.* New York: Dodd, Mead & Co., 1959.

Rossabi, Morris. *Khubilai Khan: His Life and Times.* Los Angeles: University of California Press, 1988.

Roth, Jesse, and Derek LeRoith. "Chemical Cross Talk: Why Human Cells Understand the Molecular Messages of Plants." *The Sciences,* May–June 1987, 50–55.

Rouche, Michel. "The Early Middle Ages in the West," In *A History of Private Life: From Pagan Rome to Byzantium,* edited by Paul Veyne. Translated by Arthur Goldhammer. Cambridge: Harvard University Press, 1987.

Rowse, A. L. *The Expansion of Elizabethan England.* Newport Beach, Calif.: Books on Tape, 1955.

———. *Shakespeare, the Man.* New York: Harper & Row, 1973.

Rumer, Boris. "Trouble in Samarkand." *World Monitor,* November 1988, 44–55.

Rumer, Boris, and Eugene Rumer. "Who Will Be the Next Yugoslavia?" *World Monitor,* November 1992, 36–44.

Runciman, Steven. "The First Crusade: Antioch to Ascalon." In *A History of the Crusades. See* Baldwin.

Russett, Bruce. "Peace among Democracies." *Scientific American,* November 1993, 120.

Ryan, M. J., and W. Wilczynski. "Coevolution of Sender and Receiver:

Effect on Local Mate Preference in Cricket Frogs." *Science*, 24 June 1988, 1786–87.

Sagan, Carl. *The Dragons of Eden: Speculations on the Evolution of Human Intelligence*. New York: Ballantine Books, 1977.

Sagan, Dorion. "What Narcissus Saw: The Oceanic 'I'/'Eye.' " In *The Reality Club*. See Csikszentmihalyi.

Sagan, Leonard A. "Family Ties: The Real Reason People Are Living Longer." *The Sciences*, March/April 1988, 20–29.

Sagarin, Edward, and Robert J. Kelly. "Collective and Formal Promotion of Deviance." In *The Sociology of Deviance*, edited by M. Michael Rosenberg, Robert A. Stebbins, and Allan Turowetz. New York: St. Martin's Press, 1982.

Saggs, H. W. F. *Everyday Life in Babylonia & Assyria*. New York: Dorset Press, 1965.

Sahlins, Marshall D. "Poor Man, Rich Man, Big-Man, Chief." In *Conformity and Conflict*. See Friedl.

Said, Edward W. "The Phony Islamic Threat." *New York Times Magazine*, 21 November 1993.

Salisbury, Harrison Evans. *Black Night, White Snow: Russia's Revolutions, 1905–1917*. New York: Plenum Publishing Corp., Da Capo Paperback, 1981.

———. *War between Russia and China*. New York: W. W. Norton & Co., 1969.

Sampson, Anthony. *The Seven Sisters: The Great Oil Companies and the World They Shaped*. New York: Bantam Books, 1976.

Sapolsky, Robert M. "Lessons of the Serengeti: Why Some of Us Are More Susceptible to Stress." *The Sciences*, May/June 1988, 38–42.

Sarason, I. G., B. R. Sarason, and G. R. Pierce. "Social Support, Personality, and Health." In *Topics in Health Psychology*, edited by S. Maes, C. D. Spielberger, P. B. Defares, and I. G. Sarason. New York: John Wiley & Sons, 1988.

Saulat, Sarwat. *The Life of The Prophet*. Lahore, Pakistan: Islamic Publications, 1983.

Saunders, J. J. *The History of the Mongol Conquests*. London: Routledge & Kegan Paul, 1971.

Scarisbrick, J. J. *Henry VIII*. Berkeley: University of California Press, 1968.

Schele, Linda and David Freidel. *A Forest of Kings: The Untold Story of the Ancient Maya.* New York: William Morrow and Company, 1990.

Schmandt-Besserat, Denise. "Oneness, Twoness, Threeness: How Ancient Accountants Invented Numbers." *The Sciences,* July/August 1987, 44–49.

Schmeck, Harold M., Jr. *Immunology: The Many Edged Sword.* New York: George Braziller, 1974.

Scholl-Latour, Peter. *Adventures in the East: Travels in the Land of Islam.* Translated by Ruth Hein. New York: Bantam Books, 1988.

"The Second Coming of Pol Pot: Fears of a Return to the Killing Fields." *World Press Review,* October 1988, 25–28. (First published in *Asiaweek* [Hong Kong].)

Seeley, Thomas D. *Honeybee Ecology: A Study of Adaptation in Social Life.* Princeton, N.J.: Princeton University Press, 1985.

Seeley, Thomas D., and Royce A. Levien. "A Colony of Mind: The Beehive as Thinking Machine," *The Sciences,* July/August 1987, 38–43.

Seiderman, Arthur, and Steven Schneider. *The Athletic Eye.* New York: Hearst Books, 1983.

Selye, Hans. *Stress without Distress.* New York: New American Library, 1975.

————. ed. *Selye's Guide to Stress Research.* Vol. 3. *See* Humphrey and Humphrey.

Seth, Vikram. *From Heaven Lake: Travels through Sinkiang and Tibet.* Newport Beach, Calif.: Books on Tape, 1983.

Sharabi, Hisham. "Modernity and Islamic Revival: The Critical Task of Arab Intellectuals." *Contention,* Fall 1992, 127–38.

Shaw, Stanford J. *History of the Ottoman Empire and Modern Turkey.* Vol. 1, *Empire of the Gazis: The Rise and Decline of the Ottoman Empire 1280–1808.* Cambridge: Cambridge University Press, 1976.

Shawcross, William. *The Quality of Mercy: Cambodia, Holocaust and Modern Conscience.* New York: Simon and Schuster, 1984.

Shelby, Barry. "Secessions." *World Press Review,* November 1993, 5. (Summarized from *Asiaweek* [Hong Kong].)

Shirer, William L. *20th Century Journey: A Memoir of a Life and the Times— The Start, 1904–1930.* New York: Simon and Schuster, 1976.

Shors, T. J., T. B. Seib, S. Levine, and R. F. Thompson. "Inescapable versus Escapable Shock Modulates Long-Term Potentiation in the Rat Hippocampus.'' *Science,* 14 April 1989, 224–26.

Shulman, Irving. *Valentino.* New York: Trident Press, 1967.

''Sick Men of Europe.'' *Economist,* 22 March 1986, 53.

Siddiqui, Haroon. ''The Scramble for Central Asia: A Global Contest for Hearts, Minds, Money.'' *World Press Review,* July 1992, 10. (First published in the *Toronto Star.*)

Simon, Herbert A. ''A Mechanism for Social Selection and Successful Altruism.'' *Science,* 21 December 1990, 1665–68.

Sinai, Anne, and Allen Pollack, eds. *The Syrian Arab Republic: A Handbook.* New York: American Academic Association for Peace in the Middle East, 1976.

Singer, Charles, E. J. Holmyard, A. R. Hall, and Trevor I. Williams, eds. *A History of Technology.* Vol. 5, *The Late Nineteenth Century, c. 1850 to c. 1900. See* Jarvis, ''Distribution and Utilization of Electricity.''

Smith, Adam. ''Adam Smith's Money World,'' 2 November 1987. No. 408. Educational Broadcasting Cos.

Smith, Dorothy W., and Carol P. Hanley Germain. *Care of the Adult Patient: Medical, Surgical Nursing.* 4th ed. Philadelphia: J. B. Lipincott Co., 1975.

Smith, Eric Alden, and Bruce Winterhalder. ''Natural Selection and Decision-Making: Some Fundamental Principles.'' In *Evolutionary Ecology and Human Behavior. See* Hames.

Smith, Hedrick. *The Russians.* New York: Ballantine Books, 1984.

Sondhi, Sunil. ''Losing the Competitive Edge.'' *World Press Review,* December 1988, 49. (First published in the *Times of India.*)

Southworthy, Franklin. ''The Reconstruction of Prehistoric South Asian Language Contact.'' In *The Uses of Linguistics,* edited by Edward Bendix. New York Academy of Sciences, vol. 583. New York, 1990.

Soviet Military Power, 1987. Washington, D.C.: U.S. Government Printing Office, n.d.

Sparks, John. *The Discovery of Animal Behaviour.* Boston: Little, Brown and Co., 1982.

Speer, Albert. *Inside the Third Reich—Memoirs.* Translated by Richard Winston and Clara Winston. New York: Collier Books, 1970.

Sperry, Armstrong. *Pacific Islands Speaking.* Newport Beach, Calif.: Books on Tape, 1955.

Spitz, Rene A. "Hospitalism: An Inquiry into the Genesis of Psychiatric Conditions in Early Childhood." In *The Psychoanalytic Study of the Child.* Vol. 1. New York: International Universities Press, 1945.

Spitz, Rene A., M.D., with Katherine M. Wolf. "Anaclitic Depression: An Inquiry into the Genesis of Psychiatric Conditions in Early Childhood." In *The Psychoanalytic Study of the Child.* Vol. 2. *See* Spitz, "Hospitalism."

Spradley, James P., and David W. McCurdy. *Conformity and Conflict. See* Friedl.

Spratt, H. Philip. "The Marine Steam-Engine." In *A History of Technology.* Vol. 5, *The Late Nineteenth Century, c. 1850 to c. 1900. See* Jarvis, "Distribution and Utilization of Electricity."

"Spread of Islamic Rules." *World Press Review,* November 1992, 50. (First published in *Asiaweek* [Hong Kong].)

Spuler, B. "The Disintegration of the Caliphates in the East." In *The Cambridge History of Islam.* Vol. 1, *The Central Islamic Lands. See* Holt, Lambton, and Lewis.

Starr, Chester G. *A History of the Ancient World.* New York: Oxford University Press, 1974.

Staver, Sari. "Conference Shows One Skeptic: 'It's Clear We Have a Real Syndrome.' " *American Medical News,* 26 May 1989.

Steinberg, Laurence. "Bound to Bicker; Pubescent Primates Leave Home for Good Reasons. Our Teens Stay with Us and Squabble." *Psychology Today,* September 1987, 36–39.

Stephens, Mitchell. *A History of the News: From Drum to Satellite.* New York: Viking, 1988.

Stephens, William M. *The Family in Cross-Cultural Perspective.* New York: Holt, Rinehart and Winston, 1964.

Stevens, Jane E. "Growing Rice the Old-Fashioned Way, with Computer Assist." *Technology Review,* January 1994, 16–18.

Stewart, Desmond. *T. E. Lawrence.* New York: Harper & Row, 1977.

Stimson, Henry L. "The Nurnberg Trial: Landmark in Law." *Foreign Affairs,* 1947. Quoted in Albert Speer, *Inside the Third Reich: Memoirs.* New York: Collier Books, 1970.

Stone, Lawrence. *The Family, Sex and Marriage in England, 1500–1800.* New York: Harper & Row, 1977.

Stowers, A. "The Stationary Steam Engine—1830–1900." In *A History of*

Technology. Vol. 5, *The Late Nineteenth Century, c. 1850 to c. 1900.* See Jarvis, "Distribution and Utilization of Electricity."

Strobel, Gabrielle. "Guardian Genes." *Science News,* 15 January 1994, 44–45.

Stroh, M. "Genes Determine When Cells Live or Die." *Science News,* 11 April 1992, 230.

Strong, Anna Louise. *When Serfs Stood Up.* San Francisco: Red Sun Publishers, 1976.

Stubbings, Frank H. "The Recession of Mycenaean Civilization." In *The Cambridge Ancient History.* Vol. 2, part 2, *The History of the Middle East and the Aegean Region, 1380–1000 B.C.,* edited by I. Edwards, C. Gadd, N. Hammond, and E. Sollberger. *See* Buchanan.

Sugiyama, Yukimara. "Social Organization of Hanuman Langurs." In *Social Communication among Primates. See* Altmann.

Suomi, Stephen J. and Harry F. Harlow. "Production and Alleviation of Depressive Behaviors in Monkeys." In *Psychopathology. See* Maser and Seligman. San Francisco: W. H. Freeman and Company, 1977, 167–170.

Svare, Bruce. "Steroid Use and Aggressive Behavior." *Science,* 2 December 1988, 1227.

Swaggart, Jimmy. "Rock 'n' Roll Music in the Church." *The Evangelist,* January 1987.

Szentagothai, Janos. "The 'Brain-Mind' Relation: A Pseudoproblem?" In *Mindwaves. See* Argyle.

Szulc, Tad. *Fidel: A Critical Portrait.* New York: William Morrow, 1986.

"The Tale of the Recalcitrant Imam," *The New York Times,* 25 July 1982, 12.

Tannahil, Reay. *Sex in History.* Scarborough, 1982. New York: Stein and Day, 1980.

Tanner, J. R., C. W. Previte-Orton, and Z. N. Brooke, eds. *Cambridge Medieval History.* Cambridge: Cambridge University Press, 1968.

Tanner, Nancy Makepeace. *On Becoming Human: A Model of the Transition from Ape to Human & the Reconstruction of Early Human Social Life.* New York: Cambridge University Press, 1981.

al-Tanûkhî, 'Ali al-Muhassin. "Ruminations and Reminiscences." In *The Islamic World. See* Ishaq.

Tasker, Peter. *The Japanese: A Major Exploration of Modern Japan.* New York: E. P. Dutton & Co., Truman Talley Books, 1988.

Tavris, Carol. *Anger: The Misunderstood Emotion.* New York: Simon and Schuster, 1982.

Telhami, Shibley. "Arab Public Opinion and the Gulf War." *Political Science Quarterly,* Fall 1993, 437–52.

Temple, Robert. *The Genius of China: 3,000 Years of Science, Discovery, and Invention.* New York: Simon and Schuster, 1987.

"Thabit: The Death of the Knight Rabia, Called Boy Longlocks," In *The Islamic World. See* Ishaq.

Thapar, Romila. *A History of India.* Vol. 1. 1966. Harmondsworth, Middlesex: Penguin Books, 1966.

Thomas, Antony (writer, director, and producer). *Thy Kingdom Come, Thy Will Be Done.* London: Central Television Enterprises, 1987.

Thomas, Keith. *Man and the Natural World: A History of the Modern Sensibility.* New York: Pantheon Books, 1983.

Thomas, Lewis. *Lives of a Cell: Notes of a Biology Watcher.* New York: Bantam Books, 1975.

Thomas, Lewis, and Robin Bates. *Notes of a Biology Watcher.* Produced and directed by Robin Bates. Nova, no. 818. Boston: WGBH, 1981.

Thomson, William A. R., M.D. *Black's Medical Dictionary.* Totowa, N.J.: Barnes & Noble Books, 1984.

Tiger, Lionel, and Robin Fox. *The Imperial Animal.* New York: Holt, Rinehart and Winston, 1971.

Time-Life Books, Editors of. *The Age of God Kings: Time Frame 3000–1500 BC.* Alexandria, Va.: Time-Life Books, 1987.

Time-Life Books, Editors of. *Barbarian Tides: Time Frame 1500–600 BC.* Alexandria, Va.: Time-Life Books, 1987.

Time-Life Books, Editors of. *The March of Islam.* Alexandria, Va.: Time-Life Books, 1988.

Toffler, Alvin. Introduction to *Order out of Chaos. See* Prigogine and Stengers.

Toland, John. *The Rising Sun: The Decline and Fall of the Japanese Empire.* New York: Random House, 1970.

Tomoshok, Lydia, Craig Van Dyke, M.D., and Leonard S. Zegans, M.D. *Emotions in Health and Illness: Theoretical and Research Foundations.* London: Grune & Stratton, 1983.

Trevelyan, G. M. *A Shortened History of England.* 1942. Harmondsworth, Middlesex: Penguin Books, 1959.

Troyat, Henri. *Alexander of Russia.* Newport Beach, Calif.: Books on Tape, 1980.

————. *Daily Life in Russia: Under the Last Tsar.* Newport Beach, Calif.: Books on Tape, 1959.

Tuchman, Barbara W. *A Distant Mirror: The Calamitous 14th Century.* New York: Ballantine Books, 1979.

————. *The Proud Tower: A Portrait of the World before the War, 1890–1914.* New York: Bantam Books, 1967.

Turner, Frederick Jackson. *The Frontier in American History.* New York: Henry Holt, 1920.

Ulrich, R. F., and N. H. Azrin. "Reflexive Fighting in Response to Aversive Stimulation." *Journal of the Experimental Analysis of Behavior* (October 1962): 511–20.

Uris, Leon. *The Haj.* New York: Bantam Books, 1985.

U.S. Bureau of the Census. *Statistical Abstracts of the United States: 1988.* 108th ed. Washington, D.C.: U.S. Government Printing Office, 1987.

Valero, Helena. *Yanoama: The Narrative of a White Girl Kidnapped by Amazonian Indians,* as told to Ettore Biocca. Translated by Dennis Rhodes. New York: E. P. Dutton & Co., 1970.

Varitsiotes, Ioannis M. "Security in the Mediterranean and the Balkans." *Mediterranean Quarterly,* Winter 1992, 25–34.

Voltaire [François-Marie Arouet]. *Candide.* Newport Beach, Calif.: Books on Tape.

Waite, Robert G. L. *The Psychopathic God: Adolf Hitler.* New York: New American Library, 1978.

Walder, Andrew G. "Property Rights and Stratification in Socialist Redistributive Economies." *American Sociological Review,* August 1992, 524–39.

Walker, Eric A. *The British Empire: Its Structure and Spirit, 1497–1953.* Cambridge: Harvard University Press, 1956.

Walton, Susan. "How to Watch Monkeys." *Science 86,* June 1986, 22–27.

Washburn, S. L., and D. A. Hamburg. "Aggressive Behavior in Old World Monkeys and Apes." In *Primates. See* Goodall, "Gombe Stream Chimpanzees."

Weber, Bruce H., David J. Depew, and James D. Smith, eds. *Entropy, Information, and Evolution. See* Brooks.

Weiss, Rick. "How Dare We: Scientists Seek the Sources of Risk-taking Behavior." *Science News,* 25 July 1987, 57–59.

Wells, H. G. *The Outline of History.* New York: Macmillan, 1926.

Wesson, Robert G. *Beyond Natural Selection.* Cambridge: MIT Press, 1993.

Whalen, William Joseph. *Minority Religions in America.* New York: Alba House, 1981.

What Is Clinical Ecology? Denver, Colo.: American Academy of Environmental Medicine, n.d.

Wheeler, William Morton. "The Ant Colony as an Organism." *Journal of Morphology* 22 (1911): 307–25.

White, Theodore H. *America in Search of Itself: The Making of the President, 1956–1980.* New York: Harper & Row, Cornelia and Michael Bessie Book, 1982.

Whitrow, G. J. *Einstein: The Man and His Achievement.* New York: Dover Publications, 1973.

Wicken, Jeffrey S. "Thermodynamics, Evolution and Emergence: Ingredients for a New Synthesis." In *Entropy, Information, and Evolution.* See Brooks.

Wilken, Robert R. "Marcion." In *The Encyclopedia of Religion,* edited by Mircea Eliade. New York: Macmillan, 1987.

"Will Algeria Become a Second Iran?" *World Press Review,* August 1990, 33. (First published in *Der Spiegel.*)

Williams, Rosalind. "Reindustrialization Past and Present." *Technology Review,* November/December 1982.

Williamson, James A. *The Evolution of England: A Commentary on the Facts.* Oxford: Oxford University Press, 1944.

Wilson, Edward O. *Sociobiology: The Abridged Edition.* Cambridge: Harvard University Press, Belknap Press, 1980.

———. *The Insect Societies.* Cambridge: Harvard University Press, Belknap Press 1971.

Winick, Charles. *Dictionary of Anthropology.* New York: Philosophical Library, 1956.

Wolman, Benjamin B., ed. *International Encyclopedia of Psychiatry, Psychology, Psychoanalysis & Neurology.* 12 vols. See Denber.

Wood, Michael. *In Search of the Dark Ages.* New York: Facts on File Publications, 1987.

———. *In Search of the Trojan War.* New York: New American Library, Plume Book, 1987.

Woodward, Kenneth L., with Vincent Coppola. "King of Honky-Tonk Heaven." *Newsweek,* 30 May 1983, 89–90.

The World Almanac: 1984. New York: Newspaper Enterprise Association, 1983.

Wright, Robert. "The Information Age: The Life of Meaning." *The Sciences,* May/June 1988, 10–12.

——. *Three Scientists and Their Gods: Looking for Meaning in an Age of Information.* New York: Times Books, 1988.

Wright, Robin. "Islam, Democracy and the West." *Foreign Affairs,* Summer 1992, 131–45.

Wynne-Edwards, V. C. *Animal Dispersion in Relation to Social Behavior.* New York: Hafner, 1962.

——. *Evolution through Group Selection.* Oxford: Blackwell Scientific Publications, 1986.

Year of Action. Lakemont, N.Y.: Freedom Village, 1985.

Yergin, Daniel. *The Prize: The Epic Quest for Oil, Money and Power.* New York: Simon and Schuster, 1991.

Yevtushenko, Yevgeny. "Civic Timidity Is Killing Perestroika." *World Press Review,* July 1988, 26–28. (First published in *Literaturnaya Gazeta* [Moscow].)

Yoffee, Norman, and George L. Cowgill, eds. *The Collapse of Ancient States and Civilizations. See* Bronson.

Yoshiba, Kenji. "Local and Intertroop Variability in Ecology and Social Behavior of Common Indian Langurs." In *Primates. See* Goodall, "Gombe Stream Chimpanzees."

Young, Dudley. *Origins of the Sacred: The Ecstasies of Love and War.* New York: St. Martin's Press, 1991.

Yuan, Gao. *Born Red: A Chronicle of the Cultural Revolution.* Stanford, Calif.: Stanford University Press, 1987.

Zivojinovic, Dragoljub R. "Islam in the Balkans: Origins and Contemporary Implications." *Mediterranean Quarterly,* Fall 1992, 51–61.

INDEX

ABOUT THE AUTHOR

Howard Bloom has unusual credentials for writing a book that combines science, history, politics, and mass culture. Bloom is a member of the New York Academy of Sciences, the National Association for the Advancement of Science, the American Psychological Society, the American Sociological Association, and the Academy of Political Science. He graduated magna cum laude and Phi Beta Kappa from New York University. His scientific background includes: work as a biochemistry lab assistant at one of the world's largest cancer research facilities, the Roswell Park Memorial Cancer Research Institute in Buffalo, New York; research on programmed learning at Rutgers University's Graduate School of Education; and editing foundation grant proposals for the Middlesex County Mental Health Clinic. His political experience includes writing position papers for two winning congressional candidates and co-founding Music in Action, the leading anticensorship group in the record industry during the 1980s.

In 1976, Bloom founded what Delta Airlines' in-flight magazine called "one of the most prestigious pop PR firms in the world." As president of that company until 1988, he won numerous awards and helped shape the careers of Prince, the Jacksons, Bette Midler, Run D.M.C., the Talking Heads, John Cougar Mellencamp, Billy Joel, ZZ Top, George Michael, Earth, Wind & Fire, Bob Marley, Billy Idol, and Simon and Garfunkel, among others.

Bloom has been cited or profiled in *New York Magazine,* the *Los Angeles Times,* the *New York Post, Newsday, Rolling Stone,* and numerous other

publications. Over 250 of his articles have appeared in publications rang-
ing from the *Village Voice, Omni,* and *Cosmopolitan* to the *Independent
Scholar.* He has lectured at Wesleyan University, Georgia State Univer-
sity, and New York University. He is featured in the first *Who's Who of
Science and Engineering.* And he has appeared on the "Today" show, Cable
News Network, "CBS Nightwatch," and many other television outlets.